BIOLOGICAL DNA SENSOR

The Impact of Nucleic Acids on Diseases and Vaccinology

BIOLOGICAL DNA SENSOR

SENSOR

The Impact of Nucleic Acids on Diseases and Vaccinology

KEN J. ISHII

*Vaccine Science Laboratory, Immunology Frontier Research Centre (IFReC), Osaka University, Osaka, Japan
and
Laboratory of Adjuvant Innovation, National Institute of Biomedical Innovation (NIBIO), Osaka, Japan*

CHOON KIT TANG

Vaccine Science Laboratory, Immunology Frontier Research Centre (IFReC), Osaka University, Osaka, Japan

Amsterdam • Boston • Heidelberg • London • New York • Oxford
Paris • San Diego • San Francisco • Singapore • Sydney • Tokyo
Academic Press is an imprint of Elsevier

Academic Press is an imprint of Elsevier
32 Jamestown Road, London NW1 7BY, UK
225 Wyman Street, Waltham, MA 02451, USA
525 B Street, Suite 1800, San Diego, CA 92101-4495, USA

Notice
No responsibility is assumed by the publisher for any injury and/or damage to persons
or property as a matter of products liability, negligence or otherwise, or from any use or
operation of any methods, products, instructions or ideas contained in the material herein.
Because of rapid advances in the medical sciences, in particular, independent verification
of diagnoses and drug dosages should be made.

British Library Cataloguing-in-Publication Data
A catalogue record for this book is available from the British Library

Library of Congress Cataloging-in-Publication Data
A catalog record for this book is available from the Library of Congress

ISBN: 978-0-12-404732-7

For information on all Academic Press publications visit
our website at elsevierdirect.com

Typeset by MPS Limited, Chennai, India
www.adi-mps.com

14 15 16 17   10 9 8 7 6 5 4 3 2 1

Working together
to grow libraries in
developing countries

www.elsevier.com • www.bookaid.org

CONTENTS

13. Adjuvants Targeting the DNA Sensing Pathways – Cyclic-di-GMP and other Cyclic-Dinucleotides   **313**

Rebecca Schmidt and Laurel L. Lenz

Shizuo Akira
Host Defense Laboratory, IFReC, Osaka University, Japan

Moshe Arditi
Department of Biomedical Sciences and Pediatric Infectious Diseases and Immunology, Cedars-Sinai Medical Center; Infectious and Immunologic Diseases Research Center, Cedars-Sinai Medical Center, Los Angeles, CA; David Geffen School of Medicine, University of California at Los Angeles, USA

Glen N. Barber
Department of Cell Biology and Sylvester Comprehensive Cancer Center, University of Miami Miller School of Medicine, Miami, FL, USA

Andrew G. Bowie
School of Biochemistry and Immunology, Trinity Biomedical Sciences Institute, Trinity College Dublin, Ireland

Michael P. Cancro
Department of Pathology and Laboratory Medicine, Perelman School of Medicine at the University of Pennsylvania, Philadelphia, PA, USA

Cevayir Coban
Laboratory of Malaria Immunology, WPI Immunology Frontier Research Center (IFReC), Osaka University, Osaka, Japan

Timothy R. Crother
Department of Biomedical Sciences and Pediatric Infectious Diseases and Immunology, Cedars-Sinai Medical Center; Infectious and Immunologic Diseases Research Center, Cedars-Sinai Medical Center, Los Angeles, CA; David Geffen School of Medicine, University of California at Los Angeles, USA

Christophe J. Desmet
Laboratory of Cellular and Molecular Immunology, Interdisciplinary Cluster of Applied Genoproteomics (GIGA) – Research Center and Faculty of Veterinary Medicine, University of Liège, Liège, Belgium

Katherine A. Fitzgerald
Division of Infectious Diseases and Immunology, Department of Medicine, University of Massachusetts Medical School, Worcester, MA, USA

S. Gasser
Immunology Programme and Department of Microbiology, Centre for Life Science, National University of Singapore, Singapore

S.W.S. Ho
Immunology Programme and Department of Microbiology, Centre for Life Science, National University of Singapore, Singapore

Ken J. Ishii
Laboratory of Adjuvant Innovation, National Institute of Biomedical Innovation (NIBIO), Saito-Asagi, Ibaraki City, Osaka, Japan; Vaccine Science Laboratory, Immunology Frontier Research Centre (IFReC), Osaka University, Japan

Nao Jounai
Laboratory of Adjuvant Innovation, National Institute of Biomedical Innovation, Osaka, Japan

Taro Kawai
Laboratory of Host Defense, WPI Immunology Frontier Research Center, Department of Host Defense, Research Institute for Microbial Diseases, Osaka University, Osaka, Japan

Kouji Kobiyama
Laboratory of Adjuvant Innovation, National Institute of Biomedical Innovation, Osaka, Japan

C.X. Koo
Immunology Programme and Department of Microbiology, Centre for Life Science, National University of Singapore, Singapore; Laboratory of Adjuvant Innovation, National Institute of Biomedical Innovation (NIBIO), Ibaraki, Osaka, Japan

A.R. Lam
Immunology Programme and Department of Microbiology, Centre for Life Science, National University of Singapore, Singapore

N. Le Bert
Immunology Programme and Department of Microbiology, Centre for Life Science, National University of Singapore, Singapore

Laurel L. Lenz
Integrated Department of Immunology, National Jewish Health and University of Colorado School of Medicine, Denver, CO, USA

Ann Marshak-Rothstein
Department of Medicine/Rheumatology, University of Massachusetts Medical School, Worcester, MA, USA

Jan Naujoks
Department of Internal Medicine/Infectious Diseases and Pulmonary Medicine, Charité Universitätsmedizin Berlin, Berlin, Germany

Bastian Opitz
Department of Internal Medicine/Infectious Diseases and Pulmonary Medicine, Charité Universitätsmedizin Berlin, Berlin, Germany

Søren R. Paludan
Department of Biomedicine, Aarhus Research Center for Innate Immunology, University of Aarhus, The Bartholin Building, Aarhus C, Denmark

Surya Pandey
Laboratory of Host Defense, WPI Immunology Frontier Research Center, Department of Host Defense, Research Institute for Microbial Diseases, Osaka University, Osaka, Japan

Vijay A.K. Rathinam
Division of Infectious Diseases and Immunology, Department of Medicine, University of
Massachusetts Medical School, Worcester, MA, USA

Tatsuya Saitoh
Department of Host Defense, Research Institute for Microbial Diseases, Osaka University,
Suita, Osaka, Japan; Laboratory of Host Defense, WPI Immunology Frontier Research
Center, Osaka University, Suita, Osaka, Japan

Rebecca Schmidt
Integrated Department of Immunology, National Jewish Health and University of
Colorado School of Medicine, Denver, CO, USA

Y.J. Shen
Immunology Programme and Department of Microbiology, Centre for Life Science,
National University of Singapore, Singapore

Kenichi Shimada
Department of Biomedical Sciences and Pediatric Infectious Diseases and Immunology,
Cedars-Sinai Medical Center; Infectious and Immunologic Diseases Research Center,
Cedars-Sinai Medical Center, Los Angeles, CA; David Geffen School of Medicine,
University of California at Los Angeles, USA

Fumihiko Takeshita
Laboratory of Adjuvant Innovation, National Institute of Biomedical Innovation,
Osaka, Japan

Choon Kit Tang
Laboratory of Adjuvant Innovation, National Institute of Biomedical Innovation
(NIBIO), Saito-Asagi, Ibaraki City, Osaka, Japan; Laboratory of Malaria Immunology,
WPI Immunology Frontier Research Center (IFReC), Osaka University, Osaka, Japan;
Laboratory of Vaccine Science, IFReC, Osaka University, Osaka, Japan

Miyuki Tozuka
Laboratory of Adjuvant Innovation, National Institute of Biomedical Innovation,
Osaka, Japan

Sivapriya Kailasan Vanaja
Division of Infectious Diseases and Immunology, Department of Medicine, University of
Massachusetts Medical School, Worcester, MA, USA; Department of Molecular Biology
and Microbiology, Tufts School of Medicine, Boston, MA, USA

Deoxyribonucleic acid (DNA), the fundamental molecule that packs genetic instructions for the development of living organisms, has been discovered to hold yet another vital function for our biology – inflammation. The fact that DNA is an abundant commodity in our body and causes its immunogenic property is one that cannot be taken lightly. Now, we know that DNA induced inflammation is responsible for the pathogenicity of autoimmune and infectious diseases and metabolic disorders. Furthermore, we have evidence that DNA induced inflammation is involved in the development of cancer. Therefore the awareness of the immunogenic nature of dsDNA has provided us with the opportunity to gain further insights into the workings of human diseases. DNA seemed to be an ironic choice as a trigger molecule for inflammation. After all, who would expect that self-products containing our genetic blueprint could turn against us and induce adverse inflammatory reactions upon detection by the immune system? However, on deeper reflection, it does make practical sense to have nucleic acids as signaling entities that sound off the alarm to indicate impending danger to the body. Pathogens including viruses, bacteria and parasites, like human beings, have their genetic information stored in nucleic acids. Their invasion into the host system is likely to be associated with the introduction of their genetic materials and therefore it would be the most appropriate indication of pathogenic invasion. On the other hand, the release of our own DNA into the surrounding physiological environment could also indicate cell death as a result of trauma, which may invite infection to occur and require immune defenses to be put in place. Therefore in more ways than one, DNA is an aptly chosen alarm molecule for the immune system to react to danger. Regarding the potential impact it has on various aspects of our health, we have seen an explosion of research on DNA sensing in the past eight years. This surge in publication was due in part to the discovery that DNA could be sensed not only by TLR9 in the endosomes, but also by sensors present in the cytosol. More importantly, DNA from pathogens as well as from mammalian derived sources is equally potent in inducing inflammation.

It is the purpose of this book to bring together the research on the signaling mechanism of DNA induced inflammation as well as its impact on diseases and vaccinology. The book has three sections: In Section I,

Ishii, Fitzgerald, Barber and Saitoh discuss the signaling pathway leading to DNA induced inflammation, in Section II, Kawai, Marshak-Rothstein, Opitz, Bowie, Gasser and Arditi discuss the impact of DNA inflammation on human diseases and lastly, in Section III, Coban, Desmet and Lenz comment on how the inflammatory response of DNA could influence the outcome of vaccination in DNA vaccination strategies and adjuvants targeting the DNA sensing pathway.

We are greatly indebted to the contributors whose participation and cooperation made this book possible. We thank them for their patience with our persistent requests for completion of their manuscript. We are appreciative for the editorial and technical assistance provided by Elizabeth Gibson and Mary Preap. To the publisher we are grateful for this timely opportunity to consolidate the wealth of knowledge we have acquired on DNA sensing which could facilitate future investigations. Lastly, we would like to pay tribute to Alick Issacs and his colleagues who were way ahead of their time in that they had the audacity to identify a molecule that is so fundamental to our existence, namely, DNA, as a danger signaling molecule.

The Discovery of dsDNA Immunogenicity and the Definition of DNA Sensing Pathways

Route to Discovering the Immunogenic Properties of DNA from TLR9 to Cytosolic DNA Sensors

Choon Kit Tang[1,2,3,4], Cevayir Coban[2], Shizuo Akira[3] and Ken J. Ishii[1,4]

[1]Vaccine Science Laboratory, Immunology Frontier Research Centre (IFReC), Osaka University, Japan
[2]Malaria Immunology Laboratory, IFReC, Osaka University, Japan
[3]Host Defense Laboratory, IFReC, Osaka University, Japan
[4]Laboratory of Adjuvant Innovation, National Institute of Biomedical Innovation (NIBIO), Saito-Asagi, Osaka, Japan

INTRODUCTION

The innate immune system relies on cues in the form of products released during injury or pathogenic invasion for its activation. Immunologists have come to appreciate deoxyribonucleic acid (DNA) as one of such products that signal danger to the innate immune system and induce inflammation. For example, uncleared DNA released during mechanical injury to tissues or the introduction of viral and bacterial DNA during infection was found to elicit inflammatory responses. This immune response induced by DNA is not a random process but rather it is detected by specific receptors that trigger distinct cell signaling pathways to produce inflammatory products. Our journey in understanding the mechanism of DNA sensing and inflammation began with the discovery of Toll-like receptor 9 (TLR9) in endosomal compartments and its ability to detect CpG motifs present specifically in the DNA of microbes. The immune response generated by TLR9 detection of CpG DNA is potent and capable of protection against pathogenic infection and it was also studied for applications in vaccine adjuvants. It was realized however that TLR9 could not be the sole DNA sensor present in the cell as inflammation induced by DNA persisted despite the absence of TLR9. This was not clearly understood until it was realized that DNA when directly introduced into the cytosol of cells could elicit potent type-I-interferon (type-I-IFN) responses that were independent of TLR9 but completely relied on the TBK1–IRF3 signaling axis. This discovery led to the conclusion that there are other sensors of DNA present in the cytosol and not in the endosomal compartments which could induce inflammation.

Biological DNA Sensor.
DOI: http://dx.doi.org/10.1016/B978-0-12-404732-7.00001-0

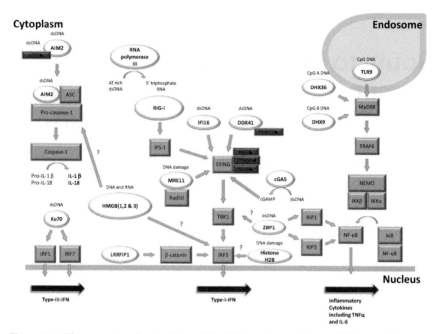

Figure 1.1 *The current understanding of the DNA sensing pathways.* Yellow oval buttons depict identified DNA sensors; blue and pink square buttons are the signaling/adaptor molecules and transcription factors involved in the particular pathway; purple tubes are ▲ positive and ◣ negative regulators of the pathways.

Since then, several cytosolic DNA sensors have been published and more continue to be reported (Figure 1.1). In this chapter, we will reflect on our early understanding of the immunogenic properties of dsDNA and give a chronological account of the journey we have taken to discover the individual cellular DNA sensors that have played important roles in mediating DNA induced inflammation.

THE IMPORTANCE OF THE INNATE IMMUNE SYSTEM

PAMPS and DAMPs

The immune responses of higher animals can be broadly categorized into those of the innate immune system and the adaptive immune system. They differ in the cell types and cytokines involved as well as the level of specificity and most importantly the responding time. Adaptive immunity provides protection with high degree of specificity and efficiency by targeting antigens present on the invading pathogen. This is put in place through a series of immunological processes including degradation of pathogens, antigen presentation and clonal expansion of T and B cells. As suggested by the multi-step process, time is required for the adaptive immune responses to be mounted. In contrast, the innate immune system is an evolutionarily older form of immunity that offers less specific but faster protection compared to adaptive immunity. Instead of reacting to specific epitopes, the innate immunity senses the small molecular motifs that are uniquely associated with pathogens such as bacterial flagellin or viral glycoproteins, which are collectively known as pathogen-associated molecular patterns (PAMPs). No further degradation or presentation to specialized cells is required for the triggering of the innate immune system by PAMPS. It is directly sensed by a class of germline-encoded receptors named pattern-recognition receptors (PRRs). So far, there are five families of PRRs identified: Toll-like receptors (TLRs), cytosolic RIG-I-like receptors (RLRs), Nod-like receptors, C-type lectins and absence inmelanoma-2 (AIM2)-like receptors. These PRRs differ in their cellular locations, type of PAMPs they recognize and the induction of downstream immune effector responses.

More recently, it has been proposed that the innate immune system should encompass the detection and reaction to another category of danger signals termed damage-associated molecular patterns (DAMPs) [1]. As the name suggests, DAMPs are non-pathogen derived self-products that are sequestered and restricted from interacting with the immune system during unprovoked physiological conditions but are exposed during cellular stress or injury. This may include intracellular proteins such as histones and heat-shock proteins that are released during necrotic or apoptotic cell death. It could also include extracellular products released after mechanical tissue injury such as heparin sulfate and hyaluronan. Irritant particles such as silica and crystals which are known to induce inflammatory responses are also considered as DAMPs. In essence, DAMPs are the agonists of sterile inflammation. Although DAMPs are categorically different from PAMPs, the patterns of inflammatory responses they induce are the same, such as infiltration of macrophages and neutrophils and the release of TNF-α and IL-1 families of cytokines. Therefore they are considered to be sensed by the same set of PRRs.

The 'shoot first ask later' nature of the innate immune system provides an effective first line of defense against pathogen invasion, but also remains as one of the greatest flaws in our immune system. In the state of chronic inflammation, when the source of the inflammatory agonists is not clear, potent cytokines and factors continually released by the innate immune system can cause widespread indiscriminate injury to the surrounding tissue, and outcomes of these processes are manifested as autoimmune diseases. On the other hand, it is also these characteristics of the innate immune response which allow adjuvants to provide the right conditions for efficacious vaccine response. In fact, the activation of the innate immune system has been viewed to be the essential first step that could pre-determine the type and magnitude of the adaptive immune response that is subsequently generated. Inflammatory cytokines induced by the innate immune system are potent immune-modulators that are capable of initiating and steering the immune response induced by vaccine preparations. For a long time, Alum has been thought to provide vaccines with adjuvant help by slowly releasing the antigen of interest which has been incorporated into its content by emulsification, a process termed depot effect. Now, we know that Alum is able to activate different arms of the innate immune system. Reports have shown that Alum can cause cell death which leads to DNA release that acts as DAMP. The release of proinflammatory cytokines by these pathways could be involved in the adjuvant activity of Alum.

Therefore the induction of the innate immunity triggered by PAMPs and DAMPs is an important process that affects many different processes in our biology including host defense, autoimmune disease and vaccine adjuvants.

Pathogen Derived Nucleic Acids are Potent Activators of the Innate Immune System

Nucleic acids have been known to be immunogenic for a long time. One of the earliest demonstrations of its immunogenic property was presented by Alick Isaacs and colleagues in 1963 [2]. At that time, interferon was recognized as an antiviral substance released by mammalian cells during viral infection, however it was not clear which component of the virus was responsible for this response [3]. Alick Isaacs and colleagues showed that nucleic acids derived from viruses including encephalomyocarditis virus and turnip yellow mosaic virus were able to induce interferon production and inhibit Chikungunya virus plaque formations when added directly to mouse and chick embryonic fibroblasts. The addition of homologous (self) DNA did not however induce strong production of interferon or give protective antiviral responses [2,4]. This was the first direct demonstration of the immunogenic properties of nucleic acids from pathogens.

Bacterial derived DNA was similarly noted to be immunogenic. In the 19th century, an innovative physician, William B. Coley, formulated a mixture composed of bacterial cell lysates (termed Coley's Toxin) for the treatment of carcinoma based on the observation that in certain cases, infection could actually be therapeutic in tumor malignancy. This form of treatment still exists in the present day, for example *Mycobacterium bovis* derived Bacillus Calmette–Guérin (BCG) is used for the therapy of bladder cancer [5]. It was not understood how this crude preparation of bacteria could have a therapeutic effect on cancer progression until the investigations made by Tokunaga and colleagues were published in 1984 [6]. In their study, they fractionated the BCG preparation and tested each of the fractions for its anti-tumor activity in guinea pigs. The anti-tumor component of BCG was narrowed to a fraction which contained mainly bacterial DNA and had no detectable cell wall components. In addition to anti-tumor activity, this fraction of BCG was found to have immunogenic properties. It was able to activate NK cell lytic activity and induce type-I-IFN response from human peripheral blood mononuclear cells (PBMCs) [7]. Subsequently, Messina et al. extended this observation and demonstrated that the immunogenic nature of DNA was not restricted to mycobacteria, but was also present in other bacterial derived DNA [8]. They showed that bacterial DNA could induce proliferation in mouse splenocytes and also directly activate B cells to proliferate and produce IgM antibodies in the absence of T cell help. This phenomenon however did not extend to mammalian derived DNA.

Another clue from early studies which suggested bacteria derived DNA is immunogenic comes from a report that investigated the generation of anti-DNA antibodies in the context of systemic lupus erythematosus (SLE). It was shown that *Escherichia coli* and *Clostridium perfringens* DNA were able to induce anti-double stranded DNA antibodies when injected into mice [9,10]. This effect was less prominent when DNA from vertebrate origin such as calf thymus, chicken blood, human placenta and salmon testes were used. These observations brought to light the immunogenic property of bacterial DNA and identified it as a source of a danger signal which forces the innate immune system to act and prepare for host defense during pathogen invasion.

ENDOSOMAL-MEMBRANE BOUND DNA SENSOR – TLR9

Toll-Like Receptors in Innate Immunity

Toll-like receptors belong to a family of transmembrane proteins that are phylogenetically conserved and play important roles in the innate immunity.

Toll-protein was first discovered in *Drosophila* (fruit fly). In addition to its role in controlling dorsal–ventral patterning during embryogenesis [11], *Drosophila*-Toll was found to be essential in host defense against microbial invasion as exemplified by the anti-fungal response against *Aspergillus fumigatus* [12]. Flies expressing a mutant non-functional form of Toll-protein were unable to secrete the anti-fungal factor drosomycin and succumbed to the fungal infection. A human homolog of *Drosophila*-Toll was later identified by Medzhitov et al. in 1997 [13]. At the same time, Rock et al. reported that they had cloned five different forms of human equivalent Toll-proteins and labeled them toll-like receptors (TLR) 1–5 [14]. Coincidentally, the first TLR identified was TLR4 in the report prepared by Rock et al.

Both *Drosophila*- and human-Toll are type-I transmembrane proteins. They are composed of a leucine rich repeat (LRR) extracellular N-terminal domain, a transmembrane region and a cytoplasmic domain that is homologous to the human IL-1 receptor (IL-1R) cytoplasmic domain (termed Toll/interleukin-1 receptor (TIR) motif in human/mammalian TLR) [15]. The signaling events triggered by human IL-1R and *Drosophila*-Toll bear many similarities. In humans, the detection of IL-1β by IL-1R triggers activation of NF-κB through signal transduction mediated by its cytoplasmic domain. The activation of NF-κB requires it to be released from IκB proteins, which inhibits NF-κB phosphorylation and translocation into the nucleus. In *Drosophila*, upon engagement and activation of the Toll protein, a signaling cascade is initiated which eventually leads to activation of the Dorsal protein by disengaging the inhibiting Cactus protein; these proteins are human equivalents of NF-κB and IκB respectively. Hence, the presence of TIR domain in human TLR and the similarities in signaling events triggered by human IL-1R and *Drosophila*-Toll indicate that human TLR might have functions in mediating the inflammatory response. Indeed, the transfection of a constitutively active form of human-Toll into THP-1 cells (human monocytic cell line) can trigger the activation of NF-κB and induce the expression of co-stimulatory molecule B7.1 and proinflammatory cytokines IL-1, IL-6 and IL-8. These responses are indicative of innate immune activity.

Although TLR4 was identified, its exact role in innate immunity was unknown until it was found to be the receptor responsible for the detection of lipopolysaccharides (LPS) from Gram-negative bacteria and the induction of inflammatory cytokines that cause sepsis. This discovery was made by studying C3H/HeJ and C57BL/10ScCr mice, which have macrophages that are strongly resistant to LPS stimulation but are highly susceptible to Gram-negative infection [16,17]. It was determined that

C3H/HeJ mice have a single missense in the TIR domain of TLR4, and C57BL/10ScCr mice do not express the TLR4 gene, which rendered these mice unresponsive to LPS stimulation. Since then, 11 mouse TLRs and 13 human TLRs based on the different ligands recognized have been identified. Four out of these identified human and mouse TLRs sense microbial nucleic acids. These TLRs have distinct specificities for the type of nucleic acids they can detect and react to. TLR3 recognizes double stranded RNA (dsRNA) while TLR7 and TLR8 recognize single stranded RNA (ssRNA). A newly identified TLR13, which exists only in mice, was found to recognize a conserved region of the bacterial 23S ribosomal RNA on to which certain antibiotics such as erythromycin bind [18]. TLR9 is the only TLR that recognizes DNA in humans and mice.

The identification of TLR and its role in the innate immune system is a significant discovery that is fundamental to our understanding of host defense mechanisms against pathogens. For their role in helping to define this area of research, Bruce A. Beutler, Jules A. Hoffmann and Ralph M. Steinman were awarded the Nobel Prize in Physiology or Medicine in 2011 [19].

TLR9 and CpG

Ten years after Tokunaga and colleagues reported on the immunogenic nature of mycobacteria derived DNA, Krieg et al. made a connection between the CpG motifs present in microbial DNA sequences and their immunogenicity [20]. CpG motifs are DNA sequences that contain a cytosine that is immediately followed by a guanine. The 'p' in the middle refers to the phosphodiester backbone of DNA. Two major differences between human and microbial DNA were reported [21,22]. Firstly, microbial DNA has four times as much CpG motif as vertebrate DNA. Secondly, the cytosines of CpG motifs found in microbes are hypomethylated compared to vertebrate DNA. True to their suspicion, Krieg et al. demonstrated that the methylation of bacterial DNA with CpG methylase completely inhibits the proliferation and the upregulation of MHC class II expression in mouse splenic derived B cells. Synthetic DNA that contains CpG motifs was also able to mimic bacterial DNA in inducing B cell proliferation and IgM production. Conversely, the removal of CpG motifs or the replacement of cytosine with 5-methylcytosine eliminated the stimulatory effect of synthetic DNA. CpG was shown to activate several immune cell types such as T cells, natural killer cells, monocytic cells and B cells. The immunogenic effect was most profound on B cells. B cells stimulated with CpG DNA were shown to induce strong IL-6, IL-12 and IFNγ secretion *in-vitro* within 1 to 2 hours [23]. An early attempt to determine the signaling pathway induced by bacterial DNA was made by Stacey et al. [24]. The authors demonstrated on macrophages that unmethylated CpG containing plasmid DNA was similar to LPS in the inflammatory signaling pathways it activates. They showed that both LPS and plasmid DNA could activate NF-κB to induce TNF-α and IL-1β responses as well as the LPS-responsive gene plasminogen activator inhibitor type-2 (PAI-2) and the integrated HIV-1 LTR. iNOS was also found to be induced by plasmid DNA, but only when the macrophages were primed with IFNγ. The awareness of the potent stimulatory effect of CpG DNA resulted in its application in various therapies including as vaccine adjuvants. Due to the cytokine milieu induced by CpG activation, it was found to trigger a strong Th1-type response when used as an adjuvant [25].

The discovery of the cellular receptor for CpG DNA came as Hemmi et al. uncovered another mammalian TLR, named TLR9, which contained a TIR motif that associates with MyD88 and TRAF6 [26]. Prior to this publication, the authors had reported that CpG responses were dependent on MyD88 and TRAF6, but TLR2 and TLR4 which signal through these molecules were not responsible for detection. This prompted Hemmi et al. to investigate whether TLR9 could be the receptor for CpG DNA. Indeed, by performing studies on TLR9 knockout mice, they confirmed that TLR9 was the cellular receptor for CpG which mediates its immunogenic effects. In the absence of TLR9, splenocytes were unable to proliferate in the presence of CpG or induce increased expression of MHC class II molecules. Similarly, peritoneal macrophages from TLR9$^{-/-}$ mice were unable to secrete detectable levels of proinflammatory cytokines such as TNF-α, IL-6 and IL-12 in response to CpG stimulation. On the contrary, TLR9$^{-/-}$ macrophages responded with a response similar to WT cells when stimulated with ligands of other TLR such as lipoprotein (TLR2/1 and TLR2/6), peptidoglycan (TLR2/2), zymosan (TLR2/6) and LPS (TLR4). TLR9 was also investigated for the role in mediating Th1 response by CpG DNA on DCs. Bone marrow derived DCs generated from TLR9$^{-/-}$ mice were unable to respond with IL-12 production or upregulate co-stimulatory molecules CD40, CD80 and CD86 when stimulated with CpG DNA. The adjuvant activity of CpG DNA was also abolished in TLR9$^{-/-}$ mice as exemplified by the lack of OVA specific IFNγ response from the splenocytes of OVA + CpG DNA immunized TLR9$^{-/-}$ mice. At the signaling level, CpG DNA failed to activate NF-κB, c-JUN N-terminal kinase (JNK) and IRAK in TLR9$^{-/-}$ macrophages. Finally, TLR9$^{-/-}$ mice were demonstrated to be resistant to the septic shock resulting from the cytokine storm induced by CpG after D-galactosamine sensitization. Latz et al. subsequently showed that CpG could indeed directly associate with TLR9 in ligand-binding studies [27]. They further demonstrated that, unlike TLR2 and TLR4, TLR9 was found to reside at the endoplasmic reticulum during the resting state. Instead of being detected on the cell surface, CpG DNA was internalized through endocytosis into the same endosomal compartments to which TLR9 were shown to migrate in conjunction with the accumulation of MyD88. All this evidence clearly demonstrates the role and specificity of TLR9 for the detection of CpG DNA in the endosomes.

The identification of TLR9 as the cellular receptor for the detection of CpG DNA remains a cornerstone in our understanding of DNA sensing. It shows that inflammation induced by DNA is not a random process, but rather it is detected by a specific receptor that induces inflammation via a distinct signaling pathway.

CYTOSOLIC DNA SENSORS
Signs that TLR9 may not be the only DNA Intracellular Sensor

The discovery of TLR9 ability to detect hypomethylated CpG-rich DNA has provided an explanation for the immuno-stimulatory property of DNA, although it is not a complete one. Numerous items of evidence have pointed to a redundancy in the system and that the detection of CpG by TLR9 could not completely account for all the inflammatory response induced by DNA. For example, it was initially thought that non-adjuvanted naked DNA vaccines were effective in mice due to the un-methylated CpG motifs present in the DNA plasmids triggering TLR9 activation. Although it was demonstrated that TLR9$^{-/-}$ DCs are defective in the production of IL-12p40 when transfected with DNA, opposing reports soon arose demonstrating that TLR9$^{-/-}$ mice responded with similar immune responses to WT mice after DNA vaccination [28–30]. This suggests that there are other danger signals present in DNA vaccines that could be detected by other innate immune sensors.

There were also reports which dispute the observation made in early studies that mammalian DNA is not immunogenic. During an investigation on the abnormal increase of MHC molecule expression in autoimmune diseases, it was found that nucleic acids released by dying cells were immunogenic and the effect could be duplicated by directly introducing dsDNA and dsRNA into the cytoplasm of cells including non-immune cells by lipofection [31]. This implied that homologous (self) and mammalian derived nucleic acids could induce inflammatory responses if delivered into the cytoplasm instead of plain addition to the culture media of *in-vitro* systems. In fact, the cytoplasmic delivery via transfection of DNA from different vertebrate sources including calf thymus, salmon sperm and genomic DNA were stimulatory. Cytosol-delivered mammalian DNA was able to upregulate MHC class I, MHC class II expression and other genes that are required for antigen processing and presentation such as LMP2 and TAP1, in mammalian non-immune cell types. Further characterization of this response revealed that the cytosolic DNA has to be double stranded and larger than 25 base pairs (bp) for it to be immunogenic; it is however not dependent on specific sequences or motifs being present. At the signaling level, cytosolic dsDNA was demonstrated to activate Stat1, Stat3, MAPK, NF-κB and IRF1. Furthermore, our group found that the immuno-stimulatory effect of mammalian derived DNA was functional and could activate the functions of antigen presenting cells (APCs), such as macrophages and DCs. As a result, C57BL/6 mice immunized with OVA mixed with dsDNA had higher levels of IgG1, IgG2a and antigen specific cytotoxic T cell activity compared to mice that were immunized with OVA alone. This adjuvant effect of DNA remains despite the treatment with methylase, therefore indicating that CpG is not involved in the stimulatory effect [32].

The immunogenic property of cytosolic dsDNA and the presence of DNA sensors other than TLR9 were also implicated in a study which investigated the lethality of mice lacking DNase II [33]. In macrophages, the primary function of DNase II is to clear redundant DNA that is taken into the cytosol during engulfment of apoptotic cells. Okabe et al. found that macrophages derived from the fetal liver of DNase II$^{-/-}$ embryos constitutively expressed a set of genes that was exclusive and not upregulated in heterozygote macrophages; these genes include IFNβ, TNF-α and CXCL10. Other interferon inducible genes including IFNγ, 2'5'-oligo(A) synthetase (OAS), IRF7 and ISG15 were also exclusively expressed in DNase II$^{-/-}$ cells, but were not restricted to macrophages and applied to other non-macrophage cell types in the fetal liver. Due to this potent induction of pro-inflammatory cytokines from the constant stimulation of uncleared DNA, DNase II deficient mice were found to die *in utero*. This lethality could be however be reversed if DNase II$^{-/-}$ mice were crossed with type-I-IFN receptor deficient mice, which confirmed that type-I-IFN production was responsible for the lethal effect. However, when DNase II$^{-/-}$ mice were crossed with TLR3$^{-/-}$, TLR9$^{-/-}$ and mice deficient for other adaptor molecules required for TLR signaling including

MyD88 and TRIF, the lethal effect of DNase II persisted. Therefore this indicated the presence of a TLR independent sensing mechanism which detects DNA and induces potent type-I-IFN responses.

This evidence challenged our conventional view on how DNA is sensed at the cellular level after the discovery of TLR9, and a few possibilities are suggested: (1) All sources of DNA and not just CpG motif containing microbial DNA maybe immunogenic; (2) DNA may not only be sensed in the endosomal compartments but also in the cytoplasm; (3) Most importantly, TLR9 may not be the only receptor that detects DNA triggering the innate immune response.

DsDNA Utilizes the TBK1-IRF3 Signaling Pathway

The next leap forward in our understanding of the immuno-stimulatory effects of DNA began with realizing that DNA of non-pathogen origins could also be immuno-stimulatory when delivered directly into the cytoplasm. Investigations in our laboratory have revealed that dsDNA, when transfected using lipofection methods, is a potent inducer of type-I-IFN and chemokinegenes such as *Cxcl10*, *Ccl5* and *Ccl2* in mouse and human dendritic cells (DCs) and stromal cells [34]. The criteria for immunogenicity were different between added CpG DNA and transfected dsDNA. Cytosolic dsDNA remained stimulatory, irrespective of its sequence, and was able to induce type-I-IFN even when it was methylated. It was also not restricted to pathogen derived DNA because calf thymus DNA and synthetic DNA which lacked the CpG motif were equally stimulatory to *E. coli* and HSV-1, when transfected. Furthermore, MEF cells which were previously demonstrated to be unresponsive to CpG because of the lack of TLR9 expression were shown to induce type-I-IFN response when transfected with DNA. The type-I-IFN stimulatory effect of dsDNA was dependent on its size. Longer dsDNA induced a stronger type-I-IFN response. Only dsDNA was stimulatory; ssDNA such as complementary DNA (cDNA) was found to be immunologically inert. The stimulatory effect was also restricted to dsDNA in the B-form (B-DNA) but not the Z-form (Z-DNA) conformation. B-DNA is a commonly occurring low energy state of dsDNA structure that adopts a right-handed helix conformation. Conversely, Z-DNA adopts a high energy left-handed helix structure that occurs less frequently in physiological states. Synthetic dsDNA made up of AT repeats poly(dA-dT)•poly(dT-dA) (poly(dA:dT) hereafter) tends to adopt the B-conformation and was able to induce potent type-I-IFN response when transfected into MEFs. On the other hand, synthetic dsDNA made up of GC repeats, poly(dG-dC)•poly(dC-dG) (poly(dG:dC) hereafter), which adopts the Z-conformation, was not stimulatory when transfected into MEFs. These secondary structures of DNA remained intact even when complexed with liposomal-based transfecting reagents [35].

We have also investigated the mechanism through which B-DNA induced a type-I-IFN response by studying whether other nucleic acid sensors and signaling molecules known at that time were involved. The TLR pathway and the RIG-I pathway which senses dsRNA in the cytoplasm were not involved, based on studies performed on MEFs that were deficient in MyD88/TRIF and RIG-I. Although the cytosolic DNA induced type-I-IFN response was not dependent on RIG-I signaling, we showed that IPS-1 was necessary for the optimal production of IFNβ response in HEK293 cells. These data were puzzling at the time of publication, but now we know that RNA polymerase III can generate RNA intermediates from the poly(dA:dT) to stimulate RIG-I. This pathway is present in human cells, but is redundant in mouse cells (discussed in a later section). Subsequently, we turned our attention to TBK1, which was previously found to be essential for the induction of type-I-IFN by viral dsRNA during infection [36–38]. TBK1 is a non-canonical IκB kinase related to IKKα and IKKβ, which are the regulators of NF-κB activation. It mediates NF-κB activation and type-I IFN responses triggered by both TLR dependent and independent pathways [39–41]. TBK1 was also shown to be dispensable for the production of IFN-α by plasmacytoid DCs (pDCs) when stimulated with CpG DNA [42]. In contrast to that observation, we showed that IFNβ and other interferon-inducible chemokine responses induced by B-DNA transfection were completely abrogated in TBK1$^{-/-}$ MEFs. TBK1 was found to be necessary for the dimerization of IRF3 that was induced by B-DNA, but it was redundant for B-DNA induced NF-κB activation. Therefore the TBK1–IRF3 signaling axis is vital to the type-I-IFN secretion induced by cytoplasmic B-DNA stimulation.

The stimulatory effect of B-DNA was also found to be functional as demonstrated by performing gene arrays on WT, TBK1$^{-/-}$, IKKi$^{-/-}$ and TBK1$^{-/-}$IKKi$^{-/-}$ double knockout MEFs, that were stimulated with poly(dA:dT). The result showed that the majority of the upregulated genes were interferon inducible genes as well as antiviral related genes such as MX1 and OASL2. For this reason, WT MEFs, but not TBK1$^{-/-}$ MEFs, were better protected from a challenge with vesicular stomatitis virus when pre-incubated with dsDNA. In addition to mounting antiviral responses,

TBK1-dependent induction of type-I-IFN response by DNA was also found to be vital to the efficacy of DNA vaccines [30]. Mice deficient for TBK1 were unable to induce increased frequency of antigen specific CD4 and CD8 T cells and IFNγ producing splenocytes, when immunized with influenza DNA vaccine.

Similar to our study, Stetson et al. showed that dsDNA from sources other than bacteria could be strongly stimulatory if delivered to the cytoplasm and that the characteristics of immune activation arising from cytoplasmic dsDNA were different from CpG DNA [43]. The authors began by demonstrating that the delivery of DNA from *Listeria monocytogenes, Legionella pneumophila* and apoptotic mammalian cells directly into the cytosol of macrophages could induce potent IFNβ and IL-6 responses. This response was independent of MyD88, TRIF and RIP2, and therefore did not involve TLR or NOD1/2 signaling. To investigate this novel pathway of DNA sensing further, they synthesized a 45 bp oligonucleotide that was without CpG motifs and named it interferon stimulatory DNA (ISD). ISD could not induce any cytokine activation in cells when added directly, but was able to trigger a strong type-I-IFN response when transfected into both immune and non-immune cell types. Unlike CpG, which was strictly stimulatory to pDCs, ISD could induce type-I-IFN across all cell types including pDC, conventional DCs, macrophages and MEFs. ISD production of IFNβ was not dependent on TLR and NOD1/2 or members of the DNA repair machinery including DNA-PK (Ku70), ATM and ATR kinase, and PI-3K. Most importantly, they showed that ISD continue to induce robust IFNβ and IL-6, despite treatment of macrophages with chloroquine, which is known to prevent endosomal acidification required for CpG DNA activation of TLR9. Therefore these studies have provided evidence that there are alternative pathways to TLR9 that detect DNA.

Identified Cytosolic DNA Sensors

After the discovery of a TLR9 independent, TBK1-IRF3 dependent pathway of DNA induced inflammation, there was a rush of publications that reported on the identification of cytosolic receptors that recognize dsDNA. It is now known that the cellular detection of DNA leading to inflammatory responses is not just plainly a dialog between TLR9 and the CpG motifs present in pathogen derived DNA. Following the discovery of dsDNA ability to induce TLR9-independent type-I-IFN response [33,34,44,45], numerous non-membrane bound cytosolic sensors of dsDNA have been identified (Figure 1.1). With the exception of AIM2 and Ku70, the induced inflammatory response common to all identified

DNA sensors thus far is the induction of type-I-IFN via the TBK1–IRF3 signaling axis. Detection of DNA by Ku70 induces type-III-IFN production, while AIM2 inflammasome activation leads to production of IL-1β and IL-18. We will be describing each of the identified cytosolic DNA sensors with a focus on the initial article that made the discovery.

Z-DNA Binding Protein-1 (ZBP-1)

ZBP-1, also known as a DNA-dependent activator of interferon regulatory factors (DAI) and DLM-1, was amongst the first of the intracellular cytosolic DNA sensors to be discovered after TLR9 [46]. Before its discovery as a cytosolic DNA sensor, ZBP-1 was recognized for its role in regulating mRNA metabolism [47] and as a tumor associated protein that is involved in host response against cellular stresses [48]. It is an IFN-inducible gene product which was initially reported to bind to Z-DNA via two binding domains. No other studies were made to study its interaction with other forms of DNA. In 2006, Takaoka and colleagues observed that the expression of ZBP-1 was strongly upregulated in L929 cells (murine fibroblast cell line) when transfected with synthetic B-DNA [49]. They showed that ZBP-1 was involved in the induction of type-I-IFN response when stimulated with B-DNA and other sources of DNA including bacteria, calf thymus and synthetic Z-DNA as indicated by upregulation of type-I-IFN production in ZBP-1 expressing L929 cells. Cells that were knockdown of ZBP-1 were unable to induce type-I-IFN secretion when stimulated with dsDNA. ZBP-1 was specific to the detection of DNA and was not involved in the type-I-IFN response induced by RNA. Due to its role in the induction of type-I-IFN upon detection of DNA, ZBP-1 was found to be involved in DNA mediated antiviral responses. ZBP-1 deficient L929 cells that were pre-stimulated with cytosolic dsDNA were found to be less resistant to encephalomyocarditis virus infection compared to WT cells. Furthermore, IFNβ expression was shown to be decreased in ZBP-1 deficient cells when infected with herpes simplex virus-1 (HSV-1) but not Newcastle disease virus (NDV), which are DNA and RNA viruses respectively. Parallel to this observation, ZBP-1 deficient L929 cells yielded higher HSV-1 viral titer than WT L929 while NDV viral titer remained the same for both WT and ZBP-1$^{-/-}$ L929 cells. Next, to demonstrate that ZBP-1 is a receptor for DNA, it was shown to directly interact with B-DNA using fluorescence resonance energy transfer (FRET) analysis and confirmed with co-immunoprecipitation assay. Microscopy experiments revealed that ZBP-1 is localized mainly in the cytoplasm.

For the signaling pathway leading to type-I-IFN response, ZBP-1 was suggested to utilize the TBK1-IRF3 signaling pathway. It was shown to induce IRF3 dimerization after B-DNA stimulation. In addition, both TBK1 and IRF3 were demonstrated to physically associate with ZBP-1 after B-DNA stimulation by co-immunoprecipitation assay. For this association between ZBP-1, IRF3 and TBK1 to occur, ZBP-1 is required to be phosphorylated at serine 352 and 353 positions. ZBP-1 was also able to mediate NF-κB activation induced by B-DNA stimulation. Detailed examinations revealed that ZBP-1 contained two receptor-interacting protein (RIP) homotypic interaction motifs (RHIMs) which recruit RHIM-containing kinases receptor protein Kinase 1 and 3 (RIP1 and RIP3) that activates NF-κB [50]. RIP1 and RIP3 are two proteins that were previously identified to be crucial for regulation of NF-κB activation by TLR3 signaling [51].

ZBP-1 Died or Dies Hard

The significance of the role of ZBP-1 in detecting and eliciting inflammatory response to DNA stimulation and in immunity against pathogens was put in doubt by subsequent reports following its discovery. Early experiments designed to study ZBP-1 and the induction of type-I-IFN by DNA were performed on ZBP-1 silenced L929 cells. However the data obtained from these experiments could not be reproduced when MEF cells were used instead [52]. Furthermore, other reports have demonstrated that ZBP-1 was not required for the induction of type-I-IFN and IL-8 by the human cell lines HEK293 and A549 cells which were transfected with dsDNA or infected with *Legionella pneumophila*. It was also proven not to be involved in the induction of type-I-IFN by *Streptococcus pneumoniae* [53], and HSV-1 [54] infection was also reported. In our laboratory, we have generated ZBP-1$^{-/-}$ in mice and studied its role in innate and adaptive immunity to dsDNA and DNA vaccines. We showed that MEFs, GM-CSF and FLT-3 ligand cultured DCs generated from ZBP-1$^{-/-}$ mice responded with normal levels of IL-6, IFNβ and other IFN-inducible genes, as WT cells, when stimulated with dsDNA. We also showed that ZBP1$^{-/-}$ mice had the same level of antibody and T cell responses as WT mice when immunized with DNA vaccine, unlike TBK1 deficient mice which were unresponsive to DNA vaccinations [30]. Despite these observations, it was reported in a separate study that ZBP-1 could be used as an adjuvant for vaccination [55]. Mice electroporated with ZBP-1 expressing plasmid had increased expression of type-I-IFN and other cytokines and co-stimulatory molecules that facilitate vaccine responses. Consequently, co-immunization of OVA and ZBP-1 expressing plasmids promotes immune responses generated against OVA. It was also able to generate potent tumor protective responses when co-immunized with plasmid expressing the tumor associated antigen Survivin. Hence, to date it is still unclear whether ZBP-1 plays an important role in DNA sensing.

Leucine-Rich Repeat (in Flightless I) Interacting Protein-1 (LRRFIP1)

Leucine rich repeats (LRR) domains are found in many functionally unrelated proteins and mediate protein–protein interactions. It is also the key motif present in the ectodomains of TLR that is responsible for the recognition of PRRs [56]. These observations point to the possibility that LRR containing proteins have functions in detecting danger signals from pathogens. In an effort to determine the LRR containing protein responsible for the induction of type-I-IFN during *Listeria monocytogenes* (an invasive DNA bacterium) infection in mouse peritoneal macrophages, Yang and colleagues used a library of siRNA that targets LRR-containing and LRR-interacting proteins [57]. Amongst the panel of siRNA used in this investigation, they found that the siRNA which targets LRRFIP1 was able to reduce the IFNβ response induced by *L. monocytogenes*. LRRFIP1 (also known as TRIP) was previously described as an HIV transactivation response RNA binding protein that interacts with the human homolog of *Drosophila* leucine-rich repeat of Flightless-I (Fli-I). Fli-I has a role in actin organization during embryogenesis and its mutation will lead to flightlessness in adult flies [58]. In humans, LRRFIP1 is localized in the cytoplasm and could bind to double-stranded but not single-stranded DNA and RNA [58,59].

Yang et al. demonstrated that in addition to *L. monocytogenes* infection, LRRFIP1 deficient macrophages could not respond with normal levels of type-I-IFN when stimulated with dsRNA poly(I:C), dsDNA poly(dG:dC) and poly(dA:dT), as well as macrophages infected with vesicular stomatitis virus (VSV) (RNA virus). For its role in detecting dsDNA, co-immunoprecipitation assay confirmed that a region in the N-terminal domain of LRRFIP1 could indeed directly interact with poly(dG:dC) and poly(dA:dT) DNA.

Surprisingly, although IFNβ response was inhibited in LRRFIP1-silenced macrophages that were infected with *L. monocytogenes*, the phosphorylation of IRF3, NF-κB, mitogen-activated protein kinase p38 and Janus Kinase (JnK) was not affected. These are common well studied transcription factors that lead to expression of proinflammatory cytokines including type-I-IFN. Instead, β-catenin was found to mediate the activation of LRRFIP1-dependent type-I-IFN responses. β-catenin is a cytosolic protein that has been linked to several biological functions including as a transcriptional co-activator in the Wnt signaling pathway that controls cell fate during embryonic development [60]. Fli-I was reported to physically associate with and regulate β-catenin dependent transcription [61].

Similar to LRRFIP1, the silencing of β-catenin in macrophages inhibited the production of IFNβ by *L. monocytogenes*. For its role in mediating type-I-IFN response induced by LRRFIP1 detection of dsDNA, β-catenin was required to physically interact with LRRFIP1 which induces phosphorylation at the Ser552 position. This process facilitates translocation of the complex into the nucleus. In the nucleus, β-catenin induced IFNβ by interacting with IRF3 and translocating to the *Ifnb1* promoter, where it recruits the histone acetyltransferase p300 and increases hyperacetylation of histones H3 and H4, which were reported to enhance IFNβ expression [62,63]. LRRFIP1 knockout mice were not available, therefore to confirm its role as a DNA sensor, the authors used β-catenin$^{-/-}$ mice instead. They concluded the study by showing that β-catenin$^{-/-}$ mice responded with lesser amounts of IFNβ in the serum when challenged with vesicular stomatitis virus (VSV) when compared to WT mice. The importance of LRRFIP1 in antiviral responses was further affirmed by a report published by a different group which showed that LRRFIP1 was recruited to endosomes containing influenza virus as well as RNA containing vesicles, and that it facilitates the induction of type-I-IFN during influenza infection [64]. The conformation changes and aggregation of LRRFIP1 upon nucleic acid binding were also reported [65]. Therefore this provided confirmation of the role of LRRFIP1 in nucleic acid sensing.

DEAD (Asp-Glu-Ala-Asp) Box Polypeptide 41 (DDX41)

The DExD/H-box helicase superfamily has had several of its members identified as cytoplasmic sensors for RNA sensing, including DDX58 [66], DDX1, DDX21 and DHX36 [67]. For this reason, Zhang et al. wanted to determine if other members of this family could function as cytosolic DNA sensors [68]. Similar to the report describing the identification of LRRFIP1, the authors used a panel of siRNA that targets each member of the DEAD box-helicase family to determine which of its member is involved in the induction of inflammatory response brought on by DNA stimulation. In this report, they used the mouse splenic myeloid DC (mDC) cell line D2SC for their screening assay. Using this methodology, the authors identified DDX41 as another member of the DExD/H-box helicase superfamily to be involved in DNA sensing. They have shown that in contrast to RIG-I and IPS-1, the silencing of DDX41 in mDC cells significantly reduced the level of IFNβ, TNF-α and IL-6 in response to poly(dA:dT) as well as poly(dG:dC). Phosphorylation of Erk1/2, p38 MAPK, TBK1, IRF3 and NF-κB was also found to be defective in these cells. DDX41 silenced mDCs had normal functions in dsRNA-dependent production of type I interferons as well as production of IL-1β by cytosolic DNA. Hence DDX41 is specific for DNA inflammatory responses and is not involved in the inflammasomes pathway of DNA sensing. In accordance with this observation, DDX41 was found to be involved in the production of proinflammatory cytokines and viral clearance during HSV-1 (DNA virus), but not influenza A (RNA virus) infection. The role of DDX41 in mediating DNA-induced inflammatory response was similarly demonstrated in mouse primary bone marrow derived dendritic cells (BMDCs) as well as human monocytes. Next, to determine if DDX41 binds directly to DNA, the authors incubated beads–linked hemagglutinin (HA)-tagged DDX41 with biotin-labeled poly(dA:dT), poly(dG:dC) or poly(I:C), and subsequently added avidin beads. DDX41 was found to bind with poly(dA:dT) and poly(dG:dC) but not poly(I:C). Using a similar approach but with plasmids expressing a truncated DDX41, it was demonstrated that the DEADc (Asp-Glu-Ala-Asp) domain of DDX41 was responsible for binding to DNA. When proteins were immunoprecipitated with anti-DDX41 antibody from resting and DNA stimulated D2SC mDCs, they was found to contain STING and TBK1. STING is an adaptor molecule that is upstream of TBK1 and IRF3 and found to have an important role in mediating type-I-IFN during viral infection [69] and stimulation with cytosolic dsDNA [70]. The association between DDX41 and STING/TBK1 suggests that it could utilize the STING-TBK-IRF3 signaling axis in mediating the type-I-IFN response during DNA sensing. Therefore the relationship between DDX41 and STING was also investigated. As expected, over-expression of DDX41 and STING in L929 cells had a synergistic effect in eliciting IFNβ response. Stimulation with DNA induces the migration of DDX41 and STING from the ER and mitochondria and co-localization into early and late endosomes. In addition STING was also shown to physically interact with DDX41 in the region between the second to fourth transmembrane domains of STING. Following this first report, two other significant publications on the role of DDX41 in DNA sensing were published. The bacterial second messenger cyclic-di-GMP (c–di–GMP) was reported to have adjuvant property which is dependent on type-I-IFN responses induced via the STING–TBK1–IRF3 signaling axis. Parvatiyar et al. showed that c–di–GMP was similar to dsDNA and was a ligand of DDX41 for the induction of type-I-IFN. However, similar to the previous study, the demonstration of the role of DDX41 as a receptor was only on knockdown cells and no data on knockout mice were available. In another study, TRIM21 was found to negatively regulate the DNA mediated type-I-IFN response induced by DDX41. It was discovered during a screen for DDX41 associating proteins in D2SC cells after DNA stimulation. When studied for its role in DNA sensing, TRIM21 displayed results which suggested opposing functions to DDX41. BMDCs derived from TRIM21$^{-/-}$ mice had higher expression of DDX41 and responded with higher IFNβ when stimulated with poly(dA:dT) and DNA viruses including vaccinia virus, adenovirus and HSV-1. These results demonstrate that DDX41 is a cytoplasmic sensor for detecting DNA and mediating type-I-IFN response; however they have not been repeated by other groups.

RNA Polymerase III

Similar to DNA, RNA is able to induce inflammation and does this via different sensors. Several members of the TLR family are known to recognize RNA. TLR7 and TLR8 detect single stranded RNA and TLR3 detects double stranded RNA. Non-membrane bound cytosolic RNA sensors have also been identified and they are categorized into the family of RIG-I-Like receptors (RLR) [71]. The members include retinoic acid-inducible gene 1 (RIG-I) [66], Melanoma Differentiation-Associated protein 5 (MDA5) [72] and LGP2 [73]. RIG-I, which is encoded by DDX58, detects the short blunt ended dsRNA as well as the 5′-triphosphate moiety of transcripts from negative-sense single stranded RNA (ssRNA) viruses. MDA5, encoded by Ifih1, recognizes and responds to long dsRNA, which is most often associated with the replication of positive-sense ssRNA viruses.

Although not directly shown to interact, DNA was found to be able to induce inflammatory responses through RIG-I mediated signaling. Two independent reports which surfaced at the same time using different approaches came to the same conclusion that dsDNA could be converted into double stranded RNA through the actions of RNA polymerase III before being detected by RIG-I to induce type-I-IFN response [74,75]. Both reports showed that transfection of B-DNA into human embryonic kidney cell lines (HEK293T) that are RIG-I and IPS-1 silenced had inhibited expression of type-I-IFN and also reduced activation of IRF3 and NF-κB phosphorylation. IPS-1, also known as MAVS, is an adaptor molecule through which RIG-I and MDA5 induce activation of IRF3 and NF-κB. At first sight, this observation seemed contradictory to the previous reports which showed that B-DNA induction of type-I-IFN from mouse antigen presenting cells and MEF cells is independent of both RIG-I and IPS-I [34,57,76]. However, further investigation revealed that RIG-I-dependent dsDNA induction of type-I-IFN could be observed only in human but not mouse cells. RIG-I and IPS-1 deficient mouse DCs had uninhibited type-I-IFN while human monocyte-derived DCs had reduced production of TNF-α and IFNα when stimulated with B-DNA [75]. This effect was most prominent in AT-rich DNA and it was dependent on the size of the DNA, where longer dsDNA induced stronger RIG-I-dependent IFNβ response. Several clues suggested that an intermediary product was responsible for the activation of RIG-I by dsDNA. Firstly, the rate at which poly(dA:dT) induces RIG-I-dependent poly(dA:dT) induced type-I-IFN in human peripheral blood mononuclear cells (PBMCs) was slower than 5′-triphosphate RNA (the native ligand of RIG-I). Secondly, nucleic acid purified from HEK293T cells stimulated with poly(dA:dT) but not from untransfected cells could induce a RIG-I-dependent IFNβ response in HEK293T cells. Lastly, the treatment of this stimulatory nucleic acid from poly(dA:dT)-stimulated HEK293T cells was inactivated by RNaseA but not DNaseI, which also suggested that the intermediary-stimulatory product is RNA-based. It is important to note that nucleic acid from poly(dA:dT)-stimulated mouse MEF cells was also able to induce RIG-I dependent type-I-IFN response when transfected into HEK293T cells, therefore indicating that mouse cells are able to generate this stimulatory RNA intermediary product, but that the RIG-I dependent pathway of detecting B-DNA is redundant in the mouse.

By treating nucleic acid purified from poly(dA:dT)-stimulated HEK293T cells with RNA alkaline phosphatase, RNA 5′-polyphosphatase, RNase III and RNase T1, Ablasser et al. came to the conclusion that the intermediary RNA product responsible for the RIG-I dependent sensing of dsDNA in human cells was a double-stranded, 5′-triphosphate RNA molecule that does not include guanosine [75]. Therefore this suggests that poly(dA:dT) acts as a template for the generation of the intermediary RNA product by a DNA-dependent RNA polymerase. RNA polymerase III which had previously been used to investigate promoter-less transcription of poly(dA:dT) [77,78] was demonstrated to be the responsible polymerase. Chiu et al. have also identified RNA polymerase III to be involved in the production of the stimulatory 5′-triphosphate RNA molecule from poly(dA:dT) by fractionating the nucleic acid purified from poly(dA:dT)-stimulated HEK293T cells and testing each fraction for its ability to stimulate type-I-IFN response. The active fraction was found to have several subunits of the RNA polymerase III. Indeed, by incubating poly(dA:dT) with RNA polymerase III in the presence of ATP and UTP, it was possible for IFNβ inducing RNA to be synthesized. With the discovery of an RIG-I dependent DNA sensing mechanism, we could now account for the RIG-I dependent induction of type-I-IFN by several DNA viruses including adenovirus [79], HSV-1 [80] and Epstein–Barr virus (EBV) [81].

PYHIN Family: AIM2 and IFI16
AIM2
Inflammasomes are multi-component protein complexes which play an important role in innate immunity. They induce the production of IL-1β and IL-18 during bacterial and viral infection. Inflammasome activation requires two signals in sequence. The first signal can be triggered by an agent that induces the production of pro-IL-1β and pro-IL-18, which is usually a TLR agonist. The second signal (such as Alum) induces the formation of the inflammasomes which activate caspase-I, a cysteine protease that cleaves and activates the proinflammatory cytokines to active forms. Inflammasomes are typically made up of an immature caspase-1, a nucleotide oligomerization domain-like receptor (NLR) protein and an adaptor protein called apoptosis-associated speck-like protein (ASC). Detection of danger signals by NLR will induce its interaction with the ASC via their PYRIN domains and this will lead to recruitment of pro-caspase-I via the caspase recruitment domain (CARD) of ASC and ultimately trigger the formation of active caspase-I [82]. Muruve et al. discovered that in addition to triggering type-I-IFN responses, DNA was also able to activate the inflammasomes to induce IL-1β and IL-18 [83]. However, the report stopped short of identifying a receptor mediating this response. Soon after this, two articles on the discovery of AIM2, which is a member of the PYHIN family of proteins, as the receptor of DNA leading to activation of inflammasomes were published [84].

Fernandes-Alnemri et al. hypothesized that the receptor responsible for this phenomenon should have a DNA binding domain as well as a pyrin-domain so that it could sense DNA and induce recruitment and activation of inflammasomes [84]. Therefore they began their investigation by searching through the NCBI database for a protein that fits these criteria. Bürckstümmer et al. took the systematic approach of using immobilized immune-stimulatory DNA (ISD) to capture DNA binding entities present in cell extracts and determined their identity by liquid chromatography–tandem mass spectrometry (LC-MSMS) [85]. At the same time, they screened for genes that were upregulated by IFNβ and cross-referenced these data with the list of DNA-binding proteins that were positively identified by LC-MSMS. Using these different approaches, both reports narrowed their search for a receptor mediating DNA induced inflammasome activation to IFI204, IFI205 and AIM2. These proteins are members of the IFN-inducible HIN-200 (PYHIN) family. Members from this family of proteins are typically made up of a pyrin domain and a HIN200 domain which binds to DNA. IFI16 and MNDA are human equivalents of IFI204 and IFI205 respectively. Of these three candidates, AIM2 seemed most likely to be the DNA sensor for inflammasome activation as it is found to reside mainly in the cytoplasm while IFI16 and MNDA are located in the nucleus. Furthermore, AIM2 was the only one capable of activating caspase-1 when transfected into a cell line expressing pro-caspase-1 and ASC. To confirm its role, AIM2 was shown to be able to directly associate with DNA, with a preference for dsDNA over ssDNA, in a pull-down assay as well as in an electrophoretic mobility shift assay. Silencing AIM2 resulted in inhibition of caspase-1 activation and IL-1β secretion from a poly(dA:dT) stimulated human monocytic leukemia cell line (THP1). Using the same cells, it was demonstrated that AIM2 induced activation of caspase-1 was dependent on the pyrin–pyrin domain interaction between AIM2 and ASC. In contrast, no interactions between ASC and IFI16 or MNDA were noticed. When THP-1 cells were stimulated with the signal 1 of inflammasome activation by phorbol 12-myristate 13-acetate (PMA), it was AIM2 but not IFI16 or MNDA that was strongly upregulated. By transfecting HEK293 cells that were unresponsive to inflammasome activation with a mixture of AIM2, ASC, caspase-1 and IL-1β expressing plasmids, Bürckstümmer et al. were able to show that IL-1β secretion could be reconstituted. AIM2 was also demonstrated to directly associate with DNA and induced the aggregation of ASC molecules which is an indication of pyroptosome formation. The role of AIM2 in DNA sensing and the activation of inflammasomes were confirmed by another group which generated AIM2$^{-/-}$ mice and proved that this molecule was necessary for the mounting of inflammasome-dependent host defense against *Fransicella tularensis*, CMV and vaccinia virus infection [86]. Therefore, AIM2 is a cytosolic DNA sensor which activates the inflammasomes to induce production of proinflammatory cytokines IL-1β and IL-18.

AIM2 Regulation by p202

p202 is a member of the PYHIN family, which IFI16 and AIM2 belong to. As opposed to other members of this family of proteins, p202 does not have a pyrin domain which recruits and activates inflammasomes, but it does possess a HIN200 domain. Roberts et al. found that p202 could bind to synthetic DNA *in-vitro* via the HIN domain and co-localize with internalized calf thymus DNA in the cytoplasm [87]. As opposed to AIM2, the deletion of p202 in primary macrophages activated higher expression levels of caspase I and III which were responsible for DNA induced cell death. As p202 had been shown to be able to form heterodimers with AIM2, the authors proposed that p202 could function to compete with AIM2 for the binding of DNA, but because it does not possess the pyrin domain there would be no inflammasome activation. Although the authors did not give direct evidence for this hypothesis, they presented results on the expression of p202 in three strains of mice, C57BL/6, BALB/C and NZB mice, and showed that it was inversely proportional to the level of AIM2-inflammasome activation in macrophages. C57BL/6 macrophages expressed no p202 and showed highest activation of caspase initiated by AIM2. Conversely, NZB macrophages expressed the highest level of p202 and showed the lowest caspase activity. NZB mice have been used as a mouse model for studying SLE; these results therefore imply that high expression levels of p202 could be one of the factors involved in increasing circulating DNA and in raising anti-DNA antibodies.

IFI16

As previously described, IFI16 was one of the PYHIN proteins identified during the screening for a receptor that is interferon inducible and contains DNA binding and PYRIN domains. Its role in DNA sensing was not further investigated in those studies as it was shown not to be involved in DNA-induced inflammasome activation; it was mainly localized in the nucleus but not the cytoplasm [84,85]. However, Unterholzner et al. reported later that IFI16 is an important DNA sensor which mediates the type-I-IFN response [88]. In their study, they used a type-I-IFN inducing 70 base pairs dsDNA from vaccinia virus (dsVACV 70mer) to screen for DNA sensors in THP-1 cells. This sequence was found to be conserved and repeated multiple times in the inverted terminal repeat region of various poxviruses. Surprisingly, this 70 bp VACV DNA induced a type-I-IFN response independent of the involvement of TLRs and other known cytosolic DNA sensors at that time; however it was dependent on TBK1 and IRF3. To determine the receptor responsible for mediating dsVACV 70mer induction of type-I-IFN, proteins isolated from pulldown assays on biotinylated dsVACV 70mer with cytosolic extracts of THP-1 cells were subjected to mass spectroscopy analysis. IFI16 was found to be one of the interacting proteins. Studies were subsequently performed to confirm whether IFI16 could indeed sense dsVACV 70mer. Firstly, it was demonstrated to be able to bind to dsDNA via its DNA binding HIN domain that is common to all the PYHIN family of proteins. Next, they showed that the knocking down of IFI16 in THP1 cells could inhibit the production of IFNβ in cells stimulated with dsVACV 70mer. In addition, silencing of IFI16 in monocytes infected with HSV-1 (DNA virus) led to significant inhibition of Cxcl10, IL-6 and TNF-α upregulation due to reduction of IRF3 and NF-κB activation. Although it was reported to exist mainly in the nucleus in previous studies, Unterholzner et al. showed that there was still detectable IFI16 present in the cytoplasm. The optimal production of type-I-IFN induced by IFI16 detection of dsVACV 70mer was found to require another cytoplasmic protein, STING. The cytoplasmic IFI16 was shown to recruit STING during stimulation with dsVACV 70mer-mediated. ASC, which was previously shown to be necessary for the activation of the inflammasomes by DNA via AIM2 inflammasomes, was not involved in the IFI16 mediated DNA sensing response. This was however disputed by a subsequent study which showed that IFI16 sensed Kaposi sarcoma-associated herpes virus (KSHV) in the nucleus and formed an inflammasome complex comprising of IFI16/ASC/caspase-1 [89]. To date, although there are no data presented on the IFI16 knockout mouse model, there are several reports on the role of IFI16 in antiviral immunity and most are on immunity against herpesvirus [90–92]. Hence, IFI16 is another member of the PYHIN family which could detect cytosolic DNA; however it induces type-I-IFN response instead of activating the inflammasomes as in the case of AIM2.

DHX36 and DHX9

Plasmacytoid DCs (pDCs) are the major cell type for producing type-I-IFN in response to viral infections. pDCs sense viral RNA and CpG DNA through their TLR7 and TLR9. It was initially thought that pDCs sense viral CpG DNA through TLR9 exclusively, as it is selectively expressed in pDCs and not mDCs. However, a few studies had shown that pDCs were able to mount type-I-IFN response against viral DNA even in the absence of TLR9, although MyD88 was still required [44,93]. This suggested that other CpG DNA receptors are present which utilize MyD88 as the adaptor molecule for downstream signaling present in pDCs. To determine the identity of these CpG DNA receptors in pDCs, Kim et al. generated biotinylated CpG-A (ODN 2216) and CpG-B (ODN 2006) and incubated them with primary human pDCs as well as pDC and B cell lines [94]. Intracellular proteins that have interacted with the biotinylated CpGs were isolated with pulldown assays and their identity was determined by liquid chromatography–mass spectrometry (LC-MS). Two members of the DExD/H box family were identified using this methodology. DHX36 was found to bind with CpG-A while DHX9 was associated with CpG-B. As other closely related members from this family of proteins (RIG-I, MDA5 and LGP2) were found to have functions in nucleic acid sensing and antiviral functions, the authors hypothesized that they may be the unidentified DNA receptors of pDCs. Indeed, the silencing of DHX36 and DHX9 in the pDC cell line Gen2.2 resulted in a decrease in IFNα and TNF-α responses to HSV-1 infection. This reduction of inflammatory response in DHX36 and DHX9 deficient Gen2.2 cells was not observed in influenza A virus challenge, therefore suggesting its specificity to DNA sensing. Inflammatory responses induced by DHX36 and DHX9 were also specific to CpG-A and CpG-B respectively. CpG-A is known to induce high IFN response via IRF7 activation while CpG-B induced lower IFN but higher TNF-α and IL-6 responses through NF-κB activation. Similar to TLR9, both DHX36 and DHX9 were shown to require MyD88 for inducing inflammatory cytokines during CpG DNA stimulation and they were also shown to physically interact with MyD88. However, unlike TLR9, DHX36 and DHX9 were found to localize in the cytoplasm instead of the endosomes. So far, these findings are not verified with studies performed on knockout mice and there is no further confirmation of DHX36 and DHX9 roles in CpG detection by other groups. However, DHX36 and DHX9 were also discovered to cooperate with DDX1/DDX21 and IPS-1 (MAVs) respectively to sense dsRNA and induce type-I-IFN in DCs [67,95].

Ku70

Ku70 is an evolutionarily conserved protein that has functions in DNA repair and maintenance [96]. It is a heterodimeric protein made up of Ku70 and Ku80 [97]. We (unpublished data) and others [43] had evaluated and shown that proteins involved in DNA repair including DNA-PK, p53, ATM and Ku70 had no effect on the type-I-IFN response induced by cytosolic dsDNA stimulation. However Zhang et al. published a short report on Ku being able to function as a cytosolic DNA sensor which induces type-III but not type-I-IFN responses via IRF1 and IRF7 signaling [98]. Type-III-IFN is a recently discovered interferon cytokine which is similar to type-I-IFN in inducing phosphorylation of STAT1, STAT2 and STAT3 and also triggers antiviral responses but binds to a different receptor from type-I-IFN [99]. In their system, Zhang et al. showed that transfection of plasmid DNA into HEK293 cells was able to upregulate secretion of IFNλ1 (a type III IFN) and relatively lower levels of IFNβ. This induction of IFNλ1 was not observed when HEK293 cells were stimulated with CpG or LPS but was limited to transfected plasmid DNA. DNA-induced upregulation of IFNλ1 was observed in HEK293 cells despite the lack of TLR7, DAI, AIM2, LRRFIP1 expression and it was also noted in RNA polymerase III-silenced cells. This suggests the presence of another DNA sensor which induces type-III instead of type-I-IFN. Both ssDNA and dsDNA were able to induce IFNλ1 responses and there is a preference for longer linearized DNA. To identify the responsible receptor, immobilized DNA beads were incubated with cytosolic extracts of HEK293 cells in the presence/absence of competing unimmo- bilized DNA, and the proteins isolated were subjected to mass spectroscopy analysis. Ku70 and Ku80 were revealed to be the two DNA-interacting pro- teins. Silencing and over-expression experiments performed on HEK293 cells confirm that Ku70 is indeed vital in the induction of IFNλ1 expression by DNA stimulation. Furthermore, IFNλ1 mRNA transcripts were found to be upregulated in splenocytes derived from WT but not in Ku70$^{-/-}$ mice. In contrast, Ku80 was demonstrated to be not involved. However, the authors did not elaborate on why Ku70 but not Ku80 was required for DNA induced IFNλ1 induction. Finally, the authors demonstrate that during DNA stimulation IRF1 and IRF7 bind to the PRDI and IRSE element of the IFNλ1 promoter region to induce the expression of IFNλ1. So far, there are no further investigations reported on the role of Ku70 as a DNA sensor, nor has it been reproduced by other independent groups. Its role in mediating anti-viral responses was not investigated either. Hence, it is unsure whether Ku70 is indeed a sensor for DNA that induces inflammatory responses.

HMGB Proteins

The High-Mobility Group Box (HMGB) family of proteins are phylo- genetically conserved and can be found across different species including plants [100]. They are abundantly and ubiquitously expressed in the body and are present in the nucleus, cytosol and extracellular environment. The primary function of this family of proteins is to maintain the health of the chromosomes and support several activities in the nucleus including DNA repair, transcription, recombination and replication. For these multiple roles in the nucleus, HMGB proteins possess a DNA-binding domain in their structure. HMGB proteins were previously indicated to have immunological functions as well [101]. They have been linked to the induction of inflammation in several autoimmune diseases including SLE [102] and arthritis [103]. A member of this family of proteins, HMGB1, was found to directly interact with members of the TLR family and influence the immune response directed to their respective ligand [104,105].

Yanai et al. [113] discovered that HMGB proteins are involved in the inflammatory response induced by cytosolic DNA sensing as well. In their investigation, HMGB proteins including HMGB1, 2 and 3 were found to be the most prominent of proteins recovered during a screen for poly(dA:dT) binding cellular components. They further demonstrate by blocking studies that HMGB1 and 3 can bind to both RNA and DNA while HMGB2 binds to DNA predominantly. By using MEFs that were deficient for HMGB expression, it was confirmed that HMGB1 was indeed required for the optimal production of inflammatory cytokines and chemokines including IFNβ, IL-6 and RANTES during stimulation with poly(dA:dT) and poly (I:C) but not LPS. HMGB2 was required for only poly(dA:dT) induced inflammation. Silencing of HMGB1, 2 and 3 simultaneously in MEFs led to an even more profound reduction in the inflammatory response induced by poly(dA:dT) and poly(I:C). In addition to cytosolic DNA induced type-I-IFN responses, HMGB proteins were also involved in the activation of inflammasomes and activation of IL-1β by cytosolic DNA. At the signaling level, the silencing on HMGB proteins resulted in the suppression of IRF3, NF-κB and ERK activation. As a result of the dependence on HMGB protein for the optimal production of inflammatory responses induced by RNA and DNA, pan-HMGB silenced MEFs had weaker resistance against vesicular stomatitis virus (VSV) and HSV-1 infection. Yanai et al. [113] further investigated the relationship between TLR9, TLR7 and HMGB proteins in the context of nucleic acid sensing. They demonstrated that conventional and plasmacytoid DCs generated from HGMB1$^{-/-}$ cells were hyporesponsive to poly(I:C), poly(U) and CpG DNA, which are ligands of TLR3, TLR7 and TLR9 respectively. As the results suggested that HMGB could detect both DNA and RNA and that it is required for optimal nucleic acid mediated TLR activation, the authors proposed HMGB to be a remnant of an evolutionarily older form of nucleic acid sensor that is promiscuous and supports the newer more discriminative TLRs.

Histone H2B

The process leading to antigen specific autoimmunity requires priming conditions that are contributed by the innate immune system. These conditions can arise during an infection or mechanical injury which triggers cell damage or death. In the clinical setting, diabetes induced by Coxsackie virus was reported to be triggered by the viral induced upregulation of immune factors such as cytokines, chemokines and MHC class II expression that may facilitate presentation of self-antigens. The induction of DNA damage to cells in *in-vitro* systems by UV radiation and treatment with DNA damaging agents such as doxorubicin have also been reported to induce type-I-IFN responses through IRF3 activation [106].

For these reasons, Kawashima et al. investigated if the induction of thyroid cell injury could similarly induce inflammation and cause dysfunctions in their role in maintaining endocrine homeostasis, which is observed in patients suffering from Graves' disease and Hashimoto's thyroiditis [107]. The authors first demonstrated that cellular stress on FRTL-5 thyroid cells induced by electric pulsing caused genomic DNA (gDNA) to be released into the cytoplasm as analyzed by agarose gel electrophoresis on the cytosol fractions. This amount was directly proportional to the magnitude of the voltage used. The release of gDNA into the cytoplasm was capable of upregulating cytokines (including IFNβ), chemokines and MHC class II molecules and it was to a level comparable to liposomal-introduction of poly(dA:dT). This implies that the introduction of foreign DNA could mimic the process of DNA damage in cells. In addition to the upregulation of immune factors, the introduction of dsDNA but not ssDNA into FRTL-5 cells impeded the uptake of radioactive iodine which indicates impaired endocrine function. This was found to be mediated by the reduction in thyroid specific genes such as Sodium Iodide Symporter (NIS) and Thyroglobulin (Tg). Lastly, the authors applied FRTL-5 cell lysates to a dsDNA affinity column and used ESI-MS/MS analysis on purified proteins to identify histone H2B as the protein that physically associates with DNA. Its role in mediating DNA induced inflammation was confirmed as histone H2B knockdown FRTL-5 cells had impaired IFNβ response when stimulated with transfected-exogenous DNA and during electro-induced cellular damage.

However as it was not a focus of this study, H2B histone was not investigated for its role in detecting viral and bacterial DNA and whether it contributes to the type-I-IFN response against infection.

MRE11

The induction of type-I-IFN during DNA damage was thought to be a defensive response against genotoxic stress. For this reason, the mechanism through which type-I-IFN is induced was speculated to involve cellular proteins that function in DNA repair such as Ku70, DNA-dependent protein kinase (DNA-PK), p53 and ataxia-telangiectasia mutated kinase (ATM). However, MEFs derived from mice deficient in the protein responded with similar levels of IFNβ response to WT MEFs, therefore confirming that these proteins were not involved in the process [43]. As such, the mechanism through which IRF3 is activated during DNA damage is still unclear.

Our group has recently identified a novel cytosolic DNA sensor, meiotic recombination homologue A (MRE11), which responds to dsDNA with type-I-IFN response and gave indications that it specifically detects DNA during cellular damage rather than DNA introduced by viruses [108]. Our investigation began with the discovery of ATM upregulation in MEFs stimulated with immune-stimulatory DNA (ISD), which was not observed in poly(I:C) stimulated cells. As ATM does not induce type-I-IFN response to DNA stimulation, this led us to investigate if components upstream of ATM activation could be involved. The MRE11-Rad50 homolog–Nijmegen breakage syndrome 1 (NBS1) complex (also known as MRN complex) functions upstream of ATM and facilitates its phosphorylation. To our surprise, the incubation of GM-DC with MIRIN, an inhibitor of MRE11, inhibited the activation of IFNβ when stimulated with cytosolic ISD but not poly(I:C). Supporting these data, we showed that the knockdown MRE11 as well as its binding protein Rad50 in GM-DC led to significant inhibition of IFNβ upregulation during ISD stimulation. However, MRE11-silenced GM-DC did not show reduction in IFNβ response when infected with HSV-1 and *Listeria monocytogenes*, which are pathogens known to induce potent type-I-IFN response through the introduction of their DNA into host cells. This observation together with its role in mediating ATM phosphorylation indicates that MRE11 has functions in mediating type-I-IFN response to DNA damage rather than during infection with DNA bearing pathogens. The ability of MRE11 to induce type-I-IFN in response to cytosolic DNA was also demonstrated in human cells. A patient with non-functional MRE11 could not respond with normal levels of IFNβ when stimulated with ISD as compared to cells reconstituted with WT normal MRE11. Further investigations revealed that MRE11 induces type-I-IFN via the STING–TBK1–IRF3 signaling axis and that MRE11 is present in the cytoplasm and nuclei. When ISD is transfected into cells, MRE11 co-localized with ISD as well as STING in cytoplasmic punctate structures. But in the absence of MRE11, STING failed to migrate away from the endoplasmic reticulum where it resides during rest, therefore this indicates that MRE11 is needed for STING activation as well as showing that MRE11 is upstream of STING. Mutation studies on MRE11 revealed that its N-terminal domain which functions in DNA cleavage was not required for the induction of type-I-IFN to DNA stimulation and that its two predicted DNA binding domains are important to the optimal induction of IFNβ upregulation. We also discovered that NBS1 was not required for the DNA sensing function of MRE11 as the mutant with deletion of a region which enables binding of MRE11 to NBS1 did not affect the type-I-IFN response induced by MRE11. Finally, we showed that DNA damage induced by cisplatin and etoposide induced the migration of STING, but this was impaired in the absence of MRE11 as represented by cells derived from the patient with mutated MRE11. This defect could however be rectified when normal WT MRE11 was reconstituted via viral transduction in the defective MRE11 expressing cells.

cGAMP as Second Messenger of the DNA Sensing and cGAS as a Cytosolic DNA Sensor

STING is an ER-bound adaptor molecule found to be essential to several of the described cytosolic DNA sensors in mediating type-I-IFN response via the TBK1–IRF3 signaling axis, but does not bind directly to DNA [69,70]. The bacterial second messenger cyclic dinucleotide c-di-GMP, which induces strong type-I-IFN response, was found to bind directly to STING instead of other DNA binding cytosolic DNA sensors to mediate type-I-IFN secretion [109]. Similarly, Wu et al. discovered a metabolite of DNA processing in eukaryotic cells, named cyclic-GMP-AMP (cGAMP), which was able to target STING directly to induce type-I-IFN [110].

To identify this STING-targeting component, the authors began by preparing lysates from STING-silenced cells transfected with dsDNA so as to obtain a source of factors which may include the component of interest. The lysate was subsequently incubated with perfringolysin-O permabilized WT cells and induced IRF3 activation. Surprisingly, this lysate continued to activate IRF3 activation in WT cells even when subjected to DNA and RNA removal by Benzonase and protein degradation by proteinease K and 95 degree heat treatment. This indicates that the STING targeting factor present in the lysate is not DNA, RNA or protein. In fact, this heat resistant factor that induces IRF3 dimerization in WT cells could be synthesized *in-vitro* by incubating DNA with STING-deficient cell lysate in the presence of ATP. This was however not possible when RNA was used instead. The identity of the STING-targeting protein was revealed by chromatography methodologies on DNA stimulated STING-silenced cell lysate. cGAMP, which is a hybrid of c-di-GMP and c-di-AMP, was found to be the heat resistant STING-targeting protein. Prior to this report, cGAMP was not studied in cells of higher orders and was only reported to function in *Vibrio cholerae* [111]. Subsequent studies performed on cGAMP proved that it could indeed induce the production of type-I-IFN when introduced into WT cells using digitonin and that it is also synthesized in the cell when transfected with DNA or infected with DNA viruses. The authors ended with a series of experiments to show that cGAMP could indeed bind with STING and that its function in mediating type-I-IFN response was dependent on STING.

The same group of researchers led by Zhijian James Chen published a separate article in the same issue of the journal, which reported on the identification of the cellular protein that was responsible for the synthesis of cyclic GMP-AMP; they have named it cyclic GMP-AMP synthetase (cGAS) [112]. cGAS was a previously uncharacterized protein which contains a nucleotidyl transferase (NTase) fold protein motif and a male abnormal 21 (Mab21) domain. These domains in cGAS are conserved across species. The role of cGAS was confirmed as the ability of various cell types to synthesize cGAMP and induce IRF3 mediated type-I-IFN was found to be directly dependent on the expression level of cGAS. It was also confirmed to be able to produce the heat resistant cGAMP product intracellularly when introduced into perfringolysine-O permeablized cells as well as *in-vitro* when cGAS was mixed with adenosine triphosphate (ATP) and guanosine triphosphate (GTP) in the presence of DNA. Similar to cGAMP stimulation, cGAS induction of type-I-IFN was dependent on STING and it was determined to function upstream of STING by using cGAS and STING knockdown cells. cGAS was also determined to be specific to DNA stimulation but not RNA. This was demonstrated when L929 cells in the absence of cGAS were found to be susceptible to HSV-1 infection but not Sendai virus infection. The authors demonstrated that cGAS is a DNA sensor by showing that cGAS activity/expression was upregulated when stimulated with mammalian derived dsDNA, and GST-fused cGAS was able to be precipitated by biotinylated ISD. Finally, they showed that cGAS is present

Table 1.1 List of currently known cellular DNA sensors

DNA Sensor	DNA Sensing Location	Activation/ Regulation	Demonstration in Anti-pathogen or DNA Damage Response	Major Immune Response	Demonstration in Knockout Mouse Model	Adaptor Molecule	Ref
TLR9	Endosome	Activation	Anti-pathogen	Type-I-IFN	Yes (whole animal)	MyD88	[26]
IFI16	Nucleus, Cytoplasm	Activation	Anti-pathogen	Type-I-IFN, and Inflammasome	No	STING, ASC	[88]
H2B	Nucleus, Cytoplasm?	Activation	DNA damage	Type-I-IFN	No	?	[107]
AIM2	Cytoplasm	Activation	Anti-pathogen	Inflammasomes	Yes (whole animal)	ASC	[84–86]
MRE11	Cytoplasm	Activation	DNA damage	Type-I-IFN	No	STING	[108]
Ku70	Nucleus?	Activation	Anti-pathogen	Type-III-IFN	Yes (in-vitro demonstration on knockout cells)	?	[98]
RNA polymerase III	Cytoplasm	Activation	Anti-pathogen	Type-I-IFN	No	RIG-I	[74,75]
DDX41	Cytoplasm	Activation	Anti-pathogen	Type-I-IFN	No	STING	[68]
DHX36 and DHX9	Cytoplasm	Activation	Anti-pathogen	Type-I-IFN	No	MyD88	[94]
cGAS	Cytoplasm	Activation	Anti-pathogen	Type-I-IFN	No	STING	[112]
ZBP1	Cytoplasm	Activation	Anti-pathogen	Type-I-IFN	Yes (whole animal; but inconclusive)	?	[46]
HMGB (1,2 and 3)	Cytoplasm	Activation	Anti-pathogen	Type-I-IFN and Inflammasome	Yes (in-vitro demonstration on knockout cells)	STING	[113]
LRRFIP1	Cytoplasm	Activation	Anti-pathogen	Type-I-IFN	No. Only on β-catenin$^{-/-}$ mice	β-catenin	[57]
p202	Cytoplasm	Regulation	–	Inhibits caspase activity – IL-1β and IL-18	No	–	[87]

in the cytoplasm and forms punctate structures that co-localized with the transfected ISD. Hence, cGAS is identified as another member of the cytoplasmic DNA sensor. In summary, the authors have made a great discovery in determining that mammalian cells are capable of producing cyclic dinucleotides, namely cGAMP, as a second messenger and that this product is synthesized by the cyclase cGMP upon the detection of cytosolic DNA. These findings have added another dimension to our understanding of inflammation induced by DNA; they have shown that metabolites derived from the cellular processes on the introduced DNA could similarly be inflammatory and target the same DNA sensing signaling pathway.

CONCLUSION

In this chapter, we have presented the highlights during our journey in uncovering the immunogenic properties of DNA. A summary of the currently known DNA sensors can be found in Table 1.1. As we will discover from the subsequent chapters of this book, cellular DNA sensing is a complicated process and many of the components involved are evolutionarily conserved and can be found across species. We are beginning to realize the impact of DNA sensing on our biology. It plays a major role in the innate immune system against pathogens, but at the same time it is involved in the pathogenesis of numerous other medical conditions. Although much has been uncovered, the exact mechanism of how DNA is detected and the signaling pathways it triggers to induce the innate immune system are still not fully understood. From this book, we hope that the collection of up-to-date findings from the respective investigators who help define this genre of research can allow us to consolidate what is known about DNA sensing and facilitate future studies.

ACKNOWLEDGMENT

This work was supported by a Health and Labour Sciences Research Grant 'Adjuvant Database Project' of the Japanese Ministry of Health, Labour and Welfare.

REFERENCES

[1] Chen GY, Nunez G. Sterile inflammation: sensing and reacting to damage. Nature Rev Immunol 2010;10(12):826–37.
[2] Isaacs A, Cox RA, Rotem Z. Foreign nucleic acids as the stimulus to make interferon. Lancet 1963;2(7299):113–6.
[3] Isaacs A. The role of interferon. Endeavour 1963;22:96–8.

[4] Rotem Z, Cox RA, Isaacs A. Inhibition of virus multiplication by foreign nucleic acid. Nature 1963;197:564–6.

[5] Alexandroff AB, Nicholson S, Patel PM, Jackson AM. Recent advances in bacillus Calmette-Guerin immunotherapy in bladder cancer. Immunotherapy 2010;2(4):551–60.

[6] Tokunaga T, Yamamoto H, Shimada S, Abe H, Fukuda T, Fujisawa Y, et al. Antitumor activity of deoxyribonucleic acid fraction from *Mycobacterium bovis* BCG. I. Isolation, physicochemical characterization, and antitumor activity. J Nat Cancer Inst 1984;72(4):955–62.

[7] Yamamoto S, Yamamoto T, Kataoka T, Kuramoto E, Yano O, Tokunaga T. Unique palindromic sequences in synthetic oligonucleotides are required to induce IFN [correction of INF] and augment IFN-mediated [correction of INF] natural killer activity. J Immunol 1992;148(12):4072–6.

[8] Messina JP, Gilkeson GS, Pisetsky DS. Stimulation of in vitro murine lymphocyte proliferation by bacterial DNA. J Immunol 1991;147(6):1759–64.

[9] Gilkeson GS, Grudier JP, Pisetsky DS. The antibody response of normal mice to immunization with single-stranded DNA of various species origin. Clin Immunol Immunopathol 1989;51(3):362–71.

[10] Gilkeson GS, Grudier JP, Karounos DG, Pisetsky DS. Induction of anti-double stranded DNA antibodies in normal mice by immunization with bacterial DNA. J Immunol 1989;142(5):1482–6.

[11] Hashimoto C, Hudson KL, Anderson KV. The Toll gene of *Drosophila*, required for dorsal-ventral embryonic polarity, appears to encode a transmembrane protein. Cell 1988;52(2):269–79.

[12] Lemaitre B, Nicolas E, Michaut L, Reichhart JM, Hoffmann JA. The dorsoventral regulatory gene cassette spatzle/Toll/cactus controls the potent antifungal response in *Drosophila* adults. Cell 1996;86(6):973–83.

[13] Medzhitov R, Preston-Hurlburt P, Janeway Jr. CA. A human homologue of the *Drosophila* Toll protein signals activation of adaptive immunity. Nature 1997;388(6640):394–7.

[14] Rock FL, Hardiman G, Timans JC, Kastelein RA, Bazan JF. A family of human receptors structurally related to *Drosophila* Toll. Proc Natl Acad Sci USA 1998;95(2):588–93.

[15] Gay NJ, Keith FJ. *Drosophila* Toll and IL-1 receptor. Nature 1991;351(6325):355–6.

[16] Poltorak A, He X, Smirnova I, Liu MY, Van Huffel C, Du X, et al. Defective LPS signaling in C3H/HeJ and C57BL/10ScCr mice: mutations in Tlr4 gene. Science 1998;282(5396):2085–8.

[17] Qureshi ST, Lariviere L, Leveque G, Clermont S, Moore KJ, Gros P, et al. Endotoxin-tolerant mice have mutations in Toll-like receptor 4 (Tlr4). J Exp Med 1999;189(4):615–25.

[18] Oldenburg M, Kruger A, Ferstl R, Kaufmann A, Nees G, Sigmund A, et al. TLR13 recognizes bacterial 23S rRNA devoid of erythromycin resistance-forming modification. Science 2012;337(6098):1111–5.

[19] Volchenkov R, Sprater F, Vogelsang P, Appel S. The 2011 Nobel Prize in physiology or medicine. Scand J Immunol 2012;75(1):1–4.

[20] Krieg AM, Yi AK, Matson S, Waldschmidt TJ, Bishop GA, Teasdale R, et al. CpG motifs in bacterial DNA trigger direct B-cell activation. Nature 1995;374(6522):546–9.

[21] Bird AP, Taggart MH, Nicholls RD, Higgs DR. Non-methylated CpG-rich islands at the human alpha-globin locus: implications for evolution of the alpha-globin pseudogene. EMBO J 1987;6(4):999–1004.

[22] Cooper DN, Gerber-Huber S. DNA methylation and CpG suppression. Cell Diff 1985;17(3):199–205.

[23] Klinman DM, Yi AK, Beaucage SL, Conover J, Krieg AM. CpG motifs present in bacteria DNA rapidly induce lymphocytes to secrete interleukin 6, interleukin 12, and interferon gamma. Proc Natl Acad Sci USA 1996;93(7):2879–83.

[24] Stacey KJ, Sweet MJ, Hume DA. Macrophages ingest and are activated by bacterial DNA. J Immunol 1996;157(5):2116–22.

[25] Jakob T, Walker PS, Krieg AM, Udey MC, Vogel JC. Activation of cutaneous dendritic cells by CpG-containing oligodeoxynucleotides: a role for dendritic cells in the augmentation of Th1 responses by immunostimulatory DNA. J Immunol 1998;161(6):3042–9.

[26] Hemmi H, Takeuchi O, Kawai T, Kaisho T, Sato S, Sanjo H, et al. A Toll-like receptor recognizes bacterial DNA. Nature 2000;408(6813):740–5.

[27] Latz E, Schoenemeyer A, Visintin A, Fitzgerald KA, Monks BG, Knetter CF, et al. TLR9 signals after translocating from the ER to CpG DNA in the lysosome. Nature Immunol 2004;5(2):190–8.

[28] Babiuk S, Mookherjee N, Pontarollo R, Griebel P, van Drunen Littel-van den Hurk S, Hecker R, et al. TLR9$^{-/-}$ and TLR9$^{+/+}$ mice display similar immune responses to a DNA vaccine. Immunology 2004;113(1):114–20.

[29] Spies B, Hochrein H, Vabulas M, Huster K, Busch DH, Schmitz F, et al. Vaccination with plasmid DNA activates dendritic cells via Toll-like receptor 9 (TLR9) but functions in TLR9-deficient mice. J Immunol 2003;171(11):5908–12.

[30] Ishii KJ, Kawagoe T, Koyama S, Matsui K, Kumar H, Kawai T, et al. TANK-binding kinase-1 delineates innate and adaptive immune responses to DNA vaccines. Nature 2008;451(7179):725–9.

[31] Suzuki K, Mori A, Ishii KJ, Saito J, Singer DS, Klinman DM, et al. Activation of target-tissue immune-recognition molecules by double-stranded polynucleotides. Proc Natl Acad Sci USA 1999;96(5):2285–90.

[32] Ishii KJ, Suzuki K, Coban C, Takeshita F, Itoh Y, Matoba H, et al. Genomic DNA released by dying cells induces the maturation of APCs. J Immunol 2001;167(5):2602–7.

[33] Okabe Y, Kawane K, Akira S, Taniguchi T, Nagata S. Toll-like receptor-independent gene induction program activated by mammalian DNA escaped from apoptotic DNA degradation. J Exp Med 2005;202(10):1333–9.

[34] Ishii KJ, Coban C, Kato H, Takahashi K, Torii Y, Takeshita F, et al. A Toll-like receptor-independent antiviral response induced by double-stranded B-form DNA. Nature Immunol 2006;7(1):40–8.

[35] Braun CS, Jas GS, Choosakoonkriang S, Koe GS, Smith JG, Middaugh CR. The structure of DNA within cationic lipid/DNA complexes. Biophys J 2003;84(2 Pt 1): 1114–23.

[36] Sharma S, tenOever BR, Grandvaux N, Zhou GP, Lin R, Hiscott J. Triggering the interferon antiviral response through an IKK-related pathway. Science 2003;300(5622):1148–51.

[37] Fitzgerald KA, McWhirter SM, Faia KL, Rowe DC, Latz E, Golenbock DT, et al. IKKepsilon and TBK1 are essential components of the IRF3 signaling pathway. Nature Immunol 2003;4(5):491–6.

[38] Hemmi H, Takeuchi O, Sato S, Yamamoto M, Kaisho T, Sanjo H, et al. The roles of two IkappaB kinase-related kinases in lipopolysaccharide and double stranded RNA signaling and viral infection. J Exp Med 2004;199(12):1641–50.

[39] Youn HS, Lee JY, Fitzgerald KA, Young HA, Akira S, Hwang DH. Specific inhibition of MyD88-independent signaling pathways of TLR3 and TLR4 by resveratrol: molecular targets are TBK1 and RIP1 in TRIF complex. J Immunol 2005;175(5):3339–46.

[40] Charrel-Dennis M, Latz E, Halmen KA, Trieu-Cuot P, Fitzgerald KA, Kasper DL, et al. TLR-independent type I interferon induction in response to an extracellular

bacterial pathogen via intracellular recognition of its DNA. Cell Host Microbe 2008;4(6):543–54.

[41] Sato S, Sugiyama M, Yamamoto M, Watanabe Y, Kawai T, Takeda K, et al. Toll/IL-1 receptor domain-containing adaptor inducing IFN-beta (TRIF) associates with TNF receptor-associated factor 6 and TANK-binding kinase 1, and activates two distinct transcription factors, NF-kappa B and IFN-regulatory factor-3, in the Toll-like receptor signaling. J Immunol 2003;171(8):4304–10.

[42] Kawai T, Sato S, Ishii KJ, Coban C, Hemmi H, Yamamoto M, et al. Interferon-alpha induction through Toll-like receptors involves a direct interaction of IRF7 with MyD88 and TRAF6. Nature Immunol 2004;5(10):1061–8.

[43] Stetson DB, Medzhitov R. Recognition of cytosolic DNA activates an IRF3-dependent innate immune response. Immunity 2006;24(1):93–103.

[44] Hochrein H, Schlatter B, O'Keeffe M, Wagner C, Schmitz F, Schiemann M, et al. Herpes simplex virus type-1 induces IFN-alpha production via Toll-like receptor 9-dependent and -independent pathways. Proc Natl Acad Sci USA 2004;101(31):11416–11421.

[45] Verthelyi D, Zeuner RA. Differential signaling by CpG DNA in DCs and B cells: not just TLR9. Trends Immunol 2003;24(10):519–22.

[46] Takaoka A, Wang Z, Choi MK, Yanai H, Negishi H, Ban T, et al. DAI (DLM-1/ZBP1) is a cytosolic DNA sensor and an activator of innate immune response. Nature 2007;448(7152):501–5.

[47] Deigendesch N, Koch-Nolte F, Rothenburg S. ZBP1 subcellular localization and association with stress granules is controlled by its Z-DNA binding domains. Nucleic Acids Res 2006;34(18):5007–20.

[48] Fu Y, Comella N, Tognazzi K, Brown LF, Dvorak HF, Kocher O. Cloning of DLM-1, a novel gene that is upregulated in activated macrophages, using RNA differential display. Gene 1999;240(1):157–63.

[49] Takaoka A, Taniguchi T. Cytosolic DNA recognition for triggering innate immune responses. Adv Drug Deliv Rev 2008;60(7):847–57.

[50] Rebsamen M, Heinz LX, Meylan E, Michallet MC, Schroder K, Hofmann K, et al. DAI/ZBP1 recruits RIP1 and RIP3 through RIP homotypic interaction motifs to activate NF-kappaB. EMBO Reports 2009;10(8):916–22.

[51] Meylan E, Burns K, Hofmann K, Blancheteau V, Martinon F, Kelliher M, et al. RIP1 is an essential mediator of Toll-like receptor 3-induced NF-kappa B activation. Nature Immunol 2004;5(5):503–7.

[52] Wang Z, Choi MK, Ban T, Yanai H, Negishi H, Lu Y, et al. Regulation of innate immune responses by DAI (DLM-1/ZBP1) and other DNA-sensing molecules. Proc Natl Acad Sci USA 2008;105(14):5477–82.

[53] Parker D, Martin FJ, Soong G, Harfenist BS, Aguilar JL, Ratner AJ, et al. Streptococcus pneumoniae DNA initiates type I interferon signaling in the respiratory tract. mBio 2011;2(3):e00016-11.

[54] Pham TH, Kwon KM, Kim YE, Kim KK, Ahn JH. DNA sensing-independent inhibition of herpes simplex virus type-1 replication by DAI/ZBP1. J Virol. 2013;87(6):3076–86.

[55] Lladser A, Mougiakakos D, Tufvesson H, Ligtenberg MA, Quest AF, Kiessling R, et al. DAI (DLM-1/ZBP1) as a genetic adjuvant for DNA vaccines that promotes effective antitumor CTL immunity. Mol Ther: J Am Soc Gene Ther 2011;19(3):594–601.

[56] Bell JK, Mullen GE, Leifer CA, Mazzoni A, Davies DR, Segal DM. Leucine-rich repeats and pathogen recognition in Toll-like receptors. Trends Immunol 2003;24(10):528–33.

[57] Yang P, An H, Liu X, Wen M, Zheng Y, Rui Y, et al. The cytosolic nucleic acid sensor LRRFIP1 mediates the production of type I interferon via a beta-catenin-dependent pathway. Nature Immunol 2010;11(6):487–94.

[58] Wilson SA, Brown EC, Kingsman AJ, Kingsman SM. TRIP: a novel double stranded RNA binding protein which interacts with the leucine rich repeat of flightless I. Nucleic Acids Res 1998;26(15):3460–7.

[59] Suriano AR, Sanford AN, Kim N, Oh M, Kennedy S, Henderson MJ, et al. GCF2/LRRFIP1 represses tumor necrosis factor alpha expression. Mol Cell Biol 2005;25(20):9073–81.

[60] Clevers H, Nusse R. Wnt/beta-catenin signaling and disease. Cell 2012;149(6):1192–205.

[61] Lee YH, Stallcup MR. Interplay of Fli-I and FLAP1 for regulation of beta-catenin dependent transcription. Nucleic Acids Res 2006;34(18):5052–9.

[62] Parekh BS, Maniatis T. Virus infection leads to localized hyperacetylation of histones H3 and H4 at the IFN-beta promoter. Mol Cell 1999;3(1):125–9.

[63] Merika M, Williams AJ, Chen G, Collins T, Thanos D. Recruitment of CBP/p300 by the IFN beta enhanceosome is required for synergistic activation of transcription. Mol Cell 1998;1(2):277–87.

[64] Bagashev A, Fitzgerald MC, Larosa DF, Rose PP, Cherry S, Johnson AC, et al. Leucine-rich repeat (in Flightless I) interacting protein-1 regulates a rapid type I interferon response. J Interfer Cytokine Res: Official J Int Soc Interfer Cytokine Res 2010;30(11):843–52.

[65] Nguyen JB, Modis Y. Crystal structure of the dimeric coiled-coil domain of the cytosolic nucleic acid sensor LRRFIP1. J Struct Biol 2013;181(1):82–8.

[66] Yoneyama M, Kikuchi M, Natsukawa T, Shinobu N, Imaizumi T, Miyagishi M, et al. The RNA helicase RIG-I has an essential function in double-stranded RNA-induced innate antiviral responses. Nature Immunol 2004;5(7):730–7.

[67] Zhang Z, Kim T, Bao M, Facchinetti V, Jung SY, Ghaffari AA, et al. DDX1, DDX21, and DHX36 helicases form a complex with the adaptor molecule TRIF to sense dsRNA in dendritic cells. Immunity 2011;34(6):866–78.

[68] Zhang Z, Yuan B, Bao M, Lu N, Kim T, Liu YJ. The helicase DDX41 senses intracellular DNA mediated by the adaptor STING in dendritic cells. Nature Immunol 2011;12(10):959–65.

[69] Ishikawa H, Barber GN. STING is an endoplasmic reticulum adaptor that facilitates innate immune signaling. Nature 2008;455(7213):674–8.

[70] Ishikawa H, Ma Z, Barber GN. STING regulates intracellular DNA-mediated, type I interferon-dependent innate immunity. Nature 2009;461(7265):788–92.

[71] Loo YM, Gale Jr. M. Immune signaling by RIG-I-like receptors. Immunity 2011;34(5):680–92.

[72] Kang DC, Gopalkrishnan RV, Wu Q, Jankowsky E, Pyle AM, Fisher PB. mda-5: An interferon-inducible putative RNA helicase with double-stranded RNA-dependent ATPase activity and melanoma growth-suppressive properties. Proc Natl Acad Sci USA 2002;99(2):637–42.

[73] Rothenfusser S, Goutagny N, DiPerna G, Gong M, Monks BG, Schoenemeyer A, et al. The RNA helicase Lgp2 inhibits TLR-independent sensing of viral replication by retinoic acid-inducible gene-I. J Immunol 2005;175(8):5260–8.

[74] Chiu YH, Macmillan JB, Chen ZJ. RNA polymerase III detects cytosolic DNA and induces type I interferons through the RIG-I pathway. Cell 2009;138(3):576–91.

[75] Ablasser A, Bauernfeind F, Hartmann G, Latz E, Fitzgerald KA, Hornung V. RIG-I-dependent sensing of poly(dA:dT) through the induction of an RNA polymerase III-transcribed RNA intermediate. Nature Immunol 2009;10(10):1065–72.

[76] Sun Q, Sun L, Liu HH, Chen X, Seth RB, Forman J, et al. The specific and essential role of MAVS in antiviral innate immune responses. Immunity 2006;24(5):633–42.

[77] Huet J, Riva M, Sentenac A, Fromageot P. Yeast RNA polymerase C and its subunits. Specific antibodies as structural and functional probes. J Biol Chem 1985;260(28):15304–15310.

[78] Zaros C, Thuriaux P. Rpc25, a conserved RNA polymerase III subunit, is critical for transcription initiation. Mol Microbiol 2005;55(1):104–14.

[79] Minamitani T, Iwakiri D, Takada K. Adenovirus virus-associated RNAs induce type I interferon expression through a RIG-I-mediated pathway. J Virol 2011; 85(8):4035–40.

[80] Xing J, Wang S, Lin R, Mossman KL, Zheng C. Herpes simplex virus 1 tegument protein US11 downmodulates the RLR signaling pathway via direct interaction with RIG-I and MDA-5. J Virol 2012;86(7):3528–40.

[81] Samanta M, Iwakiri D, Kanda T, Imaizumi T, Takada K. EB virus-encoded RNAs are recognized by RIG-I and activate signaling to induce type I IFN. EMBO J 2006;25(18):4207–14.

[82] Schattgen SA, Fitzgerald KA. The PYHIN protein family as mediators of host defenses. Immunol Rev 2011;243(1):109–18.

[83] Muruve DA, Petrilli V, Zaiss AK, White LR, Clark SA, Ross PJ, et al. The inflammasome recognizes cytosolic microbial and host DNA and triggers an innate immune response. Nature 2008;452(7183):103–7.

[84] Fernandes-Alnemri T, Yu JW, Datta P, Wu J, Alnemri ES. AIM2 activates the inflammasome and cell death in response to cytoplasmic DNA. Nature 2009;458(7237):509–13.

[85] Burckstummer T, Baumann C, Bluml S, Dixit E, Durnberger G, Jahn H, et al. An orthogonal proteomic-genomic screen identifies AIM2 as a cytoplasmic DNA sensor for the inflammasome. Nature Immunol 2009;10(3):266–72.

[86] Rathinam VA, Jiang Z, Waggoner SN, Sharma S, Cole LE, Waggoner L, et al. The AIM2 inflammasome is essential for host defense against cytosolic bacteria and DNA viruses. Nature Immunol 2010;11(5):395–402.

[87] Roberts TL, Idris A, Dunn JA, Kelly GM, Burnton CM, Hodgson S, et al. HIN-200 proteins regulate caspase activation in response to foreign cytoplasmic DNA. Science 2009;323(5917):1057–60.

[88] Unterholzner L, Keating SE, Baran M, Horan KA, Jensen SB, Sharma S, et al. IFI16 is an innate immune sensor for intracellular DNA. Nature Immunol 2010;11(11):997–1004.

[89] Kerur N, Veettil MV, Sharma-Walia N, Bottero V, Sadagopan S, Otageri P, et al. IFI16 acts as a nuclear pathogen sensor to induce the inflammasome in response to Kaposi sarcoma-associated herpesvirus infection. Cell Host Microbe 2011;9(5):363–75.

[90] Soby S, Laursen RR, Ostergaard L, Melchjorsen J. HSV-1-induced chemokine expression via IFI16-dependent and IFI16-independent pathways in human monocyte-derived macrophages. Herpesviridae 2012;3(1):6.

[91] Orzalli MH, DeLuca NA, Knipe DM. Nuclear IFI16 induction of IRF-3 signaling during herpesviral infection and degradation of IFI16 by the viral ICP0 protein. Proc Natl Acad Sci USA 2012;109(44):E3008–E3017.

[92] Johnson KE, Chikoti L, Chandran B. HSV-1 infection induces activation and subsequent inhibition of the IFI16 and NLRP3 inflammasomes. J Virol 2013;87(9):5005–18.

[93] Jung A, Kato H, Kumagai Y, Kumar H, Kawai T, Takeuchi O, et al. Lymphocytoid choriomeningitis virus activates plasmacytoid dendritic cells and induces a cytotoxic T-cell response via MyD88. J Virol 2008;82(1):196–206.

[94] Kim T, Pazhoor S, Bao M, Zhang Z, Hanabuchi S, Facchinetti V, et al. Aspartate-glutamate-alanine-histidine box motif (DEAH)/RNA helicase A helicases sense microbial DNA in human plasmacytoid dendritic cells. Proc Natl Acad Sci USA 2010;107(34):15181–15186.

[95] Zhang Z, Yuan B, Lu N, Facchinetti V, Liu YJ. DHX9 pairs with IPS-1 to sense double-stranded RNA in myeloid dendritic cells. J Immunol 2011;187(9):4501–8.

[96] Lieber MR, Ma Y, Pannicke U, Schwarz K. Mechanism and regulation of human non-homologous DNA end-joining. Nature Rev Mol Cell Biol 2003;4(9):712–20.

[97] Walker JR, Corpina RA, Goldberg J. Structure of the Ku heterodimer bound to DNA and its implications for double-strand break repair. Nature 2001;412(6847):607–14.

[98] Zhang X, Brann TW, Zhou M, Yang J, Oguariri RM, Lidie KB, et al. Cutting edge: Ku70 is a novel cytosolic DNA sensor that induces type III rather than type I IFN. J Immunol 2011;186(8):4541–5.

[99] Zhou Z, Hamming OJ, Ank N, Paludan SR, Nielsen AL, Hartmann R. Type III interferon (IFN) induces a type I IFN-like response in a restricted subset of cells through signaling pathways involving both the Jak-STAT pathway and the mitogen-activated protein kinases. J Virol 2007;81(14):7749–58.

[100] Stros M. HMGB proteins: interactions with DNA and chromatin. Biochim Biophys Acta 2010;1799(1–2):101–13.

[101] Harris HE, Andersson U, Pisetsky DS. HMGB1: a multifunctional alarmin driving autoimmune and inflammatory disease. Nature Rev Rheumatol 2012;8(4):195–202.

[102] Urbonaviciute V, Furnrohr BG, Meister S, Munoz L, Heyder P, De Marchis F, et al. Induction of inflammatory and immune responses by HMGB1-nucleosome complexes: implications for the pathogenesis of SLE. J Exp Med 2008;205(13):3007–18.

[103] Shi Y, Sandoghchian Shotorbani S, Su Z, Liu Y, Tong J, Zheng D, et al. Enhanced HMGB1 expression may contribute to Th17 cells activation in rheumatoid arthritis. Clin Develop Immunol 2012;2012:295081.

[104] Apetoh L, Ghiringhelli F, Tesniere A, Criollo A, Ortiz C, Lidereau R, et al. The interaction between HMGB1 and TLR4 dictates the outcome of anticancer chemotherapy and radiotherapy. Immunol Rev 2007;220:47–59.

[105] Tian J, Avalos AM, Mao SY, Chen B, Senthil K, Wu H, et al. Toll-like receptor 9-dependent activation by DNA-containing immune complexes is mediated by HMGB1 and RAGE. Nature Immunol 2007;8(5):487–96.

[106] Kim T, Kim TY, Song YH, Min IM, Yim J, Kim TK. Activation of interferon regulatory factor 3 in response to DNA-damaging agents. J Biol Chem 1999;274(43):30686–9.

[107] Kawashima A, Tanigawa K, Akama T, Wu H, Sue M, Yoshihara A, et al. Fragments of genomic DNA released by injured cells activate innate immunity and suppress endocrine function in the thyroid. Endocrinology 2011;152(4):1702–12.

[108] Kondo T, Kobayashi J, Saitoh T, Maruyama K, Ishii KJ, Barber GN, et al. DNA damage sensor MRE11 recognizes cytosolic double-stranded DNA and induces type I interferon by regulating STING trafficking. Proc Natl Acad Sci USA 2013;110(8):2969–74.

[109] Burdette DL, Monroe KM, Sotelo-Troha K, Iwig JS, Eckert B, Hyodo M, et al. STING is a direct innate immune sensor of cyclic di-GMP. Nature 2011;478(7370):515–8.

[110] Wu J, Sun L, Chen X, Du F, Shi H, Chen C, et al. Cyclic GMP-AMP is an endogenous second messenger in innate immune signaling by cytosolic DNA. Science 2013;339(6121):826–30.

[111] Davies BW, Bogard RW, Young TS, Mekalanos JJ. Coordinated regulation of accessory genetic elements produces cyclic di-nucleotides for *V. cholerae* virulence. Cell 2012;149(2):358–70.

[112] Sun L, Wu J, Du F, Chen X, Chen ZJ. Cyclic GMP-AMP synthase is a cytosolic DNA sensor that activates the type I interferon pathway. Science 2013;339(6121):786–91.

[113] Yanai H, Ban T, Wang Z, Choi MK, Kawamura T, Negishi H, et al. HMGB proteins function as universal sentinels for nucleic-acid-mediated innate immune responses. Nature 2009;462(7269):99–103.

CHAPTER 2

The PYHIN Family of Molecules and their Functions Sensing dsDNA

Sivapriya Kailasan Vanaja[1,2], Vijay A.K. Rathinam[1] and Katherine A. Fitzgerald[1]

[1]Division of Infectious Diseases and Immunology, Department of Medicine, University of Massachusetts Medical School, Worcester, MA, USA
[2]Department of Molecular Biology and Microbiology, Tufts School of Medicine, Boston, MA, USA

INTRODUCTION

Innate immune sensing of DNA has emerged as a central component of antimicrobial defenses and an important contributor to the development of autoimmune diseases. The PYHIN protein family is one family of proteins recently linked to the recognition of DNA of both self and microbial origin. Sensing of dsDNA via PYHIN proteins in turn initiates a range of innate immune responses. Most of the PYHIN proteins characteristically feature an N-terminal pyrin domain (PYD) and one or two DNA binding 'hematopoietic interferon-inducible nuclear proteins with a 200 amino acid repeat' (HIN-200) domains at the C-terminus [1,2]. PYHIN family members were originally classified based on their interferon-inducibility, nuclear localization and shared presence of the HIN-200 domain [3]. Although the PYHIN proteins were originally implicated in cell growth, tumor suppression and apoptosis [4], emerging data clearly position them as central players in cytosolic DNA sensing and inflammatory responses. Upon sensing foreign DNA, PYHINs typically induce two types of inflammatory responses: absent in melanoma 2 (AIM2) regulates the posttranslational proteolytic maturation of the IL-1 family of cytokines such as IL-1β and IL-18 [2,5]. On the other hand, PYHIN proteins such as interferon inducible protein 16 (IFI16) and p204 mediate the transcriptional activation of type I interferons (IFNs) [6]. Delineating the functions of hitherto uncharacterized PYHIN proteins has also revealed new functions for this family in gene regulation, pathogen detection and autoimmunity.

Biological DNA Sensor.
DOI: http://dx.doi.org/10.1016/B978-0-12-404732-7.00002-2

MEMBERS OF THE PYHIN FAMILY

The PYHIN family consists of proteins encoded by four genes in humans and 14 genes in mice [1,7]. Additional placental mammals such as rat, cow, horse, and pig also encode PYHIN genes whereas non-mammals do not express any PYHIN genes. Interestingly, marsupials such as opossum and wallaby also carry a PYHIN coding sequence indicating the existence of at least one PYHIN gene prior to the divergence of placental mammals and marsupials during evolution [1].

The PYHIN genes are located as a cluster on human and mouse chromosome 1. Evolutionary analysis indicates that this cluster is a rapidly evolving locus with a remarkable diversification of genes in different mammalian species. The human PYHIN family consists of four genes: AIM2, IFI16, IFIX, and myeloid nuclear differentiation antigen (MNDA). In contrast, 14 PYHIN genes have been identified so far in the mouse genome including p202 (p202a and p202b), p203, p204, p205, p207 (PYHIN-A), p208 (PYR-RV1), p209 (PYHIN-1), AIM2 (p210), p211 (MNDA), p212 (MNDAL), p213 (PYR-A), and p214 [1,7]. The rat genome encodes five PYHINs: Aim2, Rhin2, Rhin3, Rhin4, and Rhin5 [1]. Notably, the PYHIN genes demonstrate remarkable plasticity in evolutionary analysis and none of the members are preserved among all mammals. Similarly, orthology between different species is almost non-existent. In fact AIM2 is the only PYHIN protein that shows orthology across several different species [1,7]. Interestingly AIM2 exists as a pseudogene and therefore is non-functional in cow, sheep, dog, and elephant, which is surprising given its important role in antimicrobial defense in humans and murine models [1]. Other PYHIN genes do not show orthology across species and are more closely related to the rest of the PYHIN family members within the species. Interestingly, IFI16 and p204, which were previously considered orthologous as they both carry a single PYD and two HIN domains, were found to be non-orthologous in phylogenetic analysis. It appears that these two proteins acquired the double HIN domain structure through independent events. Furthermore, phylogenetic analysis has revealed that the AIM2 PYD and HIN domains are unique and diverge from the rest of the family whereas other members including the 13 genes in mouse emerged through duplication and rearrangement within the lineage [1,7]. It is believed that a considerable expansion of the PYHIN family occurred in a species-specific manner due to the evolutionary pressure exerted on these genes by pathogens invading each species of animals.

Most of the PYHIN proteins are primarily nuclear proteins carrying either a monopartite nuclear localization sequence (NLS), a bipartite NLS or both [3,8]. Some members carry a nuclear export sequence indicating that they may shuttle between the nucleus and the cytosol [9]. Acetylation of the NLS in IFI16 has been shown to favor its accumulation in the cytosol [10]. AIM2 and p202 do not possess NLS and therefore reside exclusively in the cytosol of resting state cells [11,12]. It has been shown that prolonged treatment with type I interferon (IFN) results in translocation of p202a to the nucleus [13]. MNDA and p206 also do not contain a recognized NLS and recently p206 was shown to localize to the cytosol [3]. The mechanism by which the proteins move from the nucleus to the cytosol remains to be clarified.

PYHIN proteins are predominantly expressed in hemopoietic tissues such as spleen, thymus, lymph nodes, peripheral blood leukocytes and bone marrow [14–18]. However, expression of each member of this family varies in a lineage- and cell type-specific manner among these tissues. Spleen is a major tissue where most of the PYHIN proteins are highly expressed. For example, AIM2 is predominantly expressed in spleen where it is detected in macrophages and dendritic cells [14]. IFI16 is typically detected in myeloid precursor cells and its expression remains strong in monocyte precursors, peripheral blood monocytes and throughout lymphoid development [3]. On the other hand, MNDA, which is myeloid restricted, is found in activated macrophages, mature granulocytes and monocytes [19,20]. p204 has a similar expression pattern to MNDA and is detected in monocytes, granulocytes and megakaryocytes [17]. p205 is found predominantly in differentiating granulocytes and macrophages [21]. p209 is found in T and B cells as well as bone marrow derived macrophages while p214 is restricted to B and T cells [1].

Recent studies have also detected expression of the PYHIN proteins in non-hemopoietic tissues. The skin is a major site of expression for most of these proteins, consistent with their emerging role in host defense. p204, p205, and p207 show their highest expression in skin. p202 is also detected in skin [1]. IFI16 is found in epithelial cells of the skin, gastrointestinal tract, urogenital tract and glands as well as in vascular endothelial cells [16,18]. In contrast, AIM2 is detected in the small intestine and testis [14]. Although p203 was originally thought to be liver-restricted, studies have since identified it in additional tissues including spleen [1,22]. p204 is highly expressed in the heart, skeletal muscles and kidney [9]. p205 is also expressed in skeletal muscle, heart and lung [21]. It is well-established that

IFN treatment markedly increases the expression of all PYHIN proteins. Interestingly, the basal expression of p207 and AIM2 is largely independent of IFN signaling in macrophages whereas all other PYHIN proteins require IFN signaling for their expression [7].

STRUCTURAL BASIS OF DNA RECOGNITION BY PYHIN PROTEINS

As mentioned earlier the presence of an N-terminal PYD and C-terminal HIN-200 domain serves as the defining structural characteristic of the PYHIN protein family. Exceptions to this conserved domain structure are p202, which lacks a PYD domain and p208, p213 and an isoform of IFIX that lack a HIN domain [3,15]. The PYD domain, also known as PAAD or DAPIN, belongs to the death domain superfamily of signaling molecules and is an alpha-helical protein–protein interaction motif that mediates the binding between PYHIN proteins and their adaptors [23–25]. Importantly, inflammasome activation by PYHIN proteins such as AIM2 and IFI16 is mediated by homotypic interactions between PYD of ASC and that of the corresponding PYHIN protein [5]. Also, proteins such as AIM2 [26], IFI16 [27] and MNDA [28] can self-associate through their PYD domains. Consistent with the findings of evolutionary studies the PYD of AIM2 has a distinct amino acid sequence compared to other PYHIN proteins, which notably hasn't resulted in a structural difference [1,7].

PYHIN proteins carry one or two copies of the HIN-200 domain, which is classified into three subclasses, 200 A, 200 B, and 200 C, based on amino acid sequence [3]. The subclass 200 C was recently designated and is found only in human AIM2 and its orthologs in mouse and rat. The 200 B domain is defined by a consensus LXCXE motif, whereas domains 200 A and 200 C carry LFCF(R/H) and LTFF(E/T), respectively, at this motif. Human PYHIN proteins MNDA and IFIX carry a single 200A domain whereas AIM2 carries a 200 C domain and IFI16 carries both 200A and 200B domains. Among the murine proteins p202 and p204 carry both 200A and 200B, whereas all others with the exception of AIM2 carry either a 200A or a 200 B domain [3,10].

The HIN-200 domain is primarily a DNA binding domain, which consists of two consecutive oligonucleotide/oligosaccharide-binding (OB) folds of ~80 residues designated as HIN-N and HIN-C subdomains [29]. OB folds are typically found in single-stranded DNA (ssDNA) binding proteins and are five-stranded beta barrel structures. Although PYHIN proteins can bind to single- or double-stranded DNA (dsDNA), the binding preferences of most of the members remain to be fully characterized. It has been clearly established that AIM2 is activated only by dsDNA [11]. Similarly p202 binds specifically to dsDNA [12] whereas IFI16 has been reported to bind to dsDNA, ssDNA as well as RNA [2,24,30]. Furthermore, IFI16 shows an interesting preference for superhelical and cruciform structures of DNA, which are frequently formed in regulating regions of several genes [31]. Therefore, this preference is believed to play a crucial role in IFI16-mediated transcription regulation and tumor suppression. A recent study by Jin et al. extensively characterized the DNA binding capacity of AIM2 and IFI16 HIN domains using X-ray crystallography [32]. The electrostatic attraction between positively charged HIN domain residues and the sugar phosphate backbone of dsDNA accounts for the non-sequence specific DNA sensing previously observed for these molecules. They also demonstrated that the PYD and HIN domains remain in an intramolecular complex prior to stimulation and that the HIN domain keeps the PYD in an inhibitory confirmation. dsDNA binding results in the release of the PYD domain from this complex and facilitates its binding with the PYD of ASC. Furthermore, the findings from this study indicated that the DNA staircase may serve as an oligomerization platform important in nucleating AIM2 inflammasome formation [32]. The lack of sequence specificity in DNA sensing by PYHIN proteins is important in host defense as it allows for a universal detection of infectious agents and host tissue damage resulting in prompt development of an effective innate immune response. Interestingly, the HIN-200 domain can also mediate protein–protein interactions through two specific motifs, a conserved MFHATVAT motif and a LXCXE pRb (retinoblastoma protein)-binding motif. These protein–protein interactions mediate self-association of p202a and transcriptional repression activities of both p202a and IFI16 [3].

MICROBIAL DNA SENSING AND SIGNALING BY PYHIN PROTEINS

Following DNA sensing PYHIN proteins have been linked to caspase-1 dependent inflammasome formation resulting in maturation of IL-1β and IL-18 as well as induction of transcriptional responses including type I IFNs. AIM2 and IFI16 have both been shown to form caspase-1 activating inflammasomes while IFI16 and p204 stimulate IFN response in a manner dependent on an endoplasmic reticulum (ER) and mitochondrial localized protein called STING [2,5,18].

Inflammasome complexes are essential mediators of cytosolic surveillance that activate a proteolytic enzyme, caspase-1. Typically an inflammasome complex contains a cytosolic receptor, which can be either a member of the NLR family (nucleotide-binding-and-oligomerization-domain and leucine-rich-repeat-containing) or PYHIN family, an adaptor protein called ASC (apoptotic spec like protein containing a CARD domain), and the procaspase-1 enzyme [5,33]. Once the inflammasome complex is assembled the proximity-induced multimerization is suggested to induce autoproteolyic cleavage of procaspase-1 leading to formation of active caspase-1, which then cleaves proforms of IL-1β and IL-18 to active IL-1β and IL-18 [5,34]. Inflammasome activation also promotes an inflammatory form of cell death known as pyroptosis [35]. All of these effector functions downstream of inflammasome activation are important in host defense. IL-1β is a potent proinflammatory cytokine that influences the development of both innate and adaptive immune responses [36]. IL-1β facilitates recruitment of inflammatory and immune cells to the site of infection and injury. It also regulates the production of IL-12 and IL-6 by dendritic cells, which are essential for differentiation of effector T cells [37,38]. Additionally, IL-1β is involved in Th17 polarization and T cell-dependent antibody production by B cells [39]. Similar to IL-1β, IL-18 is also involved in both innate and adaptive immune responses [40]. Importantly, it promotes IFN-γ production by NK cells [41]. Along with IL-12, IL-18 also primes differentiation of naïve T cells into Th1 cells [42]. Interestingly, in the presence of appropriate cytokines IL-18 can also play a role in the development of a Th2 response [43]. Given their highly proinflammatory nature, which in excess has the ability to cause collateral tissue damage, it is not surprising that production of IL-1β and IL-18 is subject to tight regulation at multiple levels.

As mentioned above, in addition to inflammasome activation, PYHINs have also been shown to mediate potent activation of type I IFNs in response to cytosolic DNA. Type I IFNs were originally identified for their pivotal role in antiviral immunity; however, recent studies have demonstrated their equally important role in anti-bacterial host responses and autoimmunity. Although TLR9 was initially thought to be the major trigger for all DNA-induced type I IFN responses, it was found later that cytosolic DNA can induce IFNs in a TLR9-independent manner [44,45]. Several recent studies have identified key new molecules and receptors involved in this cytosolic DNA-induced IFN response also known as the immune stimulatory DNA (ISD) pathway [46]. Typically following cytosolic dsDNA stimulation an ER-located adaptor protein, STING (stimulator of interferon genes), relocalizes from the ER to the Golgi and associates with a serine–threonine kinase, TBK1, to form punctuate structures [47]. This relocalization results in the activation of TBK1 and recruitment of the transcription factor IRF3 into the complex. Subsequently, IRF3 undergoes phosphorylation and nuclear translocation leading to transcriptional activation of type I IFN genes. STING is critical in this cytosolic DNA-induced IFN pathway as it specifies and promotes IRF3 phophorylation by TBK1 [47]. The mechanisms involved in activating STING were unclear until recently. Cyclic-GMP-AMP (cGAMP) synthase (cGAS), a nucleotidyltransferase, binds to cytosolic DNA and catalyzes the generation of the second messenger, cGAMP [48]. cGAMP is structurally similar to cyclic-dinucleotides, bacterial second messengers important in biofilm formation which were previously identified as microbial triggers of STING-dependent type I IFNs. Like cyclic-di-GMP and cyclic-di-AMP, cGAMP binds and activates STING leading to induction of type I IFNs [48,49]. In addition to cGAS, members of the PYHIN proteins including IFI16 and the murine PYHIN p204 as well as DDX41, a DEXDH box helicase, have also been implicated in the type I IFN response during virus infection [46].

AIM2

AIM2 was the first PYHIN protein found to be associated with inflammasome activation. Several groups had noted that dsDNA was a potent activator of inflammasome-dependent IL-1 responses; four independent studies identified AIM2 as a non-redundant sensor of cytosolic dsDNA that forms inflammasome complexes with ASC and caspase-1 [11,12,50,51]. The AIM2 inflammasome senses dsDNA of viral or bacterial origin in the cytosol [52,53]. Self-DNA can also activate AIM2, however the compartmentalization of self-DNA to the nucleus and its tight association with chromatin limit access to AIM2 and thus any unwanted inflammasome activation under steady-state-conditions. Unlike the NLR inflammasomes for which the exact ligand binding mechanisms are not well understood, it has been clearly demonstrated that AIM2 detects dsDNA by direct binding. AIM2 binds to dsDNA through its HIN-200 domain, which likely leads to its oligomerization and subsequent binding to ASC through PYD–PYD interactions (Figure 2.1) [2]. Interestingly, detection of DNA by AIM2 is not sequence specific but rather depends on the length of DNA. Any dsDNA that is greater than 80 base pairs in

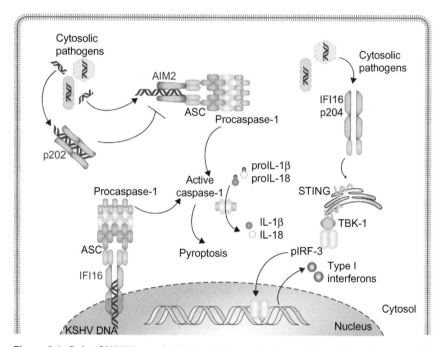

Figure 2.1 *Role of PYHIN proteins in innate immunity.* DNA released from the cytosolic bacteria and viruses is detected by AIM2, which leads to the formation of AIM2 inflammasome complexes comprising of AIM2, ASC, and procaspase-1. The recruitment of procaspase-1 to the inflammasome scaffold triggers the autoproteolytic processing and generation of active caspase-1. The active caspase-1 then catalyzes the proteolytic conversion of proIL-1β and proIL-18 to their active forms and ultimately leads to pyroptosis. IFI16 has also been reported to form an ASC-containing inflammasome complex only in response to Kaposi sarcoma associated herpes virus (KSHV). IFI16 and p204 also play a role in cytosolic DNA-driven type I interferon responses. In response to cytosolic DNA, IFI16 and p204 associate with STING, which then leads to the activation of TBK-1 and phosphorylation of IRF3. pIRF3 translocates to the nucleus to turn on the transcription of type I interferon genes. Please see color plate at the back of the book.

length can provide a scaffold facilitating robust oligomerization of AIM2 and subsequent assembly of inflammasome complexes [5].

Among all the PYHIN proteins studied to date, the role of AIM2 in host defense is well established. Several studies have demonstrated that AIM2 is essential for caspase-1 activation and production of IL-1β and IL-18 in response to DNA viruses such as vaccinia and murine cytomegalovirus (MCMV) (Figure 2.1) [11,52]. In the case of MCMV, AIM2-dependent production of IL-18 is indispensable for NK cell activation in the spleen and early virus control [52]. Herpes viruses such as MCMV typically infect cells and replicate in the nucleus. It remains unclear how AIM2 gains access to viral DNA during this life cycle. Interestingly, unlike MCMV, other herpes viruses such as HSV-1 and Kaposi's sarcoma-associated herpes virus (KSHV) activate inflammasomes in an AIM2-independent manner. This is particularly interesting in the case of HSV-1 as its DNA has been shown to accumulate in the cytosol and activate other DNA sensors such as IFI16 [6]. It is possible that HSV-1 actively subverts AIM2 dependent detection of HSV viral DNA or thwarts assembly of AIM2–ASC inflammasome complexes.

AIM2 also has a crucial role in the development of innate immune response against infections with cytosolic bacteria including *Francisella tularensis* and *Listeria monocytogenes* [52–55]. *F. tularensis* is the causative agent of tularemia, a potentially lethal disease. Following infection *F. tularensis* is released from the phagosome into the cytosol through phagosomal membrane rupture [56,57]. The live bacteria then replicate in the cytosol and this ability of cytosolic replication has been linked to the virulence of *Francisella*. The presence of *F. tularensis* in the cytosol triggers induction of IL-1β, IL-18, and pyroptosis and independent studies identified AIM2 as the sole mediator of this response [52,53]. It is suggested that bacterial DNA delivered into the cytosol during *F. tularensis* infection activates AIM2 [53]. The exact mechanism by which DNA is released into the cytosol is unknown. It has been shown that phagosomal rupture is essential for inflammasome activation during *Francisella* infection. Therefore, it is believed that some of the bacteria will be killed and degraded in the phagosome and that DNA from these degraded bacteria will escape into the cytosol through phagosomal membrane disruption caused by the live bacteria. AIM2 has also been shown to be essential for host defense against *F. tularensis in vivo*. AIM2 deficient mice typically show extreme susceptibility to *F. tularensis* with higher bacterial burden, systemic disease and mortality compared to wild type mice [53]. Notably, type I IFN signaling is essential for inflammasome activation by *Francisella* although the exact mechanism behind this regulation is not clear [58].

Unlike its non-redundant role in inflammasome response to *Francisella*, AIM2 only has a partial role in inflammasome activation by the food-borne pathogen *L. monocytogenes* [52,59]. Additional pathways such as NLRP3 and NLRC4 are also involved in this *Listeria*-mediated response [60,61]. Similar to *Francisella*, *Listeria* escapes the phagosome and replicates in the cytosol where multiple inflammasomes collaborate to detect this bacterium. It appears that defective cell wall synthesis and increased bacteriolysis make DNA more accessible to AIM2 and in doing so induce elevated AIM2 inflammasome activation during *Listeria* infection. Supporting this possibility, a study has shown that defective cell wall synthesis that occurs as a result of mutation in a gene, *lmo2473*, can lead to increased AIM2 activation [60]. Similarly, treating *Listeria*-infected macrophages with the cell wall synthesis inhibitor amoxicillin also leads to higher AIM2 activation [60]. Although AIM2 was considered to be activated only by intracellular pathogens such as DNA viruses and cytosolic bacteria, recent studies have shown that *Streptococcus pneumoniae*, an extracellular pathogen, can also induce AIM2 inflammasome activation [62,63]. Caspase-1 activation by *S. pneumoniae* was dependent on both AIM2 and NLRP3 inflammasomes [62]. The mechanism by which DNA of this extracellular bacterium gains access to the cytosol however is unclear. *S. pneumoniae* undergoes rapid death in the phagosome and it is possible that pneumolysin, a pore-forming cytotoxin secreted by *S. pneumoniae*, disrupts the phagosomal membrane through which bacterial degradation products such as DNA could access the cytosol. All these findings clearly show that AIM2 has a critical role in host resistance against both intracellular and extracellular bacteria as well as DNA viruses.

IFI16

Although IFI16 was originally described as an anti-proliferative and DNA damage response protein, recent studies have identified it as a unique PYHIN protein that can induce both type I IFNs and inflammasome activation. Unterholzner et al. identified IFI16 as a candidate cytosolic DNA sensor associated with IFN-β-inducing viral DNA [6]. Using synthetic vaccinia viral DNA motifs (VACV 70mer) this study demonstrated that IFI16 directly binds to viral DNA in the cytosol and subsequently recruits STING into the complex (Figure 2.1). Furthermore, knockdown of IFI16 resulted in inhibition of DNA-induced translocation of transcription factors, IRF3 and NF-κB, to the nucleus and reduced IFN-β activation.

While IFI16 is predominantly nuclear in most cell types, the IFN-inducing function of IFI16 in macrophages described by Unterholzner et al. was associated with a small cytosolic fraction of the protein [6]. This raised important questions about the function of IFI16 in the nucleus and the possibility of self-DNA binding by this protein. However, Kerur et al. reported that IFI16 also functions as a nuclear sensor of DNA from Kaposi sarcoma-associated herpes virus (KSHV) [64]. Interestingly, after sensing DNA of KSHV, IFI16 recruits ASC and procaspase-1 to assemble an inflammasome complex in the nucleus and perinuclear area and activates IL-1β (Figure 2.1). This was an unexpected finding because inflammasome complexes are characteristically formed in the cytosol. The IFI16 inflammasome was specifically observed in endothelial cells whereas IFI16-mediated IFN response was observed in macrophages indicating differential functions of IFI16 depending on the infected cell type [6,64]. Importantly, cell types differ with respect to permissiveness for herpes simplex virus-1 (HSV-1) infection and as a consequence detection mechanisms likely differ in different cell types. In permissive cells such as epithelial cells for example, Knipe and colleagues demonstrated that nuclear localized IFI16 recognized viral DNA and induced IRF-3 signaling [65]. IFI16 does not appear to leave the nucleus in HSV-1 infected human foreskin fibroblast (HFF) cells, although this nuclear sensing of HSV-1 still requires cytoplasmic STING signaling. Interestingly, ICP0, a viral immediate early gene, has been shown to contribute to the degradation of IFI16 in permissive cells suggesting that the virus actively subverts IFI16 signaling [65]. Another study has also shown that IFI16 induces expression of chemokines such as CCL3 during HSV-1 infection of human monocyte-derived macrophages [66]. Interestingly, IFI16 is also suggested to have a role in DNA-induced activation of human monocyte derived DCs and primary DCs [67]. Furthermore, IFI16 is implicated in restriction of human CMV replication in human embryonic lung fibroblasts [68]. All of these findings indicated that IFI16 can detect viral DNA in both the nucleus and cytosol. A multipartite NLS in IFI16 can be modified by acetylation, events which trigger the redistribution of IFI16 into the cytosol following viral DNA detection [10].

P204

As mentioned earlier, although two recent phylogenetic studies demonstrated that IFI16 and murine p204 are unrelated PYHIN proteins with distinct evolutionary profiles, Unterholzner et al. found that p204 has similar IFN-β inducing functions to IFI16. Knockdown of p204 in RAW264.7 mouse macrophages inhibited IRF3 and NF-κB activation and induction of IFN-β in response to HSV-1 (Figure 2.1) [6]. Furthermore, using in vivo siRNA knockdown Conrady et al. demonstrated that p204 was the primary sensor that augments an antiviral response against HSV-1 infection in the cornea and other clinically important HSV infection sites such as the vaginal mucosa [69]. Consequently, knockdown of p204 caused increased viral loads in HSV-1 infection [69,70].

In light of the recent evolutionary findings it is important to reana-
lyze the findings from many of the previous studies to delineate the exact
functions of IFI16 and p204. Brunette et al. revealed that p204 in fact
is not a non-redundant sensor of DNA [7]. They took into account the
high homology between p204 and other mouse PYHIN proteins and
designed highly specific siRNAs targeting p204. Knockdown of p204
using these siRNAs did not impair IFN production triggered by ISD
motif [7]. Similarly, in the study that identified cGAS as the DNA sensor that activates type I IFNs by producing cGAMP, Sun et al. found that
knockdown of p204 did not inhibit the generation of cGAMP activity
and IFN-β expression in response to herring testis DNA transfected L929
cells [48]. It is possible that the inconsistencies between recent studies and
the earlier studies are due to the differences in the types of DNA stimulus
used. Earlier studies, including the one by Unterholzner et al., used HSV-1
viral infection or stimulation with VACV 70mer whereas the studies by
Brunette et al. and Sun et al. used other types of DNA to stimulate cells
[6,7,48]. Therefore, while it remains likely that p204 has a role in antiviral
immunity, whether it serves as a sensor of cytosolic DNA during virus
infection remains unclear. The generation of p204 and cGAS knockout
mice will likely aid in clarifying these issues.

p202 (IFI202) and other PYHIN Proteins

Unlike other PYHIN proteins p202 lacks a PYD and carries two HIN-
200 domains. It has been shown to stably bind to transfected dsDNA.
Interestingly, p202 is unique in its ability to sequester dsDNA by binding
to it and preventing caspase-1 activation mediated by AIM2 (Figure 2.1)
[12]. p202 is also linked to lupus susceptibility, which is discussed in detail
in a later section of this review.

A recent study systematically analyzed the ability of all human and 13
murine PYHINs to activate inflammasome and induce STING-dependent
IFN production using reconstitution assays in HEK293T cells [7]. The
most potent STING-IFN activators included pyrin-only protein, p213
(PYR-A), p207 (PYHIN-A), p203, and p202. Although murine AIM2
failed to activate IFN, human AIM2 was found to be a robust IFN activator. Inflammasome reconstitution assays with PYHIN proteins also
revealed that murine p211 (MNDA) and p212 (MNDAL) can promote
inflammasome activation. These findings suggest a role for additional
PYHIN proteins in the interferon and inflammasome responses.

TRANSCRIPTIONAL REGULATION BY PYHIN PROTEINS

PYHIN proteins also function as regulators of transcription. As mentioned earlier, this transcriptional regulatory activity is primarily achieved through protein–protein interactions mediated by the conserved MFHATVAT motif and a LXCXE pRb-binding motif of HIN-200 domain [3]. IFI16, IFIX, p202, and p204 are some of the PYHIN family members that have been shown to regulate cell cycle transcription factors such as p53, p21, pRb, and E2F resulting in cell cycle arrest at the G1/S transition phase of cell cycle [3]. Transcriptional repression activity of IFI16 and p202 has been investigated in detail by several studies and is described below. However, the role of transcriptional regulation by PYHINs in the development of innate immunity is relatively less understood and is an area that requires further study.

IFI16 has a crucial role in transcriptional regulation, which is executed through protein–protein interactions with transcription factors or by direct interaction with DNA. One of the major interacting partners of IFI16 is the tumor suppressor p53 [71]. Typically, activation of p53, p21, and pRb results in inhibition of the cell cycle at the G1/S transition [3]. It has been suggested that IFI16 collaborates with p53 and pRb to arrest G1/S transition [72,73]. IFI16 colocalizes with p53 in the nucleoplasm and p53 has been shown to bind to the HIN-200 domain of IFI16 [71,74,75]. This IFI16–p53 binding enhances sequence-specific DNA binding by p53 leading to transcriptional activation and increased expression of p53 target genes including p21. Further confirming the inhibitory role of IFI16 in the cell cycle, siRNA-mediated knockdown of IFI16 in human fibroblasts has been shown to repress p21 expression by downregulating p53 [73]. Furthermore, IFI16 is part of BASC (BRCA1 associated surveillance protein complex), the DNA repair multiprotein complex, which assembles following UV-damage of DNA [76]. Studies suggest that IFI16 may have a role in reduced development of certain tumors and it is possible that the ability of IFI16 to sense alternative structures of DNA such as cruciform and superhelical forms contributes to DNA damage detection during the repair response [31].

Similar to IFI16, p202 also acts as a transcriptional repressor. The targets of p202-mediated transcriptional regulation include NF-κB [77], AP-1 [78], MYOD1 [79], and myogenin [2]. The HIN-200A domain of p202 interacts with p65 and p50 subunits of NF-κB to repress its activity. Interestingly, p202 can bind to transcriptionally active p65/p50 heterodimer and block its association with DNA or bind to the suppressive p50 homodimer and promote its association with DNA [77]. Similarly, p202 binds to c-Fos and c-Jun thereby blocking their interactions with DNA in a sequence-specific manner, which leads to repression of AP-1 activity [78]. p202 also inhibits the progression of the cell cycle by interacting with transcription factors such as p53, p21, pRb and E2F [80]. Overexpression of p202 leads to increased expression of p21 and potentiates pRb-mediated inhibition of cell growth. Notably, p202 can also bind to E2F1 and E2F4 thereby inhibiting transcription of E2F target genes [3].

ROLE OF PYHIN PROTEINS IN THE DEVELOPMENT OF AUTOIMMUNITY

Failure to appropriately clear apoptotic cellular debris containing DNA has been linked to the pathogenesis of autoimmune diseases [81–83]. DNA sensors including TLR9 are believed to initiate these abnormal inflammatory responses in systemic lupus erythematosus (SLE), a complex multifactorial autoimmune disease [81–84]. The role of AIM2 and IFI16 in the development of SLE is still not clear. It has been shown that expression of AIM2 is markedly low in lupus-prone NZB and B6.*Nba2* genotypes of mice compared to wild type mice [85]. Also, AIM2 deficiency has been associated with reduced expression of *Fcgr2b*, the gene encoding FcγRIIB receptor, the deficiency of which is associated with increased lupus susceptibility [86]. However, whether AIM2 has any protective or detrimental role in the development of SLE is a question that remains to be addressed. With respect to IFI16, it has been reported that compared to PBMCs from healthy individuals, those from SLE patients express higher levels of IFI16 mRNA [87]. However, the role of IFI16 in SLE remains largely unknown.

In contrast, several studies have presented compelling evidence linking the PYHIN protein p202 to murine SLE. Through gene expression analysis between B6.*Nba2* and congenic wild type mice, p202 was identified as a candidate lupus susceptibility gene [88]. Increased level of p202 in splenic B cells was associated with defective apoptosis and elevated autoantibody production. It is well established that type I IFN signaling contributes to the development of murine SLE in B6.*Nba2* mice. Consistent with this, type I IFN receptor deficient mice do not develop lupus-like disease and this lack of disease is associated with decreased expression of p202 [89]. Interestingly, studies have shown that type I IFN signaling deficiency, which results in decreased auto-antibody production, also inhibits nuclear localization of p202 [90,91]. Notably, these studies also revealed that p202 deficiency is not sufficient to cause lupus-like disease in type I IFN deficient lupus susceptible mice (B6.*Nba2*-C) [90]. Further genetic studies have indicated that epistatic interactions involving Nba2 genes such as Ifi202, AIM2, and Fcgr2b have a role in production of type I IFN and autoantibodies in B6.*Nba2* mice [85,86,92]. Interestingly, recent studies suggest that p202 contributes to lupus development through regulation of the AIM2 inflammasome [91]. Knockdown of *Ifi202* has been shown to increase cytosolic DNA-induced caspase-1 activation in BMDMs [12]. Similarly, increased expression of p202 in immune cells leads to elevated type I IFN signaling and decreased AIM2, pro–IL-1β and FcγRIIB levels [85,86]. It has also been shown that p202 can heterodimerize with other PYHIN proteins and, therefore, it is believed that p202 mediates its AIM2–inhibitory activity through binding with the AIM2 protein itself besides via sequestering DNA [93,94].

REGULATION OF PYHIN PROTEINS

As mentioned earlier IFNs dramatically regulate PYHIN gene expression. However, additional regulatory mechanisms also exist for several other PYHIN proteins especially the inflammasome associated PYHINs in line with the potential of the IL-1 family of cytokines to cause tissue damage and chronic inflammation. Not surprisingly, activation of AIM2 inflammasome is tightly controlled by various host molecules including other PYHIN proteins. Regulation of AIM2 by p202 has been described in the previous section. Another PYHIN protein that has been shown to negatively regulate AIM2 and subsequent caspase-1 activation is IFI16 [95]. In fact, AIM2 and IFI16 have been suggested to counter-regulate each other. Furthermore, it is possible that recently predicted alternatively spliced isoforms of AIM2 that lack a PYD domain can also function similar to p202 and inhibit activation of AIM2 and IFI16 [5,96].

A non-PYHIN host molecule that has been shown to inhibit AIM2 inflammasome and the DNA-induced IL-1β is cathelicidin (LL-37), an antimicrobial peptide that contributes to disease pathology in psoriatic skin lesions [97]. LL-37 forms a complex with self-DNA released from dying cells in psoriasis and accelerates intake of self-DNA resulting in activation of TLR9 in plasmacytoid DCs. It appears that the tight association of LL-37 with DNA due to the cationic charge of LL-37 or conformational changes in DNA structure sequesters the DNA and prevents sensing by the AIM2 inflammasome [97]. Additionally, studies by Panchanathan et al. have revealed a gender-dependent differential regulation of AIM2 and p202 protein expression in mice [98]. They found that treatment of mice with estrogen increases the expression of the *Ifi202* gene whereas treatment with androgen decreases its levels. In contrast, expression of the *Aim2* gene increased in response to androgen treatment and the steady state expression of *Aim2* gene is lower in immune cells of female mice compared to male mice [98]. This is an interesting observation in the context of increased susceptibility of females to SLE and the proposed protective role of AIM2 in this autoimmune disease. In addition to IFN and estrogen, Ifi202 expression is also induced in immune cells by IL-6 [99] and in B and T cells upon stimulation with anti-CD3 and anti-CD28 antibodies [100]. Interestingly, IFI16 expression has been shown to be regulated by certain cellular processes such as DNA methylation, which decreases or abolishes the expression of the *IFI16* gene in some cell types such as immortalized human diploid fibroblasts [101].

Given the the importance of inflammasomes and type I IFNs induced by PYHIN proteins in antimicrobial defenses, it is not unexpected that infectious agents have developed strategies to counter the functions of PYHIN proteins. MviN, a putative lipid II phospholipid translocase expressed by *F. tularensis*, has been shown to limit AIM2 inflammasome activation [102]. Another *Francisella* protein, RipA, also suppresses inflammasome activation, however the exact mechanism by which these proteins exert their inhibitory effect is not clear [103]. As mentioned above, the herpes viral protein ICP0 targets IFI16 for degradation thereby limiting protective type I IFN responses [65]. Several additional viral proteins have also been shown to bind to various PYHIN proteins and it is possible that they also regulate the function of PYHIN proteins. One such protein is latency associated nuclear antigen (LANA) of KSHV, which bound to MNDA in yeast two-hybrid assays [104]. Similarly, E1A from adenovirus interacts with p202 and rescues p202-mediated suppression of E2F transcriptional activity [105]. Furthermore, human CMV protein pUL83 binds to IFI16 [106]. Further studies are required to determine the downstream consequences of these viral protein–PYHIN interactions and to evaluate their role in the development of an inflammatory response.

CONCLUSIONS

Members of the PYHIN protein family have clearly emerged as important factors which contribute to innate immune sensing of DNA, and this family is therefore important in both antimicrobial defenses and the development of autoimmune diseases such as SLE. The PYHIN protein family consists of a continuously expanding list of predominantly nuclear proteins that are expressed mainly in hematopoietic tissues. They typically elicit two important effector responses: inflammasome activation and induction of type I IFNs. Although certain members of the family such as AIM2 are better characterized, the biological significance of other members particularly those linked to DNA sensing is still being investigated. Their astounding evolutionary diversity with little preservation among different species indicates that each member of the PYHIN family has a specific biological function. Extensive studies that involve generation of knockout mice for these proteins are required to address these important questions.

REFERENCES

[1] Cridland JA, Curley EZ, Wykes MN, Schroder K, Sweet MJ, Roberts TL, et al. The mammalian PYHIN gene family: phylogeny, evolution and expression. BMC Evolut Biol 2012;12:140.

[2] Schattgen SA, Fitzgerald KA. The PYHIN protein family as mediators of host defenses. Immunol Rev 2011;243(1):109–18.

[3] Ludlow LE, Johnstone RW, Clarke CJ. The HIN-200 family: more than interferon-inducible genes?. Exp Cell Res 2005;308(1):1–17.

[4] Asefa B, Klarmann KD, Copeland NG, Gilbert DJ, Jenkins NA, Keller JR. The interferon-inducible p200 family of proteins: a perspective on their roles in cell cycle regulation and differentiation. Blood Cells Mols Dis 2004;32(1):155–67.

[5] Rathinam VA, Vanaja SK, Fitzgerald KA. Regulation of inflammasome signaling. Nat Immunol 2012;13(4):333–42.

[6] Unterholzner L, Keating SE, Baran M, Horan KA, Jensen SB, Sharma S, et al. IFI16 is an innate immune sensor for intracellular DNA. Nat Immunol 2010;11(11):997–1004.

[7] Brunette RL, Young JM, Whitley DG, Brodsky IE, Malik HS, Stetson DB. Extensive evolutionary and functional diversity among mammalian AIM2-like receptors. J Exp Med 2012;209(11):1969–83.

[8] Briggs LJ, Johnstone RW, Elliot RM, Xiao CY, Dawson M, Trapani JA, et al. Novel properties of the protein kinase CK2-site-regulated nuclear-localization sequence of the interferon-induced nuclear factor IFI 16. Biochem J 2001;353(Pt 1):69–77.

[9] Liu C, Wang H, Zhao Z, Yu S, Lu YB, Meyer J, et al. MyoD-dependent induction during myoblast differentiation of p204, a protein also inducible by interferon. Mol Cell Biol 2000;20(18):7024–36.

[10] Li T, Diner BA, Chen J, Cristea IM. Acetylation modulates cellular distribution and DNA sensing ability of interferon-inducible protein IFI16. Proc Natl Acad Sci USA 2012;109(26):10558–63.

[11] Hornung V, Ablasser A, Charrel-Dennis M, Bauernfeind F, Horvath G, Caffrey DR, et al. AIM2 recognizes cytosolic dsDNA and forms a caspase-1-activating inflammasome with ASC. Nature 2009;458(7237):514–8.

[12] Roberts TL, Idris A, Dunn JA, Kelly GM, Burnton CM, Hodgson S, et al. HIN-200 proteins regulate caspase activation in response to foreign cytoplasmic DNA. Science 2009;323(5917):1057–60.

[13] Choubey D, Lengyel P. Interferon action: cytoplasmic and nuclear localization of the interferon-inducible 52-kD protein that is encoded by the Ifi 200 gene from the gene 200 cluster. J Interfer Res 1993;13(1):43–52.

[14] DeYoung KL, Ray ME, Su YA, Anzick SL, Johnstone RW, Trapani JA, et al. Cloning a novel member of the human interferon-inducible gene family associated with control of tumorigenicity in a model of human melanoma. Oncogene 1997;15(4):453–7.

[15] Ding Y, Wang L, Su LK, Frey JA, Shao R, Hunt KK, et al. Antitumor activity of IFIX, a novel interferon-inducible HIN-200 gene, in breast cancer. Oncogene 2004;23(26):4556–66.

[16] Gariglio M, Azzimonti B, Pagano M, Palestro G, De Andrea M, Valente G, et al. Immunohistochemical expression analysis of the human interferon-inducible gene IFI16, a member of the HIN200 family, not restricted to hematopoietic cells. J Interf Cytok Res: Off J Int Soc Interf Cytok Res 2002;22(7):815–21.

[17] Gariglio M, De Andrea M, Lembo M, Ravotto M, Zappador C, Valente G, et al. The murine homolog of the HIN 200 family, Ifi 204, is constitutively expressed in myeloid cells and selectively induced in the monocyte/macrophage lineage. J Leucocyte Biol 1998;64(5):608–14.

[18] Wei W, Clarke CJ, Somers GR, Cresswell KS, Loveland KA, Trapani JA, et al. Expression of IFI 16 in epithelial cells and lymphoid tissues. Histochem Cell Biol 2003;119(1):45–54.

[19] Miranda RN, Briggs RC, Shults K, Kinney MC, Jensen RA, Cousar JB. Immunocytochemical analysis of MNDA in tissue sections and sorted normal bone marrow cells documents expression only in maturing normal and neoplastic myelomonocytic cells and a subset of normal and neoplastic B lymphocytes. Hum Pathol 1999;30(9):1040–9.

[20] Dawson MJ, Trapani JA, Briggs RC, Nicholl JK, Sutherland GR, Baker E. The closely linked genes encoding the myeloid nuclear differentiation antigen (MNDA) and IFI16 exhibit contrasting haemopoietic expression. Immunogenetics 1995;41(1):40–3.

[21] Weiler SR, Gooya JM, Ortiz M, Tsai S, Collins SJ, Keller JR. D3: a gene induced during myeloid cell differentiation of Linlo c-Kit+ Sca-1(+) progenitor cells. Blood 1999;93(2):527–36.

[22] Zhang K, Kagan D, DuBois W, Robinson R, Bliskovsky V, Vass WC, et al. Mndal, a new interferon-inducible family member, is highly polymorphic, suppresses cell growth, and may modify plasmacytoma susceptibility. Blood 2009;114(14):2952–60.

[23] Park HH, Lo YC, Lin SC, Wang L, Yang JK, Wu H. The death domain superfamily in intracellular signaling of apoptosis and inflammation. Ann Rev Immunol 2007;25:561–86.

[24] Pawlowski K, Pio F, Chu Z, Reed JC, Godzik A. PAAD – a new protein domain associated with apoptosis, cancer and autoimmune diseases. Trends Biochem Sci 2001;26(2):85–7.

[25] Staub E, Dahl E, Rosenthal A. The DAPIN family: a novel domain links apoptotic and interferon response proteins. Trends Biochem Sci 2001;26(2):83–5.

[26] Cresswell KS, Clarke CJ, Jackson JT, Darcy PK, Trapani JA, Johnstone RW. Biochemical and growth regulatory activities of the HIN-200 family member and putative tumor suppressor protein, AIM2. Biochem Biophys Res Commun 2005;326(2):417–24.

[27] Johnstone RW, Kershaw MH, Trapani JA. Isotypic variants of the interferon-inducible transcriptional repressor IFI 16 arise through differential mRNA splicing. Biochemistry 1998;37(34):11924–31.

[28] Xie J, Briggs JA, Briggs RC. MNDA dimerizes through a complex motif involving an N-terminal basic region. FEBS letters 1997;408(2):151–5.

[29] Albrecht M, Choubey D, Lengauer T. The HIN domain of IFI-200 proteins consists of two OB folds. Biochem Biophys Res Commun 2005;327(3):679–87.

[30] Yan H, Dalal K, Hon BK, Youkharibache P, Lau D, Pio F. RPA nucleic acid-binding properties of IFI16-HIN200. Biochim Biophys Acta 2008;1784(7–8):1087–97.

[31] Brazda V, Coufal J, Liao JC, Arrowsmith CH. Preferential binding of IFI16 protein to cruciform structure and superhelical DNA. Biochem Biophys Res Commun 2012;422(4):716–20.

[32] Jin T, Perry A, Jiang J, Smith P, Curry JA, Unterholzner L, et al. Structures of the HIN domain:DNA complexes reveal ligand binding and activation mechanisms of the AIM2 inflammasome and IFI16 receptor. Immunity 2012;36(4):561–71.

[33] Martinon F, Burns K, Tschopp J. The inflammasome: a molecular platform triggering activation of inflammatory caspases and processing of proIL-beta. Mol Cell 2002;10(2):417–26.

[34] Bryan NB, Dorfleutner A, Rojanasakul Y, Stehlik C. Activation of inflammasomes requires intracellular redistribution of the apoptotic speck-like protein containing a caspase recruitment domain. J Immunol 2009;182(5):3173–82.

[35] Schroder K, Tschopp J. The inflammasomes. Cell 2010;140(6):821–32.

[36] Dinarello CA. IL-1: discoveries, controversies and future directions. Eur J Immunol 2010;40(3):599–606.

[37] Luft T, Jefford M, Luetjens P, Hochrein H, Masterman KA, Maliszewski C, et al. IL-1 beta enhances CD40 ligand-mediated cytokine secretion by human dendritic cells (DC): a mechanism for T cell-independent DC activation. J Immunol 2002;168(2):713–22.

[38] Rathinam VA, Fitzgerald KA. Inflammasomes and anti-viral immunity. J Clin Immunol 2010;30(5):632–7.

[39] Acosta-Rodriguez EV, Napolitani G, Lanzavecchia A, Sallusto F. Interleukins 1beta and 6 but not transforming growth factor-beta are essential for the differentiation of interleukin 17-producing human T helper cells. Nat Immunol 2007;8(9):942–9.

[40] Dupaul-Chicoine J, Yeretssian G, Doiron K, Bergstrom KS, McIntire CR, LeBlanc PM, et al. Control of intestinal homeostasis, colitis, and colitis-associated colorectal cancer by the inflammatory caspases. Immunity 2010;32(3):367–78.

[41] Chaix J, Tessmer MS, Hoebe K, Fuseri N, Ryffel B, Dalod M, et al. Cutting edge: Priming of NK cells by IL-18. J Immunol 2008;181(3):1627–31.

[42] Micallef MJ, Ohtsuki T, Kohno K, Tanabe F, Ushio S, Namba M, et al. Interferon-gamma-inducing factor enhances T helper 1 cytokine production by stimulated human T cells: synergism with interleukin-12 for interferon-gamma production. Eur J Immunol 1996;26(7):1647–51.

[43] Nakanishi K, Yoshimoto T, Tsutsui H, Okamura H. Interleukin-18 is a unique cytokine that stimulates both Th1 and Th2 responses depending on its cytokine milieu. Cytokine Growth Fact Rev 2001;12(1):53–72.

[44] Stetson DB, Medzhitov R. Recognition of cytosolic DNA activates an IRF3-dependent innate immune response. Immunity 2006;24(1):93–103.

[45] Ishii KJ, Coban C, Kato H, Takahashi K, Torii Y, Takeshita F, et al. A Toll-like receptor-independent antiviral response induced by double-stranded B-form DNA. Nat Immunol 2006;7(1):40–8.

[46] Rathinam VA, Fitzgerald KA. Cytosolic surveillance and antiviral immunity. Curr Opinion Virol 2011;1(6):455–62.

[47] Ishikawa H, Ma Z, Barber GN. STING regulates intracellular DNA-mediated, type I interferon-dependent innate immunity. Nature 2009;461(7265):788–92.

[48] Sun L, Wu J, Du F, Chen X, Chen ZJ. Cyclic GMP-AMP synthase is a cytosolic DNA sensor that activates the type I interferon pathway. Science 2012;339(6121):786–91.

[49] Wu J, Sun L, Chen X, Du F, Shi H, Chen C, et al. Cyclic GMP-AMP is an endogenous second messenger in innate immune signaling by cytosolic DNA. Science 2012;339(6121):826–30.

[50] Fernandes-Alnemri T, Yu JW, Datta P, Wu J, Alnemri ES. AIM2 activates the inflammasome and cell death in response to cytoplasmic DNA. Nature 2009;458(7237):509–13.

[51] Burckstummer T, Baumann C, Bluml S, Dixit E, Durnberger G, Jahn H, et al. An orthogonal proteomic-genomic screen identifies AIM2 as a cytoplasmic DNA sensor for the inflammasome. Nat Immunol 2009;10(3):266–72.

[52] Rathinam VA, Jiang Z, Waggoner SN, Sharma S, Cole LE, Waggoner L, et al. The AIM2 inflammasome is essential for host defense against cytosolic bacteria and DNA viruses. Nat Immunol 2010;11(5):395–402.

[53] Fernandes-Alnemri T, Yu JW, Juliana C, Solorzano L, Kang S, Wu J, et al. The AIM2 inflammasome is critical for innate immunity to *Francisella tularensis*. Nat Immunol 2010;11(5):385–93.

[54] Jones JW, Kayagaki N, Broz P, Henry T, Newton K, O'Rourke K, et al. Absent in melanoma 2 is required for innate immune recognition of *Francisella tularensis*. Proc Natl Acad Sci USA 2010;107(21):9771–6.

[55] Warren SE, Armstrong A, Hamilton MK, Mao DP, Leaf IA, Miao EA, et al. Cutting edge: Cytosolic bacterial DNA activates the inflammasome via Aim2. J Immunol 2010;185(2):818–21.

[56] Clemens DL, Lee BY, Horwitz MA. *Francisella tularensis* phagosomal escape does not require acidification of the phagosome. Infect Immun 2009;77(5):1757–73.

[57] Clemens DL, Lee BY, Horwitz MA. Virulent and avirulent strains of *Francisella tularensis* prevent acidification and maturation of their phagosomes and escape into the cytoplasm in human macrophages. Infect Immun 2004;72(6):3204–17.

[58] Henry T, Brotcke A, Weiss DS, Thompson LJ, Monack DM. Type I interferon signaling is required for activation of the inflammasome during *Francisella* infection. J Exp Med 2007;204(5):987–94.

[59] Kim S, Bauernfeind F, Ablasser A, Hartmann G, Fitzgerald KA, Latz E, et al. *Listeria monocytogenes* is sensed by the NLRP3 and AIM2 inflammasome. Eur J Immunol 2010;40(6):1545–51.

[60] Sauer JD, Witte CE, Zemansky J, Hanson B, Lauer P, Portnoy DA. *Listeria monocytogenes* triggers AIM2-mediated pyroptosis upon infrequent bacteriolysis in the macrophage cytosol. Cell Host Microbe 2010;7(5):412–9.

[61] Wu J, Fernandes-Alnemri T, Alnemri ES. Involvement of the AIM2, NLRC4, and NLRP3 inflammasomes in caspase-1 activation by *Listeria monocytogenes*. J Clin Immunol 2010;30(5):693–702.

[62] Fang R, Tsuchiya K, Kawamura I, Shen Y, Hara H, Sakai S, et al. Critical roles of ASC inflammasomes in caspase-1 activation and host innate resistance to *Streptococcus pneumoniae* infection. J Immunol 2011;187(9):4890–9.

[63] Koppe U, Hogner K, Doehn JM, Muller HC, Witzenrath M, Gutbier B, et al. *Streptococcus pneumoniae* stimulates a STING- and IFN regulatory factor 3-dependent type I IFN production in macrophages, which regulates RANTES production in macrophages, cocultured alveolar epithelial cells, and mouse lungs. J Immunol 2012;188(2):811–7.

[64] Kerur N, Veettil MV, Sharma-Walia N, Bottero V, Sadagopan S, Otageri P, et al. IFI16 acts as a nuclear pathogen sensor to induce the inflammasome in response to Kaposi sarcoma-associated herpesvirus infection. Cell Host Microbe 2011;9(5):363–75.

[65] Orzalli MH, DeLuca NA, Knipe DM. Nuclear IFI16 induction of IRF-3 signaling during herpesviral infection and degradation of IFI16 by the viral ICP0 protein. Proc Natl Acad Sci USA 2012;109(44):E3008–E3017.

[66] Soby S, Laursen RR, Ostergaard L, Melchjorsen J. HSV-1-induced chemokine expression via IFI16-dependent and IFI16-independent pathways in human monocyte-derived macrophages. Herpesviridae 2012;3(1):6.

[67] Kis-Toth K, Szanto A, Thai TH, Tsokos GC. Cytosolic DNA-activated human dendritic cells are potent activators of the adaptive immune response. J Immunol 2011;187(3):1222–34.

[68] Gariano GR, Dell'Oste V, Bronzini M, Gatti D, Luganini A, De Andrea M, et al. The intracellular DNA sensor IFI16 gene acts as restriction factor for human cytomegalovirus replication. PLoS pathogens 2012;8(1):e1002498.

[69] Conrady CD, Zheng M, Fitzgerald KA, Liu C, Carr DJ. Resistance to HSV-1 infection in the epithelium resides with the novel innate sensor, IFI-16. Mucosal Immunol 2012;5(2):173–83.

[70] Conrady CD, Zheng M, Mandal NA, van Rooijen N, Carr DJ. IFN-alpha-driven CCL2 production recruits inflammatory monocytes to infection site in mice. Mucosal Immunol 2013;6(1):45–55.

[71] Johnstone RW, Wei W, Greenway A, Trapani JA. Functional interaction between p53 and the interferon-inducible nucleoprotein IFI 16. Oncogene 2000;19(52):6033–42.

[72] Raffaella R, Gioia D, De Andrea M, Cappello P, Giovarelli M, Marconi P, et al. The interferon-inducible IFI16 gene inhibits tube morphogenesis and proliferation of primary, but not HPV16 E6/E7-immortalized human endothelial cells. Exp Cell Res 2004;293(2):331–45.

[73] Xin H, Curry J, Johnstone RW, Nickoloff BJ, Choubey D. Role of IFI 16, a member of the interferon-inducible p200-protein family, in prostate epithelial cellular senescence. Oncogene 2003;22(31):4831–40.

[74] Fujiuchi N, Aglipay JA, Ohtsuka T, Maehara N, Sahin F, Su GH, et al. Requirement of IFI16 for the maximal activation of p53 induced by ionizing radiation. J Biol Chem 2004;279(19):20339–44.

[75] Liao JC, Lam R, Brazda V, Duan S, Ravichandran M, Ma J, et al. Interferon-inducible protein 16: insight into the interaction with tumor suppressor p53. Structure 2011;19(3):418–29.

[76] Johnstone RW, Kerry JA, Trapani JA. The human interferon-inducible protein, IFI 16, is a repressor of transcription. J Biol Chem 1998;273(27):17172–7.

[77] Ma XY, Wang H, Ding B, Zhong H, Ghosh S, Lengyel P. The interferon-inducible p202a protein modulates NF-kappaB activity by inhibiting the binding to DNA of p50/p65 heterodimers and p65 homodimers while enhancing the binding of p50 homodimers. J Biol Chem 2003;278(25):23008–19.

[78] Min W, Ghosh S, Lengyel P. The interferon-inducible p202 protein as a modulator of transcription: inhibition of NF-kappa B, c-Fos, and c-Jun activities. Mol Cell Biol 1996;16(1):359–68.

[79] Datta B, Min W, Burma S, Lengyel P. Increase in p202 expression during skeletal muscle differentiation: inhibition of MyoD protein expression and activity by p202. Mol Cell Biol 1998;18(2):1074–83.

[80] Datta B, Li B, Choubey D, Nallur G, Lengyel P. p202, an interferon-inducible modulator of transcription, inhibits transcriptional activation by the p53 tumor suppressor protein, and a segment from the p53-binding protein 1 that binds to p202 overcomes this inhibition. J Biol Chem 1996;271(44):27544–55.

[81] Munoz LE, Janko C, Schulze C, Schorn C, Sarter K, Schett G, et al. Autoimmunity and chronic inflammation - two clearance-related steps in the etiopathogenesis of SLE. Autoimmun Rev 2010;10(1):38–42.

[82] Baccala R, Gonzalez-Quintial R, Lawson BR, Stern ME, Kono DH, Beutler B, et al. Sensors of the innate immune system: their mode of action. Nature Rev Rheumatol 2009;5(8):448–56.

[83] Baccala R, Hoebe K, Kono DH, Beutler B, Theofilopoulos AN. TLR-dependent and TLR-independent pathways of type I interferon induction in systemic autoimmunity. Nature medicine 2007;13(5):543–51.

[84] Crispin JC, Liossis SN, Kis-Toth K, Lieberman LA, Kyttaris VC, Juang YT, et al. Pathogenesis of human systemic lupus erythematosus: recent advances. Trends Mol Med 2010;16(2):47–57.

[85] Panchanathan R, Duan X, Shen H, Rathinam VA, Erickson LD, Fitzgerald KA, et al. Aim2 deficiency stimulates the expression of IFN-inducible Ifi202, a lupus susceptibility murine gene within the Nba2 autoimmune susceptibility locus. J Immunol 2010;185(12):7385–93.

[86] Panchanathan R, Shen H, Duan X, Rathinam VA, Erickson LD, Fitzgerald KA, et al. Aim2 deficiency in mice suppresses the expression of the inhibitory Fcgamma receptor (FcgammaRIIB) through the induction of the IFN-inducible p202, a lupus susceptibility protein. J Immunol 2011;186(12):6762–70.

[87] Kimkong I, Avihingsanon Y, Hirankarn N. Expression profile of HIN200 in leukocytes and renal biopsy of SLE patients by real-time RT-PCR. Lupus 2009;18(12):1066–72.

[88] Rozzo SJ, Allard JD, Choubey D, Vyse TJ, Izui S, Peltz G, et al. Evidence for an interferon-inducible gene, Ifi202, in the susceptibility to systemic lupus. Immunity 2001;15(3):435–43.

[89] Jorgensen TN, Roper E, Thurman JM, Marrack P, Kotzin BL. Type I interferon signaling is involved in the spontaneous development of lupus-like disease in B6.Nba2 and (B6.Nba2 x NZW)F(1) mice. Genes Immun 2007;8(8):653–62.

[90] Jorgensen TN, Alfaro J, Enriquez HL, Jiang C, Loo WM, Atencio S, et al. Development of murine lupus involves the combined genetic contribution of the SLAM and FcgammaR intervals within the Nba2 autoimmune susceptibility locus. J Immunol 2010;184(2):775–86.

[91] Choubey D. DNA-responsive inflammasomes and their regulators in autoimmunity. Clin Immunol 2012;142(3):223–31.

[92] Bolland S, Yim YS, Tus K, Wakeland EK, Ravetch JV. Genetic modifiers of systemic lupus erythematosus in FcgammaRIIB(−/−) mice. J Exp Med 2002;195(9):1167–74.

[93] Choubey D. P202: an interferon-inducible negative regulator of cell growth. J Biol Regulat Homeostat Agents 2000;14(3):187–92.

[94] Choubey D, Kotzin BL. Interferon-inducible p202 in the susceptibility to systemic lupus. Front Biosci: A J Virtual Libr 2002;7:e252–62.

[95] Veeranki S, Duan X, Panchanathan R, Liu H, Choubey D. IFI16 protein mediates the anti-inflammatory actions of the type-I interferons through suppression of activation of caspase-1 by inflammasomes. PloS one 2011;6(10):e27040.

[96] Choubey D, Duan X, Dickerson E, Ponomareva L, Panchanathan R, Shen H, et al. Interferon-inducible p200-family proteins as novel sensors of cytoplasmic DNA: role in inflammation and autoimmunity. J Interf Cytok Res: Off J Int Soc Interf Cytok Res 2010;30(6):371–80.

[97] Dombrowski Y, Peric M, Koglin S, Kammerbauer C, Goss C, Anz D, et al. Cytosolic DNA triggers inflammasome activation in keratinocytes in psoriatic lesions. Sci Translat Med 2011;3(82):82ra38.

[98] Panchanathan R, Duan X, Arumugam M, Shen H, Liu H, Choubey D. Cell type and gender-dependent differential regulation of the p202 and Aim2 proteins: implications for the regulation of innate immune responses in SLE. Mol Immunol 2011;49(1–2):273–80.

[99] Pramanik R, Jorgensen TN, Xin H, Kotzin BL, Choubey D. Interleukin-6 induces expression of Ifi202, an interferon-inducible candidate gene for lupus susceptibility. J Biol Chem 2004;279(16):16121–7.

[100] Chen J, Panchanathan R, Choubey D. Stimulation of T cells up-regulates expression of Ifi202, an interferon-inducible lupus susceptibility gene, through activation of JNK/c-Jun pathway. Immunol Lett 2008;118(1):13–20.

[101] Xin H, Pereira-Smith OM, Choubey D. Role of IFI 16 in cellular senescence of human fibroblasts. Oncogene 2004;23(37):6209–17.

[102] Ulland TK, Buchan BW, Ketterer MR, Fernandes-Alnemri T, Meyerholz DK, Apicella MA, et al. Cutting edge: mutation of *Francisella tularensis* mviN leads to increased macrophage absent in melanoma 2 inflammasome activation and a loss of virulence. J Immunol 2010;185(5):2670–4.

[103] Huang MT, Mortensen BL, Taxman DJ, Craven RR, Taft-Benz S, Kijek TM, et al. Deletion of ripA alleviates suppression of the inflammasome and MAPK by Francisella tularensis. J Immunol 2010;185(9):5476–85.

[104] Fukushi M, Higuchi M, Oie M, Tetsuka T, Kasolo F, Ichiyama K, et al. Latency-associated nuclear antigen of Kaposi's sarcoma-associated herpesvirus interacts with human myeloid cell nuclear differentiation antigen induced by interferon alpha. Virus Gene 2003;27(3):237–47.

[105] Xin H, D'Souza S, Fang L, Lengyel P, Choubey D. p202, an interferon-inducible negative regulator of cell growth, is a target of the adenovirus E1A protein. Oncogene 2001;20(47):6828–39.

[106] Cristea IM, Moorman NJ, Terhune SS, Cuevas CD, O'Keefe ES, Rout MP, et al. Human cytomegalovirus pUL83 stimulates activity of the viral immediate-early promoter through its interaction with the cellular IFI16 protein. J Virol 2010;84(15):7803–14.

CHAPTER 3

Cytosolic DNA-Sensing and the STING Pathway

Glen N. Barber
Department of Cell Biology and Sylvester Comprehensive Cancer Center, University of Miami Miller School of Medicine, Miami, FL, USA

INTRODUCTION

Invading microbes are detected by cellular sensors, the consequences of which result in the production of potent anti-pathogen proteins such as type I interferon (IFN) as well as other cytokines capable of stimulating the adaptive immune response. Examples comprise the RIG-I-like helicase (RLH) and the Toll-like receptor (TLR) families which recognize non-self-pathogen derived molecules (PAMPs) including bacterial lipopolysaccharides as well as nucleic acids [1–4]. In addition, an endoplasmic reticulum (ER) associated transmembrane protein referred to as STING (for stimulator of interferon genes) was established as being essential for triggering the production of innate immune proteins in response to the sensing of cytosolic DNA. Such DNA can be 'self'-DNA produced from necrotic or apoptotic cells, or the actual genomes of DNA pathogens that become exposed following infection [5–6]. Moreover, while STING appears essential for controlling innate signaling events triggered by DNA microbes, chronic STING activation also appears to be responsible for certain inflammatory diseases manifested by 'self'-DNA. Thus, understanding STING function may lead to the design of new compounds that may facilitate vaccine development, or conversely that may provide new therapies for the treatment of inflammatory disease.

Microbial-derived nucleic acids, typically constituting the genomes of pathogens (referred to as PAMPs, pathogen associated molecular patterns, or DAMPS, for danger associated molecular patterns) are robust stimulators of host-defense related innate immune responses [1–3]. Indeed, a number of cellular sensors have evolved, in different cell-types, to distinguish pathogen-associated RNA or DNA following infection. However, while significant progress has been made in determining how the cell recognizes

Biological DNA Sensor.
DOI: http://dx.doi.org/10.1016/B978-0-12-404732-7.00003-4

microbial, especially viral-derived RNA, species [3,4], it has been less clear how the cell triggers innate immune signaling in response to intracellular DNA, such as that of DNA viruses like herpes simplex virus (HSV) or bacteria including *Listeria monocytogenes*. The triggering of these defense pathways is essential for protecting us from lethal disease. In addition, microbial derived or endogenous self-DNA may be responsible for stimulating our own innate immune pathways and proinflammatory gene induction which can trigger autoimmune disorders [5,6]. DNA sensing pathways are also critical for initiating innate and adaptive immune pathways mediated by plasmid-based vaccines and perhaps gene therapy procedures [7,8]. Comprehending the mechanisms of DNA-dependent innate signaling may therefore lead to the development of more effective DNA vaccine and gene therapeutic strategies [8]. Within the last decade, considerable progress has been made in understanding the mechanisms of cytosolic DNA signaling. The most well characterized DNA sensor is now known to be a member of the Toll-like pathway, referred to as Toll-like receptor 9 (TLR9). TLR9 has been shown to be important for recognizing pathogen derived non-self CpG species [9,10]. Conversely, a component of the Nod-like receptor family, known as AIM2, has been shown to be important for cytosolic DNA-mediated inflammasome-dependent innate signaling [11–14]. However, less was known about non-CpG intracellular dsDNA-mediated innate immune signaling processes, such as those critical for producing type I IFN. Then, in 2008, the identification of a molecule referred to as STING (for stimulator of interferon genes) was reported [15]. It is now known that STING is evolutionarily conserved and controls a novel TLR-independent, cytosolic DNA activated innate immune signaling pathway [15,16]. Here, we describe the discovery and importance of STING in intracellular DNA signaling, the understanding of which may lead to an increase in our knowledge of the causes of inflammatory disease and may generate new therapies to help combat such disorders.

THE TOLL-LIKE PATHWAY: TLR9 AND CPG DNA

TLR9 was one of the first sensors found to play a role in the triggering of innate immune signaling in response to foreign DNA [9]. There are approximately 13 TLRs in humans and mice, which sense a variety of microbial PAMPs (pathogen associated molecule patterns) including lipopolysaccharides (TLR4) which are common on bacterial cell walls, viral dsRNA (TLR3), viral ssRNA (TLR7/8), as well as viral or bacterial unmethylated DNA referred to as CpG DNA (TLR9) [1–3]. TLR9 is mainly expressed in pDCs and B cells and recognizes CpG (cytidine-phosphate-guanosine) DNA typically found in bacteria and viruses, but atypical in vertebrates. TLR9 contains a leucine rich repeat \(LRR) motif, a Toll/IL-1R (TIR) homology domain and is a member of the type I integral membrane glycoprotein family. Inactive TLR9 is principally found in the endoplasmic reticulum (ER) of resting pDCs. The receptor traffics to lysosomal and endosomal organelles from the ER and after cleavage becomes active [17,18]. CpG DNA that is internalized by pDCs traffics to and binds to TLR9 that has also localized from the ER to these regions. TLR9 interacts with the TIR and death domain (DD) containing adaptor myeloid differentiation primary response gene 88 (MyD88) which in turn associates with IRAK-1 (IL-1R associated kinase 1) and IRAK-4 as well as select TRAFs (tumor necrosis factor (TNF) receptor associated factors) to activate the transcription factors IRF-7 (interferon regulatory factor 7), MAPK (mitogen activated protein kinase) and NF-κB (nuclear factor kappa-light-chain-enhancer of activated B cells), collectively required for the transcription of IFNβ.

Genetically engineered mice that lack TLR9 are viable. This enabled the importance of TLR9 to be evaluated in host defense against pathogen infection. Such studies indicated that TLR9 could stimulate type I IFN production following infection with DNA viruses such as herpes simplex virus [19,20]. Further reports indicated that TLR9 may play a role in stimulating adaptive immune responses in response to plasmid based immunization protocols [21]. In this light, the use of TLR ligands has been used to improve the immunogenicity of a variety of antigens used in vaccine strategies. For example, synthetic oligodeoxynucleotide (ODN) ligands, such as CpG ODN for TLR9, have been reported to be potent adjuvants which enhance immunization events such as T and B cell activation [10]. Aside from exhibiting strong adjuvant properties, CpG monotherapy has also been demonstrated to be effective against a wide range of viral, bacterial and parasitic microbes [8,10]. In addition, TLR9 agonists have been utilized as treatments for hematologic malignancies, melanoma and glioblastoma. However, through these studies, it became apparent that transfected DNA, DNA based pathogens, as well as plasmid DNA based immunization regimes were still able to mount immune responses in cells or mice that did not express TLR9 [21,22]. Therefore, while TLR9 signaling certainly comprises a key element of our innate immune responses, evidence suggested the existence of additional DNA sensors, especially in non–pDCs [23]. Subsequently, in 2008, the identification of STING was reported and shown to play a key role in facilitating innate immunity in response to DNA pathogens and cytosolic DNA, as we describe next.

THE STING PATHWAY AND INTRACELLULAR DNA SIGNALING

It was apparent that a TLR9–independent cytosolic DNA signaling pathway existed in a variety of cells [24]. Consequently, STING was discovered by screening a cDNA expression library and identifying molecules that could activate the IFNβ promoter when transfected in 293T cells [15,16]. STING turned out to be a 398 amino acid protein in humans (378 in mouse) that comprises several transmembrane regions in its N-terminal region and which exists as a dimer in the endoplasmic reticulum (ER) of numerous cell types (endothelial cells, macrophages, T-cells, DCs). Overexpression of STING, also known as MITA/MPYS/ERIS, can robustly activate the transcription factors NF-κB as well as IRF3/7 to stimulate type I IFN production and numerous other cytokines [25–27]. STING appears to be evolutionarily conserved and similar homologs have been noted to exist in *Drosophila*. In attempts to further characterize STING, yeast two hybrid experiments using STING as bait were used to isolate associating proteins. This study indicated that STING interacted with a member of the ER translocon system, known as TRAPβ (translocon associated protein beta) [15,16]. The TRAP complex is made up of four subunits (α-Δ) and is tightly linked with the Sec61 complex, itself made up of three subunits (SEC61α, SEC61β and SEC61γ) [28–30]. Proteins intended for secretion or for membrane incorporation are shunted into the translocon by the SRP (signal recognition particle) associated with translating ribosome. The SRP comprises several proteins and an RNA molecule (7SL) and recognizes a signal sequence at the N-terminus of the nascent protein as it is translated [31]. Peptide synthesis is halted while the SRP attaches to the signal recognition receptor (SRR) attached to the translocon. As the peptide enters the translocon, the SRP and SRR detach and translation continues. Thus, peptide synthesis and translocation are tightly associated and important for ensuring proper protein folding, glycosylation, secretion, or delivery to the appropriate membrane compartment. It is considered that up to 30% of cellular proteins may be directed into the translocon for posttranslational modification [31]. The translocon has also been indicated to play an important role in

ER–associated degradation (ERAD), where misfolded or aberrantly gly-cosylated proteins are re-directed for destruction, following ubquitination, by the 26S proteosomal pathway (retrotranslocation) [30]. It remains to be determined, however, whether STING plays a role in regulating these pathways.

A key question that remains to be fully resolved is whether STING is the actual sensor of cytosolic DNA, or whether accessory molecules facili-tate this process. A recent report indicated that purified STING was able to directly associate with dsDNA, implying that STING could indeed play a role in DNA sensing [32]. However, yet another report has indicated a role for a cGAMP synthase (cyclic guanosine monophosphate-adenosine monophosphate-cyclic GMP-AMP) in the activation of STING. In this situation cGAS was reported to bind to cytosolic DNA to catalyze cGAMP synthesis from ATP and GTP [33]. cGAMP was then able to bind to and activate STING. The interest in cyclic-dinucleotide activa-tion of STING came following studies with intracellular bacteria such as *Listeria monocytogenes*. Bacteria that enter non-phagocytic cells via the clathrin-mediated pathway can leak bacterial DNA or second messenger dinucleotides and are able to activate innate signaling processes [34]. A study first implicating dinucleotides in regulating STING activity involved using a forward genetic mutagenesis screen to generate mice that failed to produce type I IFN in response to *Listeria*. This screen identified STING as the basis for the defect [35]. The mouse, referred to as Goldenticket, had a single point mutation (T596A) in STING which impeded signaling [35]. A subsequent study by the same group confirmed that STING is able to directly associate with cyclic dinucleotides (cyclic di-GMP) to trigger STING function [36]. Thus, bacteria may directly secrete cyclic dinucleo-tides that activate STING while DNA may either bind to STING directly or trigger the production of cGAMPs via the molecule cGAS. Some con-fusion is apparent in these models, however, since cyclic dinucleotides only appear to robustly activate murine STING and not human STING [37]. This has been explained by providing evidence that cGAS produces non-canonical cyclic dinucleotides that contain a single $2'-5'$-phosphodi-ester bond, which are responsible for stimulating human STING [38]. In contrast, dsDNA greater than 45 base pairs are able to activate STING in both human and mouse cells [32]. A further observation is that overex-pression of STING is also able to trigger the induction of the type IIFN promoter in 293T cells, which do not express cGAS [15]. In addition to these observations, a number of type I IFN-inducing compounds such as

DMXAA (5,6-dimethylxanthenone-4-acetic acid) or CMA (10-carboxy-methyl-9-acridanone) were found to also function by binding to and triggering murine STING activity similar to cyclic-dinucelotides [37,39]. However, these compounds also failed to bind human STING to activate signaling. A number of groups have now determined the crystal structure of cyclic-dinucleotides binding to STING, or at least the C-terminal region of STING since the full length protein was not able to be purified for such studies. These studies confirmed that STING exists as a dimer, in a V-shaped structure, in an inactive state [40–44]. Cyclic-dinucleotides bind at the groove between two STING molecules and perhaps alter the confirmation to initiate STING function. Amino acid variations between human and mouse STING have been reported to explain why these compounds associate with one species of STING and not another. In summary, it's apparent that human STING has altered its ability to bind to cyclic-dinucleotides, although perhaps this pathway remains important in protecting other types of mammals. The full understanding of how DNA activates STING thus remains to be fully clarified.

Alternatively, other DNA sensors that feed into the STING pathway have also been reported, but evidence is still lacking that they are bonafide DNA sensors important for innate immune signaling and for the triggering of type I IFN production [2]. Following infection, it is unclear how pathogen-derived DNA activates the 'sensor'. However, it is plausible that, after cellular invasion, the DNA from the microbe 'leaks' out into the cytosol and encounters molecules such as STING and/or cGAS residing in the ER and cytosol respectively. DNA viruses, such as the polyomavirus SV40, are known to traffic to the ER compartment where viral disassembly may occur [45]. SV40 entry involves caveolin-mediated endocytosis, while other viruses utilize macropinocytosis, or caveolin- or clathrin-dependent entry. It is further possible that endosomes/lysosomes carrying viruses become associated with the ER, where pathogen DNA could activate STING-dependent signaling. What is clear is that in the presence of intracellular DNA, STING rapidly traffics from the ER region through to the Golgi to reside in a distinct perinuclear endosome. TBK1 was observed to similarly traffic to these regions in a STING-dependent manner [16]. Presumably, STING escorts TBK1 to endosomal compartments to associate with and activate IRF3 and IRF7 which translocate into the nucleus to activate innate immune gene transcription. STING is also known to stimulate NF-κB signaling, although the mechanisms remain to be clarified

[15]. However, it is apparent that STING trafficking involves autophagy and perhaps members of the exocyst family since autophagy-related gene 9a (Atg9a) has been reported to facilitate STING action [46]. The process of utilizing autophagy is unique in innate signaling processes and is not thought to govern the control of TLR-signaling. Following autophagy, STING related autophagosomes likely fuse with endosomes and lysosomes containing transcription factors. STING has been reported to be ubiquitinated (lysine 63-linked) by interferon-inducible tripartite-motif (TRIM) 56, TRIM32 as well as TRIM21 and rapidly undergoes phosphorylation and degradation after the delivery of TBK1 [47–49]. These procedures almost certainly provide a mechanism to prevent STING from overstimulating innate immune gene transcription for reasons that will become clear, as described below. Thus, intracellular dsDNA induces autophagy and the trafficking of STING/TBK1 through the Golgi to endosomal compartments which harbor the IRF and NF-kB family. These transcription factors become activated and numerous immune-related genes are induced such as type I IFN, a variety of cytokines and chemokines such as CXCL10, as well as members of the IFIT family [32]. STING is then degraded and the delivery of TBK1 halted. These events ensure the transient production of host defense genes that exert direct antimicrobial effects, as well as stimulating the adaptive immune response. These events additionally ensure that the chronic production of cytokines is prevented, thus avoiding the possibility of inflammatory disease.

The importance of STING in host defense has been emphasized through the generation of STING deficient mice which are viable but sensitive to infection by a number of microbes. For example, HSV-1 or Gram-positive *Listeria monocytogenes*, or a variety of dsDNA species including viral or bacterial genomes or plasmids, fail to trigger the production of type I IFN in STING deficient MEFs, macrophages or conventional DCs [15,16]. The DNA pathogens *Chlamydia muridarum* and *trachomatis* have similarly been shown to activate the STING pathway [50]. *Mycobacterium tuberculosis* activates STING-dependent autophagy which leads to the lysosomal destruction of these microbes [51,52]. Our studies additionally indicated that STING was important for both intracellular dsDNA-mediated and HSV-1 activated type I IFN production in pDCs, although the presence of TLR9 in these cells was found to substitute, partially, for the loss of STING in response to *Listeria* infection [16]. STING deficient animals succumb to lethal HSV-1 infection due to a lack of type I IFN production required to protect the host [16]. The production of IL1β in STING deficient macrophages following HSV1 and intracellular *Listeria* infection did not appear to be affected by loss of STING, confirming that this pathway is controlled by AIM2 (see below) [16]. Thus, STING functions independently of AIM2 and is not essential for inflammasome activation.

DNA-dependent innate immune signaling pathways are important for plasmid DNA-based immune responses. Previous work indicated the importance of TBK-1 in facilitating these processes [7]. TBK1 dependent pathways are triggered by intracellular plasmid B-form DNA which stimulates immune responses by activating type I IFN- and NF-κB-dependent pathways, similar to agonists of the TLR9 pathway [7,24]. Since we observed that STING facilitates TBK-1 function we investigated whether STING was important for the adjuvant effects of plasmid DNA and observed that STING was indeed important for effective plasmid DNA and vaccinia virus stimulated adaptive immune responses [16]. Thus, the regulation of STING controlled pathways by novel adjuvants may lead to the development of safe, effective vaccines.

It is also apparent that STING may not only be important for dsDNA-dependent or cyclic dinucleotide-dependent innate immune signaling, but may also be critical for facilitating innate immune responses by negative-stranded and positive-stranded RNA viruses such as vesicular stomatitis virus (VSV), Sendai virus (SV) or Dengue virus [15,16,53]. MEFs lacking STING were defective in VSV- and SV-dependent type I IFN production. STING-deficient mice were also extremely sensitive to VSV infection. However, synthetic dsRNA (polyIC) was not affected in its ability to produce IFN, in the absence of STING [15]. This suggested that STING may play an important role in RIG-I mediated signaling, which senses RNA viruses such as VSV and SV, but not MDA5, responsible for facilitating polyIC signaling. RIG-I signaling is also known to be dependent on IPS-1, a mitochondrial protein [32,53,54]. It remains unclear how STING regulates the replication of RNA viruses. Plausibly, STING may exhibit multiple functions in the cell, and while DNA-mediated signaling may affect the translation or posttranslational modification of viral proteins, this may require efficient translocon function. STING has been found to exhibit homology with flavivirus proteins (Dengue virus, yellow fever virus and hepatitis C virus; NS4) viral proteins that are known to reside in the ER. Moreover, both Dengue virus and HCV have been reported to target STING for repression [53,55].

Thus, flaviviruses, and possibly other viruses, can stop STING function and inactivate host defense.

STING AND INFLAMMATORY DISEASE

Type I IFN is an essential factor for host defense against invading pathogens, but inappropriate production of type I IFN leads to autoimmune diseases such as SLE [5]. In eukaryotes, localization of self-DNA is restricted to the cell nucleus and mitochondria, thereby sequestering self-DNA from cytoplasmic DNA sensing mechanisms which may activate proinflammatory cytokine pathways. Cellular DNases eliminate aberrant self-DNA found in apoptotic bodies, extracellular space, cytosol, and endosomes. Several studies have shown that defective clearance of self-DNA leads to inappropriate activation of type I IFN production through a TLR-independent innate immune signaling, which is tightly linked to autoimmune diseases [5,56–58]. For example, DNase I deficiency or mutations are associated with lupus-like syndrome in mice and humans. In addition, DNase II deficient mice have shown the accumulation of incompletely digested DNA, which causes TLR-independent type I IFN production, inflammatory responses and early death, which are linked to autoimmune disease like chronic polyarthritis [58]. Crossing susceptible mice with mice deficient in the type I IFN receptor abrogated lethality, indicating the importance of excessive IFN production in pathogenesis. DNAseII-mediated lethality and the triggering of IFN production were noted to be TLR-independent. However, crossing STING-deficient mice with DNaseII heterozygote mice led to the production of relatively healthy STING $^{-/-}$ DNaseII $^{-/-}$ mice that did not exhibit any signs of polyarthritis [6]. Thus, the STING pathway appears responsible for DNAseII-mediated embryonic lethal disease.

In addition, several studies have reported that Trex1, 3′-repair exonuclease 1, regulates DNA homeostasis and its deficiency is linked to autoimmune diseases [59]. Mutations in the human Trex1 gene cause SLE and AGS (Aicardi–Goutieres syndrome) [59,60]. Moreover, Trex1-deficient mice develop lethal autoimmunity after approximately 8 weeks via elevated production of type I IFNs and auto-antibodies [61]. It was known that genetic ablation of IRF3 or IFNR rescued Trex1-deficient mice from early death, suggesting IRF3-dependent IFN production was linked to autoimmune symptoms in Trex1-deficient mice [62,63]. Collectively, these observations suggest that Trex1 is required for the prevention of autoimmunity that would otherwise develop during the activation of the cytosolic DNA-mediated innate immune signaling by cell-intrinsic substrates. It is plausible that Trex1 may play a role in negatively regulating the STING pathway [64]. For example, a report indicated that reverse transcribed human immunodeficiency virus type I (HIV1) DNA could activate STING to suppress viral replication. HIV was found to use TREX1 to rapidly digest excess viral DNA to avoid STING activation. Accordingly, the crossing of STING $^{-/-}$ mice with TREX1 $^{-/-}$ mice also eliminated TREX1-mediated disease and lethality [65]. Thus, quite remarkably, the STING pathway appears to regulate inflammatory responses mediated through two pathways, DNaseII and Trex1 (DNaseIII). However, while DNaseII-mediated lethality is known to be caused by incompletely digested apoptotic DNA, it is not known what causes TREX1-mediated lethality, although it could plausibly involve self-DNA.

OTHER SENSORS OF INTRACELLULAR DNA

The NLR pathway has also been shown to be important in sensing cytosolic DNA but triggers inflammasome-dependent innate immune signaling. This involves caspase-1 dependent processing of cytokines such as IL-1β, a key proinflammatory mediator that can stimulate recruitment of macrophages and DCs to sites of infection or injury. Indeed, a number of groups identified AIM2, a HIN-200 family member, as the cytoplasmic dsDNA sensor that activates ASC/caspase 1-mediated secretion of IL-1β [12–14]. Intracellular DNA species bind to dsDNA through the AIM2s HIN-200 domain which is independent of NLRP3. This causes the N-terminal pyrin domains (PYDs) to recruit ASC and activate caspase-1. Importantly, AIM2 inflammasome is essential for caspase-1 activation, but completely dispensable for type I IFN production in response to cytosolic dsDNA. This underscores that cytosolic DNA-mediated AIM2 inflammasome-dependent signaling is distinct from type I IFN-dependent innate signaling, which is largely dependent on STING. Recent studies have also demonstrated the existence of a novel cytosolic DNA sensing mechanism that is able to induce type I IFN in a DNA-dependent RNA polymerase III manner. Transfected poly(dA:dT) (synthetic AT-rich dsDNA), but not other types of DNA including poly(dG:dC) (synthetic GC-rich dsDNA), calf thymus DNA, PCR fragments, or plasmid DNA, have been found capable of producing dsRNA containing 5′triphosphate ends and to activate RIG-I/IPS-1 dependent innate signaling [54,66]. Alternate molecules that have been reported to play a role in the sensing of intracellular DNA to produce IFN include DAI (DNA-dependent activator of IRFs, also referred to as Z DNA-binding protein-1 (ZBP-1) [67]. However, cells derived from DAI deficient mice induced type I IFN normally in response to cytosolic DNA [7]. DAI deficient mice were also found to elicit a normal immune response to DNA based vaccines. These data would suggest that DAI is not solely responsible for the recognition of foreign DNA. Other molecules reported to play a role in sensing intracellular DNA species include the high mobility group box (HMGB) proteins [68]. Cells lacking HMGB 1 exhibited defects in both intracellular DNA and poly-IC mediated IFN production, but not LPS. Cells lacking HMG2 exhibited defects in IFN production only in response to intracellular DNA. Cells where HMGB1-3 was suppressed exhibited defects in AIM2-dependent IL-β production. It was thus suggested that HMBGs are required for full-blown activation of innate immune responses by cytosolic nucleic acids (i.e. regulate the RLR, TLR, AIM2 and STING pathways). The physiological importance of the HMGBs in innate signaling awaits clarification as does how these proteins may interact with the various downstream signaling molecules. Yet another molecule that may be involved in sensing DNA is IFI16, which is an INF-inducible protein and member of the PYHIN protein family that contains a pryin domain and two DNA binding HIN domains. RNAi studies have indicated that IFI16 depletion reduces the induction of IFN by synthetic DNA and HSV-1 and is dependent on STING. DDX41 is yet another molecule considered to play a role in STING function [47]. Further analysis of animals lacking IFI16 or DDX41 will further shed light on the importance of these molecules in innate signaling pathways.

CONCLUSION

STING appears to be essential for the recognition of intracellular DNA species and for the stimulation of innate immune signaling processes. STING may directly interact with DNA or with cyclic dinucleotides generated directly from bacteria or via cGAS. Chronic STING activation also appears to play a key role in inflammatory disease. Plausibly, this effect is caused by self-DNA. The rapid advance in our understanding of cytosolic DNA signaling pathways involving STING may lead to the development of novel adjuvants that may be useful in vaccine formulations designed to combat infectious disease and cancer. Conversely, compounds that inhibit STING function may prove to be potent therapeutics for the treatment of inflammatory disease.

REFERENCES

[1] Blasius AL, Beutler B. Intracellular toll-like receptors. Immunity 2010;32(3):305–15. [Epub 2010/03/30].

[2] Iwasaki A, Medzhitov R. Regulation of adaptive immunity by the innate immune system. Science 2010;327(5963):291–5. [Epub 2010/01/16].

[3] Kawai T, Akira S. The roles of TLRs, RLRs and NLRs in pathogen recognition. Internat Immunol 2009;21(4):317–37. [Epub 2009/02/28].

[4] Yoneyama M, Fujita T. RNA recognition and signal transduction by RIG-I-like receptors. Immunological reviews 2009;227(1):54–65. [Epub 2009/01/06].

[5] Nagata S, Hanayama R, Kawane K. Autoimmunity and the clearance of dead cells. Cell 2010;140(5):619–30. [Epub 2010/03/10].

[6] Ahn J, Gutman D, Saijo S, Barber GN. STING manifests self DNA-dependent inflammatory disease. Proc Nat Acad Sci USA 2012;109(47):19386–91. [Epub 2012/11/08].

[7] Ishii KJ, Kawagoe T, Koyama S, Matsui K, Kumar H, Kawai T, et al. TANK-binding kinase-1 delineates innate and adaptive immune responses to DNA vaccines. Nature 2008;451(7179):725–9. [Epub 2008/02/08].

[8] Koyama S, Coban C, Aoshi T, Horii T, Akira S, Ishii KJ. Innate immune control of nucleic acid-based vaccine immunogenicity. Expert Rev Vaccines 2009;8(8):1099–107. [Epub 2009/07/25].

[9] Hemmi H, Takeuchi O, Kawai T, Kaisho T, Sato S, Sanjo H, et al. A Toll-like receptor recognizes bacterial DNA. Nature 2000;408(6813):740–5. [Epub 2000/12/29].

[10] Wilson HL, Dar A, Napper SK, Marianela Lopez A, Babiuk LA, Mutwiri GK. Immune mechanisms and therapeutic potential of CpG oligodeoxynucleotides. Internat Rev Immunol 2006;25(3-4):183–213. [Epub 2006/07/05].

[11] Burckstummer T, Baumann C, Bluml S, Dixit E, Durnberger G, Jahn H, et al. An orthogonal proteomic-genomic screen identifies AIM2 as a cytoplasmic DNA sensor for the inflammasome. Nature Immunol 2009;10(3):266–72. [Epub 2009/01/23].

[12] Hornung V, Ablasser A, Charrel-Dennis M, Bauernfeind F, Horvath G, Caffrey DR, et al. AIM2 recognizes cytosolic dsDNA and forms a caspase-1-activating inflammasome with ASC. Nature 2009;458(7237):514–8. [Epub 2009/01/23].

[13] Roberts TL, Idris A, Dunn JA, Kelly GM, Burnton CM, Hodgson S, et al. HIN-200 proteins regulate caspase activation in response to foreign cytoplasmic DNA. Science 2009;323(5917):1057–60. [Epub 2009/01/10].

[14] Fernandes-Alnemri T, Yu JW, Datta P, Wu J, Alnemri ES. AIM2 activates the inflammasome and cell death in response to cytoplasmic DNA. Nature 2009;458(7237):509–13. [Epub 2009/01/23].

[15] Ishikawa H, Barber GN. STING is an endoplasmic reticulum adaptor that facilitates innate immune signalling. Nature 2008;455(7213):674–8. [Epub 2008/08/30].

[16] Ishikawa H, Ma Z, Barber GN. STING regulates intracellular DNA-mediated, type I interferon-dependent innate immunity. Nature 2009;461(7265):788–92. [Epub 2009/09/25].

[17] Kim YM, Brinkmann MM, Paquet ME, Ploegh HL. UNC93B1 delivers nucleotide-sensing toll-like receptors to endolysosomes. Nature 2008;452(7184):234–8. [Epub 2008/02/29].

[18] Ewald SE, Lee BL, Lau L, Wickliffe KE, Shi GP, Chapman HA, et al. The ectodomain of Toll-like receptor 9 is cleaved to generate a functional receptor. Nature 2008;456(7222):658–62. [Epub 2008/09/30].

[19] Lund J, Sato A, Akira S, Medzhitov R, Iwasaki A. Toll-like receptor 9-mediated recognition of herpes simplex virus-2 by plasmacytoid dendritic cells. J Exp Med 2003;198(3):513–20. [Epub 2003/08/06].

[20] Hochrein H, Schlatter B, O'Keeffe M, Wagner C, Schmitz F, Schiemann M, et al. Herpes simplex virus type-1 induces IFN-alpha production via Toll-like receptor 9-dependent and -independent pathways. Proc Nat Acad Sci USA 2004;101(31): 11416–21. [Epub 2004/07/24].

[21] Spies B, Hochrein H, Vabulas M, Huster K, Busch DH, Schmitz F, et al. Vaccination with plasmid DNA activates dendritic cells via Toll-like receptor 9 (TLR9) but functions in TLR9-deficient mice. J Immunol 2003;171(11):5908–12. [Epub 2003/11/25].

[22] Heit A, Maurer T, Hochrein H, Bauer S, Huster KM, Busch DH, et al. Cutting edge: Toll-like receptor 9 expression is not required for CpG DNA-aided cross-presentation of DNA-conjugated antigens but essential for cross-priming of CD8 T cells. J Immunol 2003;170(6):2802–5. [Epub 2003/03/11].

[23] Stetson DB, Medzhitov R. Recognition of cytosolic DNA activates an IRF3-dependent innate immune response. Immunity 2006;24(1):93–103. [Epub 2006/01/18].

[24] Ishii KJ, Coban C, Kato H, Takahashi K, Torii Y, Takeshita F, et al. A Toll-like receptor-independent antiviral response induced by double-stranded B-form DNA. Nature Immunol 2006;7(1):40–8. [Epub 2005/11/16].

[25] Zhong B, Yang Y, Li S, Wang YY, Li Y, Diao F, et al. The adaptor protein MITA links virus-sensing receptors to IRF3 transcription factor activation. Immunity 2008;29(4):538–50. [Epub 2008/09/27].

[26] Sun W, Li Y, Chen L, Chen H, You F, Zhou X, et al. ERIS, an endoplasmic reticulum IFN stimulator, activates innate immune signaling through dimerization. Proc Nat Acad Sci USA 2009;106(21):8653–8. [Epub 2009/05/13].

[27] Jin L, Waterman PM, Jonscher KR, Short CM, Reisdorph NA, Cambier JC. MPYS a novel membrane tetraspanner, is associated with major histocompatibility complex class II and mediates transduction of apoptotic signals. Mol Cell Biol 2008;28(16):5014–26. [Epub 2008/06/19].

[28] Hartmann E, Gorlich D, Kostka S, Otto A, Kraft R, Knespel S, et al. A tetrameric complex of membrane proteins in the endoplasmic reticulum. Eur J Biochem/FEBS 1993;214(2):375–81. [Epub 1993/06/01].

[29] Menetret JF, Hegde RS, Aguiar M, Gygi SP, Park E, Rapoport TA, et al. Single copies of Sec61 and TRAP associate with a nontranslating mammalian ribosome. Structure 2008;16(7):1126–37. [Epub 2008/07/10].

[30] Menetret JF, Hegde RS, Heinrich SU, Chandramouli P, Ludtke SJ, Rapoport TA, et al. Architecture of the ribosome-channel complex derived from native membranes. J Mol Biol 2005;348(2):445–57. [Epub 2005/04/07].

[31] Akopian D, Shen K, Zhang X, Shan SO. Signal recognition particle: an essential protein-targeting machine. Ann Rev Biochem 2013;82:693–721. [Epub 2013/02/19].

[32] Abe T, Harashima A, Xia T, Konno H, Konno K, Morales A, et al. STING recognition of cytoplasmic DNA instigates cellular defense. Mol Cell 2013;50(1):5–15. [Epub 2013/03/13].

[33] Sun L, Wu J, Du F, Chen X, Chen ZJ. Cyclic GMP-AMP synthase is a cytosolic DNA sensor that activates the type I interferon pathway. Science 2013;339(6121):786–91. [Epub 2012/12/22].

[34] Woodward JJ, Iavarone AT, Portnoy DA. c-di-AMP secreted by intracellular *Listeria monocytogenes* activates a host type I interferon response. Science 2010;328(5986):1703–5. [Epub 2010/05/29].

[35] Sauer JD, Sotelo-Troha K, von Moltke J, Monroe KM, Rae CS, Brubaker SW, et al. The N-ethyl-N-nitrosourea-induced Goldenticket mouse mutant reveals an essential function of Sting in the in vivo interferon response to *Listeria monocytogenes* and cyclic dinucleotides. Infect Immun 2011;79(2):688–94. [Epub 2010/11/26].

[36] Burdette DL, Monroe KM, Sotelo-Troha K, Iwig JS, Eckert B, Hyodo M, et al. STING is a direct innate immune sensor of cyclic di-GMP. Nature 2011;478(7370):515–8. [Epub 2011/09/29].

[37] Conlon J, Burdette DL, Sharma S, Bhat N, Thompson M, Jiang Z, et al. Mouse, but not human STING, binds and signals in response to the vascular disrupting agent 5,6-dimethylxanthenone-4-acetic acid. J Immunol 2013;190(10):5216–25. [Epub 2013/04/16].

[38] Diner EJ, Burdette DL, Wilson SC, Monroe KM, Kellenberger CA, Hyodo M, et al. The innate immune DNA sensor cGAS produces a noncanonical cyclic dinucleotide that activates human STING. Cell Rep 2013 [Epub 2013/05/28].

[39] Cavlar T, Deimling T, Ablasser A, Hopfner KP, Hornung V. Species-specific detection of the antiviral small-molecule compound CMA by STING. EMBO J 2013;32(10):1440–50. [Epub 2013/04/23].

[40] Chin KH, Tu ZL, Su YC, Yu YJ, Chen HC, Lo YC, et al. Novel c-di-GMP recognition modes of the mouse innate immune adaptor protein STING. Acta Crystallogr Section D, Biol Crystallogr 2013;69(Pt 3):352–66. [Epub 2013/03/23].

[41] Huang YH, Liu XY, Du XX, Jiang ZF, Su XD. The structural basis for the sensing and binding of cyclic di-GMP by STING. Nature Struct Mol Biol 2012;19(7):728–30. [Epub 2012/06/26].

[42] Ouyang S, Song X, Wang Y, Ru H, Shaw N, Jiang Y, et al. Structural analysis of the STING adaptor protein reveals a hydrophobic dimer interface and mode of cyclic di-GMP binding. Immunity 2012;36(6):1073–86. [Epub 2012/05/15].

[43] Shu C, Yi G, Watts T, Kao CC, Li P. Structure of STING bound to cyclic di-GMP reveals the mechanism of cyclic dinucleotide recognition by the immune system. Nature Struct Mol Biol 2012;19(7):722–4. [Epub 2012/06/26].

[44] Su YC, Tu ZL, Yang CY, Chin KH, Chuah ML, Liang ZX, et al. Crystallization studies of the murine c-di-GMP sensor protein STING. Acta Crystallogr Section F, Struct Biol Crystallization Commun 2012;68(Pt 8):906–10. [Epub 2012/08/08].

[45] Spooner RA, Smith DC, Easton AJ, Roberts LM, Lord JM. Retrograde transport pathways utilised by viruses and protein toxins. Virol J 2006;3:26. [Epub 2006/04/11].

[46] Saitoh T, Fujita N, Hayashi T, Takahara K, Satoh T, Lee H, et al. Atg9a controls dsDNA-driven dynamic translocation of STING and the innate immune response. Proc Nat Acad Sci USA 2009;106(49):20842–6. [Epub 2009/11/21].

[47] Zhang J, Hu MM, Wang YY, Shu HB. TRIM32 protein modulates type I interferon induction and cellular antiviral response by targeting MITA/STING protein for K63-linked ubiquitination. J Biol Chem 2012;287(34):28646–55. [Epub 2012/06/30].

[48] Zhang Z, Bao M, Lu N, Weng L, Yuan B, Liu YJ. The E3 ubiquitin ligase TRIM21 negatively regulates the innate immune response to intracellular double-stranded DNA. Nature Immunol 2013;14(2):172–8. [Epub 2012/12/12].

[49] Tsuchida T, Zou J, Saitoh T, Kumar H, Abe T, Matsuura Y, et al. The ubiquitin ligase TRIM56 regulates innate immune responses to intracellular double-stranded DNA. Immunity 2010;33(5):765–76. [Epub 2010/11/16].

[50] Barker JR, Koestler BJ, Carpenter VK, Burdette DL, Waters CM, Vance RE, et al. STING-dependent recognition of cyclic di-AMP mediates type i interferon responses during Chlamydia trachomatis infection. mBio 2013;4(3) [Epub 2013/05/02].

[51] Manzanillo PS, Shiloh MU, Portnoy DA, Cox JS. Mycobacterium tuberculosis activates the DNA-dependent cytosolic surveillance pathway within macrophages. Cell Host and Microbe 2012;11(5):469–80. [Epub 2012/05/23].

[52] Watson RO, Manzanillo PS, Cox JS. Extracellular M. tuberculosis DNA targets bacteria for autophagy by activating the host DNA-sensing pathway. Cell 2012;150(4):803–15. [Epub 2012/08/21].

[53] Aguirre S, Maestre AM, Pagni S, Patel JR, Savage T, Gutman D, et al. DENV inhibits type I IFN production in infected cells by cleaving human STING. PLoS Pathogens 2012;8(10):e1002934. [Epub 2012/10/12].

[54] Ablasser A, Bauernfeind F, Hartmann G, Latz E, Fitzgerald KA, Hornung V. RIG-I-dependent sensing of poly(dA:dT) through the induction of an RNA polymerase III-transcribed RNA intermediate. Nature Immunol 2009;10(10):1065–72. [Epub 2009/07/18].

[55] Ding Q, Cao X, Lu J, Huang B, Liu YJ, Kato N, et al. Hepatitis C virus NS4B blocks the interaction of STING and TBK1 to evade host innate immunity. J Hepatol 2013;59(1):52–8. [Epub 2013/04/02].

[56] Kawane K, Ohtani M, Miwa K, Kizawa T, Kanbara Y, Yoshioka Y, et al. Chronic polyarthritis caused by mammalian DNA that escapes from degradation in macrophages. Nature 2006;443(7114):998–1002. [Epub 2006/10/27].

[57] Okabe Y, Kawane K, Akira S, Taniguchi T, Nagata S. Toll-like receptor-independent gene induction program activated by mammalian DNA escaped from apoptotic DNA degradation. J Exp Med 2005;202(10):1333–9. [Epub 2005/11/23].

[58] Yoshida H, Okabe Y, Kawane K, Fukuyama H, Nagata S. Lethal anemia caused by interferon-beta produced in mouse embryos carrying undigested DNA. Nature Immunol 2005;6(1):49–56. [Epub 2004/11/30].

[59] Crow YJ, Hayward BE, Parmar R, Robins P, Leitch A, Ali M, et al. Mutations in the gene encoding the 3′-5′ DNA exonuclease TREX1 cause Aicardi-Goutieres syndrome at the AGS1 locus. Nature Genet 2006;38(8):917–20. [Epub 2006/07/18].

[60] Lee-Kirsch MA, Gong M, Chowdhury D, Senenko L, Engel K, Lee YA, et al. Mutations in the gene encoding the 3′-5′ DNA exonuclease TREX1 are associated with systemic lupus erythematosus. Nature Genet 2007;39(9):1065–7. [Epub 2007/07/31].

[61] Stetson DB, Ko JS, Heidmann T, Medzhitov R. Trex1 prevents cell-intrinsic initiation of autoimmunity. Cell 2008;134(4):587–98. [Epub 2008/08/30].

[62] Yang YG, Lindahl T, Barnes DE. Trex1 exonuclease degrades ssDNA to prevent chronic checkpoint activation and autoimmune disease. Cell 2007;131(5):873–86. [Epub 2007/11/30].

[63] Morita M, Stamp G, Robins P, Dulic A, Rosewell I, Hrivnak G, et al. Gene-targeted mice lacking the Trex1 (DNase III) 3′→5′ DNA exonuclease develop inflammatory myocarditis. Mol Cell Biol 2004;24(15):6719–27. [Epub 2004/07/16].

[64] Yan N, Regalado-Magdos AD, Stiggelbout B, Lee-Kirsch MA, Lieberman J. The cytosolic exonuclease TREX1 inhibits the innate immune response to human immunodeficiency virus type 1. Nature Immunol 2010;11(11):1005–13. [Epub 2010/09/28].

[65] Gall A, Treuting P, Elkon KB, Loo YM, Gale Jr. M, Barber GN, et al. Autoimmunity initiates in nonhematopoietic cells and progresses via lymphocytes in an interferon-dependent autoimmune disease. Immunity 2012;36(1):120–31. [Epub 2012/01/31].

[66] Chiu YH, Macmillan JB, Chen ZJ. RNA polymerase III detects cytosolic DNA and induces type I interferons through the RIG-I pathway. Cell 2009;138(3):576–91. [Epub 2009/07/28].

[67] Takaoka A, Wang Z, Choi MK, Yanai H, Negishi H, Ban T, et al. DAI (DLM-1/ZBP1) is a cytosolic DNA sensor and an activator of innate immune response. Nature 2007;448(7152):501–5. [Epub 2007/07/10].

[68] Yanai H, Ban T, Wang Z, Choi MK, Kawamura T, Negishi H, et al. HMGB proteins function as universal sentinels for nucleic-acid-mediated innate immune responses. Nature 2009;462(7269):99–103. [Epub 2009/11/06].

CHAPTER 4

Regulation of Intracellular dsDNA-Induced Innate Immune Responses by Autophagy-Related Proteins

Tatsuya Saitoh[1,2]

[1]Laboratory of Host Defense, WPI Immunology Frontier Research Center, Osaka University, 3-1 Yamadaoka, Suita, Japan
[2]Department of Host Defense, Research Institute for Microbial Diseases, Osaka University, 3-1 Yamadaoka, Suita, Japan

PART I. AUTOPHAGY-RELATED PROTEINS AND STING-DEPENDENT IFN RESPONSES

STING Mediates dsDNA-Induced Expression of Type I IFNs

Innate immunity is triggered by the engagement of pattern-recognition receptors (PRRs), the sensors for pathogen-associated molecular patterns (PAMPs) [1–3]. After sensing PAMPs, PRRs stimulate the activation of transcription factors leading to the expression of inflammatory cytokines and type I interferons (IFNs), which are required for protection of a host against microbial infection. One such PAMP, microbial double-stranded (ds)DNA, derived from bacteria or DNA viruses, induces the expression of type I IFNs and autophagy, resulting in the potent activation of innate immunity [4–6]. Synthesized dsDNA is also a potent stimulator of innate immunity and is used as a vaccine for the induction of efficient acquired immune responses [7]. Recently the regulatory mechanisms involved in dsDNA-induced signaling pathways have been identified (Figure 4.1). Stimulator of interferon genes (STING), also called TMEM173/MPYS/ MITA/ERIS, is a multi-spanning membrane protein and identified as an activator of the IFN-β promoter [8–11]. STING is capable of sensing cytosolic dsDNA [12]. After dsDNA stimulation, TANK-binding kinase 1 (TBK1), a serine/threonine kinase, assembles with STING and induces phosphorylation of the transcription factor interferon regulatory factor 3 (IRF3) [13,14]. Phosphorylated IRF3 forms a homo-dimer and induces IFN-stimulation responsive element-dependent expression of type I IFNs

Biological DNA Sensor.
DOI: http://dx.doi.org/10.1016/B978-0-12-404732-7.00004-6

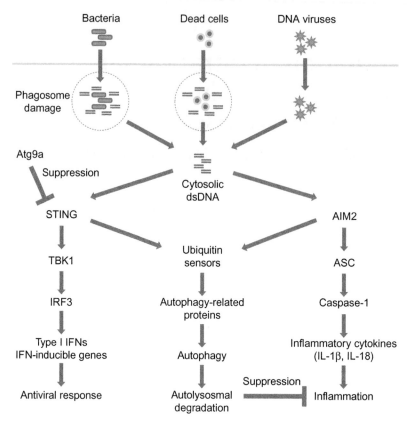

Figure 4.1 *Cytosolic dsDNA induces STING-dependent and AIM2-dependent innate immune responses.* STING mediates dsDNA-induced expression of type I IFNs and eliminates invading mycobacteria. Autophagy is involved in the STING-dependent bactericidal response. Atg9a negatively regulates the STING-dependent IFN response. AIM2 mediates dsDNA-induced production of inflammatory cytokines IL-1β and IL-18. After dsDNA stimulation, autophagy is induced in an AIM2-dependent manner and causes degradation of activated AIM2-inflammasomes, which limits the immune response.

and IFN–inducible genes. Hence, STING mediates dsDNA–induced expression of type I IFNs and protects hosts from lethal infection of herpes simplex virus 1.

Translocation and assembly of the essential signal transducers, STING and TBK1, is required for innate immune responses to dsDNA. After dsDNA stimulation, STING translocates from the endoplasmic reticulum (ER) to the Golgi apparatus, finally reaching the cytoplasmic punctate structures to associate with TBK1 [13,14]. Sec5 (also called EXOC2), a component of the exocyst complex involved in targeting exocytic vesicles to the plasma membrane, co-localizes with STING to regulate STING movement and STING-dependent induction of type I IFNs. Because the addition of an ER-retention signal to the C-terminus of STING or treatment with Brefeldin A (an inhibitor of protein transport from the ER to the Golgi apparatus) dampens its ability to stimulate IRF-3, STING translocation from ER to the Golgi apparatus is a critical process for dsDNA-induced innate immune responses. Hence, dsDNA activates membrane trafficking machinery that mediates the dynamic translocation of STING, leading to the efficient expression of type I IFNs and IFN-inducible genes.

Autophagy-Related Protein Regulates Membrane Trafficking and Induces Autophagosome Formation

Autophagy is a system that delivers cytoplasmic constituents into lysosomes for degradation (Figure 4.2) [15,16]. A number of essential components of autophagic machinery, autophagy-related proteins, have been identified by yeast genetic screening. Autophagy-related proteins are highly conserved and mammalian counterparts such as ULK1/2 (mammalian Atg1), Atg2a/b, Atg3, Atg4a/b/c/d, Atg5, Beclin-1 (mammalian Atg6), Atg7, microtubule-associated protein 1 light chain 3 (LC3, mammalian Atg8), Atg9a/b, Atg10, Atg12, Atg13, Atg14, Atg16L1/L2, FIP200 (mammalian Atg17) and WIPIs (mammalian Atg18) have been reported [17–30]. The cooperation of autophagy-related proteins mediates the membrane trafficking required for generation of isolation membrane, a source membrane of the autophagosome, from ER–mitochondria contact sites and the subsequent formation of the autophagosome [15,16,31]. Autophagy enables the re-use of intracellular constituents and supplies an amino acid pool during periods of neonatal starvation [15,16]. Indeed, mice deficient in Atg3, Atg5, Atg7, Atg9a or Atg16L1 die within 1 day of delivery, indicating that autophagy is essential for survival during neonatal starvation [14,20,32–34]. Autophagy mediates the clearance of damaged organelles, long-lived proteins, insoluble protein aggregates and invading pathogens to maintain cellular homeostasis and limit pathogen replication [35–38]. Furthermore, autophagy-related proteins mediate autophagy-independent immune responses. After type II IFN stimulation, Atg5, Atg7 and Atg16L1 mediate recruitment of IFN-inducible p47 GTPase IIGP1 (also called Irga6) to pathogen-containing vacuoles by regulating membrane trafficking, resulting in the elimination of invading pathogens [39–41]. In addition, Atg5, Atg7, Atg4b, and LC3 are involved in the polarized secretion of

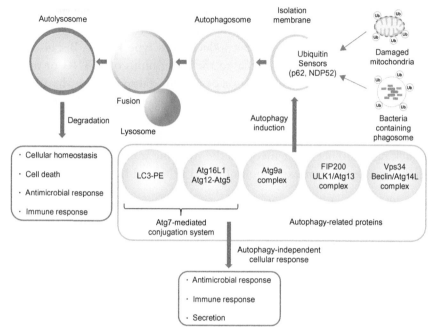

Figure 4.2 *Autophagy-related proteins.* Autophagy, an intracellular clearance system, is induced after sensing various types of stresses such as the accumulation of damaged organelles and invasion by microbes. This system enables the elimination of unfavorable substances, thus contributing to the maintenance of cellular homeostasis and the prevention of cell death. Autophagy-related proteins are recruited to the isolation membrane, a source membrane of autophagosomes, and drive the membrane trafficking necessary for the generation of autophagosomes. Autophagosomes then fuse with lysosomes to become autolysosomes that degrade engulfed constituents. In mammalian cells, some autophagy-related proteins regulate autophagy-independent immune responses by regulating membrane trafficking.

lysosomal contents into the extracellular space by directing lysosomes to fuse with the plasma membrane, and regulate bone resorption by osteoclasts [42]. Hence, autophagy-related proteins play multiple roles in the regulation of cellular homeostasis, cell death, antimicrobial responses and immune responses.

Autophagy-Related Proteins Regulate STING-Mediated Expression of Type I IFNs

The importance of autophagy machinery in STING-dependent innate immune responses is highlighted by the co-localization of STING with LC3, an autophagy-related protein, on cytoplasmic punctate structures [14].

STING also co-localizes with Atg9a in the Golgi apparatus. Electron microscopy studies showed that synthetic dsDNA-induced puncta did not have the morphological characteristics of autophagosomes, but were unique membrane-bound compartments, suggesting an unexpected function for either Atg9a or LC3 in the regulation of STING-mediated innate immune responses. Atg9a is the sole multi-spanning membrane protein in mammals known to be an autophagy-related protein, and localizes to the Golgi apparatus and late endosomes. Atg9a does not reside at one site, but cycles vigorously between these organelles under nutrient-starvation conditions, regulating the transport of source membrane to generate autophagosomes. The loss of Atg9a significantly enhances the dsDNA-induced assembly of STING and TBK1 on cytoplasmic punctate structures, resulting in the aberrant activation of IRF3-dependent innate immune responses (Figure 4.3). However, Atg7-deficiency does not alter the localization of

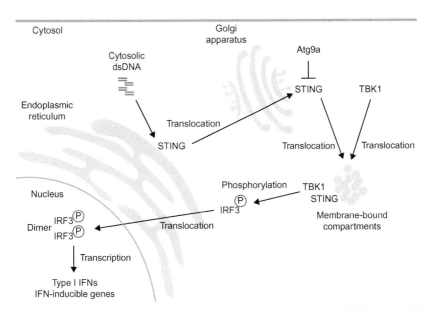

Figure 4.3 *Loss of Atg9a augments dsDNA-induced movement of STING and STING-dependent expression of type I IFNs.* After stimulation by dsDNA, STING translocates from the endoplasmic reticulum to the Golgi apparatus, finally reaching the cytoplasmic punctate compartments to assemble with the IRF3 kinase, TBK1. TBK1 then induces the nuclear translocation of IRF3 by inducing phosphorylation, resulting in the expression of type I IFNs and IFN-inducible genes. STING co-localizes with Atg9a on the Golgi apparatus after dsDNA stimulation. Atg9a-deficiency enhances the dsDNA-induced assembly of STING and TBK1, leading to aberrant expression of type I IFNs and IFN-inducible genes.

STING or the subsequent induction of type I IFNs by dsDNA. Thus, Atg9a-dependent membrane trafficking, but not autophagy, disrupts the assembly of STING and TBK1 after stimulation by dsDNA. Further studies are required to clarify the unique function of Atg9a in the regulation of membrane trafficking involved in STING movement and in the regulation of innate immune responses to microbes such as DNA viruses.

PART II. AUTOPHAGY AND STING-DEPENDENT BACTERICIDAL RESPONSES

Autophagy Functions as a Host Defense System Against Invading Mycobacteria

Mammalian cells sterilize the cytosol by autophagic elimination of invading bacteria in autolysosomes (Figure 4.4) [43,44]. Upon infection, pathogenic bacteria enter phagosomes and disrupt their vacuolar integrity. Ubiquitinated proteins accumulate on the surface of the damaged phagosomes and act as an alarm system that monitors bacterial invasion. Ubiquitin sensors, p62/SQSTM1 and NDP52, recognize ubiquitinated proteins on damaged phagosomes by their ubiquitin binding domain and recruit autophagy-related proteins such as LC3 and Beclin-1 to the damaged phagosomes [45–48]. Bacteria-containing damaged phagosomes are selectively degraded by autophagy, also called xenophagy. In addition to autophagy-related proteins and ubiquitin sensors, TBK1 and its binding partner optineurin are required for the maintenance of vacuolar integrity of bacteria-containing phagosomes. TBK1 assembles with NDP52 on damaged phagosomes and phosphorylates optineurin to recruit LC3, independently of its role in the expression of type I IFNs [49]. The NDP52–TBK1–optineurin signaling axis mediates the recruitment of core autophagy machinery to bacteria-containing phagosomes. Thus, autophagy mediates a host defense response to vacuolar pathogens.

STING Mediates Autophagy-Dependent Elimination of Invading Mycobacteria

Recently, a novel function of STING in the elimination of mycobacteria was reported (Figure 4.4) [6]. In macrophages, the induction of autophagy can suppress replication of *Mycobacterium tuberculosis*, a vacuolar pathogen. *M. tuberculosis* infection causes phagosomal permeabilization mediated by the ESX-1 secretion system. Disintegration of the phagosomal membrane results in release of bacterial extracellular dsDNA into macrophage

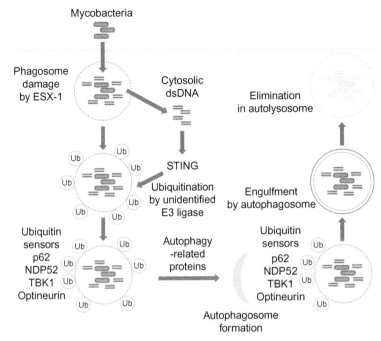

Figure 4.4 *STING mediates autophagy-dependent elimination of M. tuberculosis.* *M. tuberculosis* causes phagosomal permeabilization by the ESX-1 secretion system. Damage to phagosomal membranes results in the release of bacterial extracellular dsDNA into the cytoplasm. Recognition of dsDNA by the STING-dependent cytosolic pathway triggers the marking of bacteria with ubiquitin. Ubiquitin sensors p62 and NDP52 are recruited to bacteria-containing vacuoles, which then assemble with dsDNA-responsive kinase TBK1. TBK1 phosphorylates optineurin to enhance recruitment of LC3 to the bacteria-containing vacuoles and assists in the delivery of *M. tuberculosis* to lysosomes. STING stimulates the ubiquitin sensors–TBK1–optineurin signaling axis in response to dsDNA and mediates autophagic elimination of *M. tuberculosis*.

cytoplasm. Importantly, STING recognizes bacterial dsDNA in cytosol and promotes the marking of damaged phagosomes with ubiquitin. Ubiquitin sensors p62 and NDP52 detect accumulated ubiquitinated proteins and promote the recruitment of autophagy-related proteins to *M. tuberculosis*-containing phagosomes. STING also mediates recruitment of TBK1, a DNA-responsive kinase, to phagosomes containing *M. tuberculosis*. Thus, STING promotes p62, NDP52 and TBK1 on damaged phagosomes to induce autophagic elimination of *M. tuberculosis*. Furthermore, mice with phagocytes incapable of delivering *M. tuberculosis* to the autophagosomes are highly susceptible to infection. Thus, STING links dsDNA sensing, innate immunity and autophagy as well as playing a key role in host defense responses to bacterial infection. Very recently, MRE11 was identified as a cytosolic dsDNA sensor that induces STING movement and subsequent expression of type I IFNs [50]. It would be interesting to assess whether MRE11 is involved in STING-dependent marking of damaged phagosomes with ubiquitin and the subsequent induction of autophagic elimination of *M. tuberculosis*.

PART III. AUTOPHAGY AND AIM2-DEPENDENT INFLAMMATORY RESPONSES

Inflammasomes Mediate Inflammatory Innate Immune Responses

PRRs detect microbial components, environmental irritants and host-derived stimulatory factors to induce the expression of inflammatory cytokines [1,2]. Among PRRs, Nod-like receptors (NLRs) and AIM2 form a molecular platform, the so-called inflammasome, with the adaptor protein ASC and the protease caspase-1 [51]. Activation of inflammasomes triggers caspase-1-dependent processing of immature pro-interleukin (IL)-1β and pro-IL-18, leading to the generation and production of mature bioactive IL-1β and IL-18. Inflammasomes also mediate the induction of cell death, called pyroptosis, in a caspase-1-dependent manner. NLRC4, a member of the NLR family, detects microbial proteins such as flagellin and mediates host defense responses to Gram-negative bacteria [52–53]. NLRP3, another member of the NLR family, senses organelle damage caused by microbial infection, and mediates host defense responses to varieties of viruses such as influenza A virus, adenovirus and encephalomyocarditis virus [54–56]. Although microbe-derived cytosolic dsDNA induces the production of IL-1/IL-18 and cell death in an ASC- and caspase-1-dependent manner, NLR family members are not involved in this response [57]. Recently, AIM2 was identified as a sensor for cytosolic dsDNA (Figure 4.1) [58–60]. AIM2 recognizes dsDNA in a HIN200 domain and signals to ASC through a PYRIN domain. The AIM2-inflammasome mediates production of IL-1/IL-18 and induction of cell death by microbe-derived cytosolic dsDNA, and is critically involved in host defense responses to microbes, such as *Francisella tularensis*, *M. tuberculosis* and mouse cytomegalovirus [61–63]. However, aberrant accumulation of host-derived genomic dsDNA often activates the AIM2-inflammasome and causes persistent induction of inflammatory responses, resulting in the development of severe inflammatory diseases, such as systemic lupus erythematosus

(SLE) and psoriasis [64–66]. IFI202 is another member of the HIN200 domain-containing protein family and negatively regulates dsDNA-induced inflammatory responses by interfering with AIM2 recognition of dsDNA [67]. Consequently, IFI202 is regarded as a candidate susceptibility gene for SLE [68–69]. Therefore, the study of inflammasomes is important not only for a better understanding of host defense responses, but also for the development of effective therapeutic treatments for autoimmune diseases.

Autophagy-Deficiency Enhances Production of Proinflammatory Cytokine IL-1β

Genome-wide association studies have identified Atg16L1 as a candidate susceptibility gene for Crohn's disease [70–73]. Atg16L1 forms a high molecular weight protein complex with Atg12–Atg5 conjugates, and the self-multimerization of Atg16L1 via its coiled-coil domain mediates the recruitment of this conjugate into an isolation membrane [74]. The Atg12–Atg5/Atg16L1 complex then recruits an Atg3–LC3 intermediate to the isolation membrane, which defines the site at which LC3 is conjugated to phosphatidylethanolamine (PE). LC3–PE mediates membrane tethering and hemifusion, and promotes the formation of autophagosomes [75]. Hence, Atg16L1 is an essential component of autophagy during nutrient-rich and nutrient-starved conditions.

The importance of autophagy in the regulation of inflammatory responses has attracted attention from researchers in the immunology field [76–78]. Commensal bacteria are thought to be a major causative agent of bowel disease [51]. When intestinal epithelial cells are damaged, commensal bacteria are able to pass through the epithelial layer and stimulate PRRs, resulting in the induction of intestinal inflammation. Toll-like receptors (TLRs), one family of pattern recognition receptors, detect bacterial components such as lipopolysaccharide (LPS), and induce the production of inflammatory cytokines and chemokines by phagocytes [1,2]. Inflammatory cytokines and chemokines induce the infiltration of activated lymphocytes into the intestinal tissues and cause severe inflammation. Recent studies highlight a role for Atg16L1 in TLR-mediated inflammatory responses (Figure 4.5) [34]. Macrophages lacking Atg16L1 produce high amounts of IL-1β and IL-18 in response to LPS, a TLR4 ligand. Macrophages from Atg7-deficient mice also show enhanced IL-1β production. Thus, autophagy, but not the unique function of autophagy-related protein, is involved in the regulation of IL-1β and IL-18

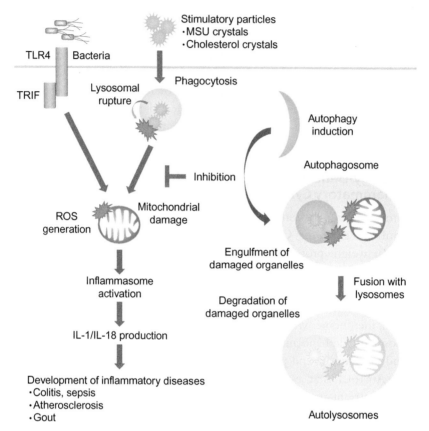

Figure 4.5 Autophagy-deficiency enhances production of inflammatory cytokines IL-1β and IL-18. Stimulatory metabolites such as monosodium urate (MSU) crystals cause rupture of the lysosomal membrane, resulting in mitochondrial damage. Lipopolysaccharide (LPS) engages TLR4 to cause damage to mitochondria via Toll/IL-1 receptor domain-containing adaptor inducing IFN-β (TRIF), an adaptor protein. Reactive oxygen species (ROS) generated by damaged mitochondria activate the NLRP3-inflammasome leading to IL-1β and IL-18 inflammatory cytokine production. Autophagy eliminates damaged mitochondria and suppresses activation of the NLRP3-inflammasome.

production. Atg16L1–deficiency does not enhance production of IL-1β after engagement of other TLR family members, except TLR3. Toll/IL-1 receptor domain-containing adaptor inducing IFN-β (TRIF), an adapter protein involved in TLR3 and TLR4 signaling pathways, mediates IL-1β production in Atg16L1-deficient macrophages. Reactive oxygen species (ROS) play an important role in the production of IL-1β induced by ATP, uric acid crystals, silicas and asbestos. Enhanced IL-1β production by

Atg16L1–deficient macrophages is also dependent on ROS and is inhibited by ROS scavengers. In LPS-stimulated macrophages, Atg16L1–deficiency results in TRIF-dependent generation of ROS, leading to the caspase-1-dependent production of IL-1β and IL-18. Mice with hematopoietic cells lacking Atg16L1 are susceptible to dextran sulfate sodium-induced colitis, which is alleviated by the injection of anti-IL-1β and anti-IL-18 blocking antibodies. Thus, the loss of Atg16L1, an essential component of the autophagic machinery, causes the excessive production of inflammatory cytokines in response to microbial components such as LPS.

Autophagy Limits dsDNA-Induced Activation of AIM2-Inflammasome

Accumulating evidence demonstrates that autophagy significantly contributes to control of inflammasome activation. The NLRP3-inflammasome is activated after stimulation with environmental irritants and host-derived stimulatory factors, such as asbestos, ATP, cholesterol crystals, ceramide, free fatty acids, human islet polypeptides, monosodium urate crystals and silica [79–86]. Macrophages lacking autophagy-related proteins or treated with a chemical inhibitor of autophagy produce high amounts of IL-1β in response to various NLRP3-inflammasome inducers (Figure 4.5) [34,85,87–91]. Enhanced LPS-induced IL-1β production in autophagy-deficiency depends on the NLRP3-inflammasome [87]. However, the chemical inducers of autophagy suppress activation of the NLRP3-inflammasome and dampen inflammation. ROS generated by damaged mitochondria is critically involved in activation of the NLRP3-inflammasome [87–88,92]. Autophagy-deficiency disrupts turnover of mitochondria generation and elimination, resulting in the accumulation of old mitochondria. Old mitochondria are susceptible to membrane damage and generate high amounts of ROS in response to various stresses. Loss of autophagy enhances the generation of mitochondrial ROS, resulting in massive activation of the NLRP3-inflammasome. Thus, inhibition of autophagy causes the development of NLRP3-related inflammatory disorders, whereas induction of autophagy ameliorates the symptoms of NLRP3-related inflammatory disorders.

Autophagy also regulates activation of the AIM2-inflammasome in a mouse macrophage cell line and human primary monocytes/macrophages (Figure 4.6) [90]. Engagement of AIM2 with cytosolic dsDNA induces IL-1β/IL-18 production and the formation of autophagosomes. The induction of autophagy depends on the presence of AIM2, but not

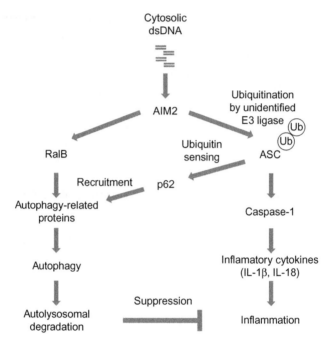

Figure 4.6 *Autophagy mediates degradation of ASC and suppresses activation of the AIM2-inflammasome.* Engagement of AIM2 by dsDNA triggers formation of the inflammasome with the adaptor ASC or the protease capase-1 and the subsequent production of IL-1β. The engagement of AIM2 by dsDNA also triggers activation of the G protein RalB leading to the induction of autophagy. ASC or capase-1 is dispensable for autophagy induction. After dsDNA stimulation, ASC is ubiquitinated and is recognized by the autophagic adaptor p62, which assists delivery of the AIM2-inflammasome to autolysosomes. Autophagy induces selective degradation of the activated AIM2-inflammasome. Inhibition of autophagy promotes production of IL-1β by the AIM2-inflammasome, whereas stimulating autophagy suppresses it.

on ASC or caspase-1. AIM2 engagement causes activation of the G protein RalB, a Ras-related small GTPase that regulates membrane traffic [93], to induce autophagy, although the molecular mechanism of signal transduction from AIM2 to RalB requires further clarification. After dsDNA stimulation, ASC undergoes ubiquitination by an unidentified ubiquitin E3 ligase. A ubiquitin sensor, p62, targets ubiquitinated ASC to recruit autophagy-related proteins such as beclin-1 and LC3, and mediates delivery of the activated AIM2-inflammasome to autolysosomes by autophagosomes. Thus, inhibition of autophagy potentiates activation of the AIM2-inflammasome, whereas induction of autophagy limits it. Because persistent activation of the AIM2-inflammasome is implicated in the development of autoimmune diseases, it would be interesting to assess whether reduction of autophagy activity is linked to the elevated production of inflammatory cytokines in patients suffering from autoimmune diseases. Further studies are required to elucidate the importance of autophagy in regulation of dsDNA-induced inflammatory responses mediated by the AIM2-inflammasome.

CONCLUSION

Autophagy-related proteins are involved in regulation of dsDNA-induced innate immune responses. Atg9a regulates membrane trafficking involved in the movement of STING and dsDNA-induced production of type I IFNs. Autophagosomes target and engulf bacteria-containing vacuoles in a STING-dependent manner and mediate autolysosomal degradation of invading bacteria. Autophagosomes also target and engulf ubiquitinated ASC and mediate autolysosomal degradation of the AIM2-inflammasome to suppress dsDNA-induced production of inflammatory cytokines. However, it is important to note that these studies mainly utilized an *in vitro* approach. Thus, the contribution of autophagy-related proteins to the regulation of dsDNA-induced innate immune responses *in vivo* remains unclear. Further studies using conditional knockout mice are required to clarify the role of autophagy-related proteins in the regulation of dsDNA-induced innate immune responses.

DNA vaccination is a therapeutic method for protecting an organism against microbial infection and cancers [94,95]. The injection of genetically engineered dsDNA encoding a protein antigen can efficiently stimulate both humoral and cellular immune responses to the protein antigen. Upon DNA vaccination, induction of TBK1-dependent innate immunity greatly promotes subsequent induction of acquired immunity [96]. STING is responsible for TBK1 activation by DNA vaccine. DNA vaccination causes activation of the STING–TBK1 signaling axis to induce acquired immune responses, resulting in the generation of antigen-specific T cells and production of antigen-specific antibody [13]. Cytosolic dsDNA stimulation also causes activation of the AIM2-inflammasome to induce the release of HMGB1, a mediator of inflammatory responses [97]. HMGB1 promotes the production of inflammatory cytokines by stimulating TLR as well as RAGE, a receptor for HMGB1 [98–99]. Indeed, HMGB1 acts as an adjuvant and enhances the effect of DNA vaccination [100–103]. The AIM2-inflammasome might mediate DNA vaccine-induced antigen-specific immune responses by inducing HMGB1 release. Because autophagy-related proteins are involved in the regulation of both STING- and AIM2-dependent innate immune responses, an effect of DNA vaccination might be boosted by modulation of the function of autophagy-related proteins. In future studies, identification of chemical compounds and bioactive factors that can modulate the functions of autophagy-related proteins will be important for improved effectiveness of DNA vaccine.

REFERENCES

[1] Kawai T, Akira S. The roles of TLRs, RLRs and NLRs in pathogen recognition. Int Immunol 2009;21:317–37.

[2] Takeuchi O, Akira S. Pattern recognition receptors and inflammation. Cell 2010;140:805–20.

[3] Barber GN. Cytoplasmic DNA innate immune pathways. Immunol Rev 2011;243:99–108.

[4] Ishii KJ, Coban C, Kato H, et al. A Toll-like receptor-independent antiviral response induced by double-stranded B-form DNA. Nat Immunol 2006;7:40–8.

[5] Stetson DB, Medzhitov R. Recognition of cytosolic DNA activates an IRF3-dependent innate immune response. Immunity 2006;24:93–103.

[6] Watson RO, Manzanillo PS, Cox JS. Extracellular *M. tuberculosis* DNA targets bacteria for autophagy by activating the host DNA-sensing pathway. Cell 2012;150:803–15.

[7] Aoshi T, Koyama S, Kobiyama K, et al. Innate and adaptive immune responses to viral infection and vaccination. Curr Opin Virol 2011;1:226–32.

[8] Ishikawa H, Barber GN. STING is an endoplasmic reticulum adaptor that facilitates innate immune signalling. Nature 2008;455:674–8.

[9] Jin L, Waterman PM, Jonscher KR, et al. MPYS, a novel membrane tetraspanner, is associated with major histocompatibility complex class II and mediates transduction of apoptotic signals. Mol Cell Biol 2008;28:5014–26.

[10] Zhong B, Yang Y, Li S, et al. The adaptor protein MITA links virus-sensing receptors to IRF3 transcription factor activation. Immunity 2008;29:538–50.

[11] Sun W, Li Y, Chen L, et al. ERIS, an endoplasmic reticulum IFN stimulator, activates innate immune signaling through dimerization. Proc Natl Acad Sci USA 2009;106:8653–8.

[12] Abe T, Harashima A, Xia T, et al. STING recognition of cytoplasmic DNA instigates cellular defense. Mol Cell 2013;50:5–15.

[13] Ishikawa H, Ma Z, Barber GN. STING regulates intracellular DNA-mediated, type I interferon-dependent innate immunity. Nature 2009;461:788–92.

[14] Saitoh T, Fujita N, Hayashi T, et al. Atg9a controls dsDNA-driven dynamic translocation of STING and the innate immune response. Proc Natl Acad Sci USA 2009;106:20842–6.

[15] Mizushima N, Komatsu M. Autophagy: renovation of cells and tissues. Cell 2011;147:728–41.

[16] Mizushima N, Yoshimori T, Ohsumi Y. The role of Atg proteins in autophagosome formation. Annu Rev Cell Dev Biol 2011;27:107–32.

[17] Young AR, Chan EY, Hu XW, et al. Starvation and ULK1-dependent cycling of mammalian Atg9 between the TGN and endosomes. J Cell Sci 2006;119:3888–900.

[18] Hara T, Takamura A, Kishi C, et al. FIP200, a ULK-interacting protein, is required for autophagosome formation in mammalian cells. J Cell Biol 2008;181:497–510.

[19] Velikkakath AK, Nishimura T, Oita E, et al. Mammalian Atg2 proteins are essential for autophagosome formation and important for regulation of size and distribution of lipid droplets. Mol Biol Cell 2012;23:896–909.

[20] Sou YS, Waguri S, Iwata J, et al. The Atg8 conjugation system is indispensable for proper development of autophagic isolation membranes in mice. Mol Biol Cell 2008;19:4762–75.

[21] Fujita N, Hayashi-Nishino M, Fukumoto H, et al. An Atg4B mutant hampers the lipidation of LC3 paralogues and causes defects in autophagosome closure. Mol Biol Cell 2008;19:4651–9.

[22] Mizushima N, Yamamoto A, Hatano M, et al. Dissection of autophagosome formation using Apg5-deficient mouse embryonic stem cells. J Cell Biol 2001;152:657–68.

[23] Liang XH, Jackson S, Seaman M, et al. Induction of autophagy and inhibition of tumorigenesis by beclin 1. Nature 1999;402:672–6.

[24] Tanida I, Tanida-Miyake E, Ueno T, et al. The human homolog of *Saccharomyces cerevisiae* Apg7p is a protein-activating enzyme for multiple substrates including human Apg12p, GATE-16, GABARAP, and MAP-LC3. J Biol Chem 2001;276:1701–6.

[25] Kabeya Y, Mizushima N, Ueno T, et al. LC3, a mammalian homologue of yeast Apg8p, is localized in autophagosome membranes after processing. EMBO J 2000;19:5720–8.

[26] Young AR, Chan EY, Hu XW, et al. Starvation and ULK1-dependent cycling of mammalian Atg9 between the TGN and endosomes. J Cell Sci 2006;119:3888–900.

[27] Hosokawa N, Hara T, Kaizuka T, et al. Nutrient-dependent mTORC1 association with the ULK1-Atg13-FIP200 complex required for autophagy. Mol Biol Cell 2009;20:1981–91.

[28] Itakura E, Kishi C, Inoue K, et al. Beclin 1 forms two distinct phosphatidylinositol 3-kinase complexes with mammalian Atg14 and UVRAG. Mol Biol Cell 2008;19:5360–72.

[29] Mizushima N, Kuma A, Kobayashi Y, et al. Mouse Apg16L, a novel WD-repeat protein, targets to the autophagic isolation membrane with the Apg12-Apg5 conjugate. J Cell Sci 2003;116:1679–88.

[30] Proikas-Cezanne T, Waddell S, Gaugel A, et al. WIPI-1alpha (WIPI49), a member of the novel 7-bladed WIPI protein family, is aberrantly expressed in human cancer and is linked to starvation-induced autophagy. Oncogene 2004;23:9314–25.

[31] Hamasaki M, Furuta N, Matsuda A, et al. Autophagosomes form at ER-mitochondria contact sites. Nature 2013;495:389–93.

[32] Kuma A, Hatano M, Matsui M, et al. The role of autophagy during the early neonatal starvation period. Nature 2004;432:1032–6.

[33] Komatsu M, Waguri S, Ueno T, et al. Impairment of starvation-induced and constitutive autophagy in Atg7-deficient mice. J Cell Biol 2005;169:425–34.

[34] Saitoh T, Fujita N, Jang MH, et al. Loss of the autophagy protein Atg16L1 enhances endotoxin-induced IL-1beta production. Nature 2008;456:264–8.

[35] Nakagawa I, Amano A, Mizushima N, et al. Autophagy defends cells against invading group A *Streptococcus*. Science 2004;306:1037–40.

[36] Ogawa M, Yoshimori T, Suzuki T, et al. Escape of intracellular *Shigella* from autophagy. Science 2005;307:727–31.

[37] Gutierrez MG, Master SS, Singh SB, et al. Autophagy is a defense mechanism inhibiting BCG and *Mycobacterium tuberculosis* survival in infected macrophages. Cell 2004;119:753–66.

[38] Narendra D, Tanaka A, Suen DF, et al. Parkin is recruited selectively to impaired mitochondria and promotes their autophagy. J Cell Biol 2008;183:795–803.

[39] Zhao Z, Fux B, Goodwin M, et al. Autophagosome-independent essential function for the autophagy protein Atg5 in cellular immunity to intracellular pathogens. Cell Host Microbe 2008;4:458–69.

[40] Cadwell K, Patel KK, Maloney NS, et al. Virus-plus-susceptibility gene interaction determines Crohn's disease gene Atg16L1 phenotypes in intestine. Cell 2010;141:1135–45.

[41] Hwang S, Maloney NS, Bruinsma MW, et al. Nondegradative role of Atg5-Atg12/Atg16L1 autophagy protein complex in antiviral activity of interferon gamma. Cell Host Microbe 2012;11:397–409.

[42] DeSelm CJ, Miller BC, Zou W, et al. Autophagy proteins regulate the secretory component of osteoclastic bone resorption. Dev Cell 2011;21:966–74.

[43] Fujita N, Yoshimori T. Ubiquitination-mediated autophagy against invading bacteria. Curr Opin Cell Biol 2011;23:492–7.

[44] Deretic V. Autophagy in immunity and cell-autonomous defense against intracellular microbes. Immunol Rev 2011;240:92–104.

[45] Yoshikawa Y, Ogawa M, Hain T, et al. *Listeria monocytogenes* ActA-mediated escape from autophagic recognition. Nat Cell Biol 2009;11:1233–40.

[46] Zheng YT, Shahnazari S, Brech A, et al. The adaptor protein p62/SQSTM1 targets invading bacteria to the autophagy pathway. J Immunol 2009;183:5909–16.

[47] Ponpuak M, Davis AS, Roberts EA, et al. Delivery of cytosolic components by autophagic adaptor protein p62 endows autophagosomes with unique antimicrobial properties. Immunity 2010;32:329–41.

[48] Thurston TL, Ryzhakov G, Bloor S, et al. The TBK1 adaptor and autophagy receptor NDP52 restricts the proliferation of ubiquitin-coated bacteria. Nat Immunol 2009;10:1215–21.

[49] Wild P, Farhan H, McEwan DG, et al. Phosphorylation of the autophagy receptor optineurin restricts *Salmonella* growth. Science 2011;333:228–33.

[50] Kondo T, Kobayashi J, Saitoh T, et al. DNA damage sensor MRE11 recognizes cytosolic double-stranded DNA and induces type I interferon by regulating STING trafficking. Proc Natl Acad Sci USA 2013;110:2969–74.

[51] Strowig T, Henao-Mejia J, Elinav E, et al. Inflammasomes in health and disease. Nature 2012;481:278–86.

[52] Mariathasan S, Newton K, Monack DM, et al. Differential activation of the inflammasome by caspase-1 adaptors ASC and Ipaf. Nature 2004;430:213–8.

[53] Franchi L, Amer A, Body-Malapel M, et al. Cytosolic flagellin requires Ipaf for activation of caspase-1 and interleukin 1beta in salmonella-infected macrophages. Nat Immunol 2006;7:576–82.

[54] Ichinohe T, Pang IK, Iwasaki A. Influenza virus activates inflammasomes via its intracellular M2 ion channel. Nat Immunol 2010;11:404–10.

[55] Barlan AU, Griffin TM, McGuire KA, et al. Adenovirus membrane penetration activates the NLRP3 inflammasome. J Virol 2011;85:146–55.

[56] Ito M, Yanagi Y, Ichinohe T. Encephalomyocarditis virus viroporin 2B activates NLRP3 inflammasome. PLoS Pathog 2012;8:e1002857.

[57] Muruve DA, Pétrilli V, Zaiss AK, et al. The inflammasome recognizes cytosolic microbial and host DNA and triggers an innate immune response. Nature 2008;452:103–7.

[58] Hornung V, Ablasser A, Charrel-Dennis M, et al. AIM2 recognizes cytosolic dsDNA and forms a caspase-1-activating inflammasome with ASC. Nature 2009;458:514–8.

[59] Fernandes-Alnemri T, Yu JW, Datta P, et al. AIM2 activates the inflammasome and cell death in response to cytoplasmic DNA. Nature 2009;458:509–13.

[60] Bürckstümmer T, Baumann C, Blüml S, et al. An orthogonal proteomic-genomic screen identifies AIM2 as a cytoplasmic DNA sensor for the inflammasome. Nat Immunol 2009;10:266–72.

[61] Rathinam VA, Jiang Z, Waggoner SN, et al. The AIM2 inflammasome is essential for host defense against cytosolic bacteria and DNA viruses. Nat Immunol 2010;11:395–402.

[62] Fernandes-Alnemri T, Yu JW, Juliana C, et al. The AIM2 inflammasome is critical for innate immunity to *Francisella tularensis*. Nat Immunol 2010;11:385–93.

[63] Saiga H, Kitada S, Shimada Y, et al. Critical role of AIM2 in *Mycobacterium tuberculosis* infection. Int Immunol 2012;24:637–44.

[64] Zhang W, Cai Y, Xu W, et al. AIM2 facilitates the apoptotic DNA-induced systemic lupus erythematosus via arbitrating macrophage functional maturation. J Clin Immunol 2013;33:925–37.

[65] Choubey D. Interferon-inducible Ifi200-family genes as modifiers of lupus susceptibility. Immunol Lett 2012;147:10–17.

[66] Dombrowski Y, Peric M, Koglin S, et al. Cytosolic DNA triggers inflammasome activation in keratinocytes in psoriatic lesions. Sci Transl Med 2011;3:82ra38.

[67] Roberts TL, Idris A, Dunn JA, et al. HIN-200 proteins regulate caspase activation in response to foreign cytoplasmic DNA. Science 2009;323:1057–60.

[68] Haywood ME, Rose SJ, Horswell S. Overlapping BXSB congenic intervals, in combination with microarray gene expression, reveal novel lupus candidate genes. Genes Immun 2006;7:250–63.

[69] Rozzo SJ, Allard JD, Choubey D, et al. Evidence for an interferon-inducible gene, Ifi202, in the susceptibility to systemic lupus. Immunity 2001;15:435–43.

[70] Hampe J, Franke A, Rosenstiel P, et al. A genome-wide association scan of nonsynonymous SNPs identifies a susceptibility variant for Crohn disease in ATG16L1. Nat Genet 2007;39:207–11.

[71] Rioux JD, Xavier RJ, Taylor KD, et al. Genome-wide association study identifies new susceptibility loci for Crohn disease and implicates autophagy in disease pathogenesis. Nat Genet 2007;39:596–604.

[72] Cummings JR, Cooney R, Pathan S, et al. Confirmation of the role of ATG16L1 as a Crohn's disease susceptibility gene. Inflamm Bowel Dis 2007;13:941–6.

[73] Prescott NJ, Fisher SA, Franke A, et al. A nonsynonymous SNP in ATG16L1 predisposes to ileal Crohn's disease and is independent of CARD15 and IBD5. Gastroenterology 2007;132:1665–71.

[74] Fujita N, Itoh T, Omori H, et al. The Atg16L complex specifies the site of LC3 lipidation for membrane biogenesis in autophagy. Mol Biol Cell 2008;19:2092–100.

[75] Nakatogawa H, Ichimura Y, Ohsumi Y. Atg8 a ubiquitin-like protein required for autophagosome formation, mediates membrane tethering and hemifusion. Cell 2007;130:165–78.

[76] Saitoh T, Akira S. Regulation of innate immune responses by autophagy-related proteins. J Cell Biol 2010;189:925–35.

[77] Stappenbeck TS, Rioux JD, Mizoguchi A, et al. Crohn disease: a current perspective on genetics, autophagy and immunity. Autophagy 2011;7:355–74.

[78] Naser SA, Arce M, Khaja A, et al. Role of ATG16L, NOD2 and IL23R in Crohn's disease pathogenesis. World J Gastroenterol 2012;18:412–24.

[79] Martinon F, Pétrilli V, Mayor A, et al. Gout-associated uric acid crystals activate the NALP3 inflammasome. Nature 2006;440:237–41.

[80] Mariathasan S, Weiss DS, Newton K, et al. Cryopyrin activates the inflammasome in response to toxins and ATP. Nature 2006;440:228–32.

[81] Dostert C, Pétrilli V, Van Bruggen R, et al. Innate immune activation through Nalp3 inflammasome sensing of asbestos and silica. Science 2008;320:674–7.

[82] Hornung V, Bauernfeind F, Halle A, et al. Silica crystals and aluminum salts activate the NALP3 inflammasome through phagosomal destabilization. Nat Immunol 2008;9:847–56.

[83] Duewell P, Kono H, Rayner KJ, et al. NLRP3 inflammasomes are required for atherogenesis and activated by cholesterol crystals. Nature 2010;464:1357–61.

[84] Masters SL, Dunne A, Subramanian SL, et al. Activation of the NLRP3 inflammasome by islet amyloid polypeptide provides a mechanism for enhanced IL-1β in type 2 diabetes. Nat Immunol 2010;11:897–904.

[85] Wen H, Gris D, Lei Y, et al. Fatty acid-induced NLRP3-ASC inflammasome activation interferes with insulin signaling. Nat Immunol 2011;12:408–15.

[86] Vandanmagsar B, Youm YH, Ravussin A, et al. The NLRP3 inflammasome instigates obesity-induced inflammation and insulin resistance. Nat Med 2011;17:179–88.

[87] Zhou R, Yazdi AS, Menu P, et al. A role for mitochondria in NLRP3 inflammasome activation. Nature 2011;469:221–5.

[88] Nakahira K, Haspel JA, Rathinam VA, et al. Autophagy proteins regulate innate immune responses by inhibiting the release of mitochondrial DNA mediated by the NALP3 inflammasome. Nat Immunol 2011;12:222–30.

[89] Harris J, Hartman M, Roche C, et al. Autophagy controls IL-1beta secretion by targeting pro-IL-1beta for degradation. J Biol Chem 2011;286:9587–97.

[90] Shi CS, Shenderov K, Huang NN, Kabat J, et al. Activation of autophagy by inflammatory signals limits IL-1β production by targeting ubiquitinated inflammasomes for destruction. Nat Immunol 2012;13:255–63.

[91] Razani B, Feng C, Coleman T, et al. Autophagy links inflammasomes to atherosclerotic progression. Cell Metab 2012;15:534–44.

[92] Shimada K, Crother TR, Karlin J, et al. Oxidized mitochondrial DNA activates the NLRP3 inflammasome during apoptosis. Immunity 2012;36:401–14.

[93] Bodemann BO, Orvedahl A, Cheng T, et al. RalB and the exocyst mediate the cellular starvation response by direct activation of autophagosome assembly. Cell 2011;144:253–67.

[94] Desmet CJ, Ishii KJ. Nucleic acid sensing at the interface between innate and adaptive immunity in vaccination. Nat Rev Immunol 2012;12:479–91.

[95] Coban C, Kobiyama K, Aoshi T, et al. Novel strategies to improve DNA vaccine immunogenicity. Curr Gene Ther 2011;11:479–84.

[96] Ishii KJ, Kawagoe T, Koyama S, et al. TANK-binding kinase-1 delineates innate and adaptive immune responses to DNA vaccines. Nature 2008;451:725–9.

[97] Lu B, Nakamura T, Inouye K, et al. Novel role of PKR in inflammasome activation and HMGB1 release. Nature 2012;488:670–4.

[98] Yang H, Hreggvidsdottir HS, Palmblad K, et al. A critical cysteine is required for HMGB1 binding to Toll-like receptor 4 and activation of macrophage cytokine release. Proc Natl Acad Sci USA 2010;107:11942–7.

[99] Tian J, Avalos AM, Mao SY, et al. Toll-like receptor 9-dependent activation by DNA-containing immune complexes is mediated by HMGB1 and RAGE. Nat Immunol 2007;8:487–96.

[100] Rovere-Querini P, Capobianco A, Scaffidi P, et al. HMGB1 is an endogenous immune adjuvant released by necrotic cells. EMBO Rep 2004;5:825–30.

[101] Muthumani G, Laddy DJ, Sundaram SG, et al. Co-immunization with an optimized plasmid-encoded immune stimulatory interleukin, high-mobility group box 1 protein, results in enhanced interferon-gamma secretion by antigen-specific CD8 T cells. Immunology 2009;128:e612–20.

[102] Kimura R, Shiibashi R, Suzuki M, et al. Enhancement of antibody response by high mobility group box protein-1-based DNA immunization. J Immunol Methods 2010;361:21–30.

[103] Fagone P, Shedlock DJ, Bao H, et al. Molecular adjuvant HMGB1 enhances anti-influenza immunity during DNA vaccination. Gene Ther 2011;18:1070–7.

dsDNA and Its Part in Modulating Human Diseases

CHAPTER 5

Host DNA Induced Inflammation and Autoimmune Diseases

Surya Pandey and Taro Kawai

Laboratory of Host Defense, WPI Immunology Frontier Research Center, Department of Host Defense, Research Institute for Microbial Diseases, Osaka University, Osaka, Japan

INTRODUCTION

During the past decade, there has been significant advancement in our understanding of how infectious pathogens are recognized initially by the innate immune system. The innate immune system is armed with an arsenal of germline encoded pattern recognition receptors (PRRs) to safeguard the host from invading pathogens through recognition of conserved structures or pathogen associated molecular patterns (PAMPs) in pathogens, and to initiate protective immune responses. A diverse array of PAMPs are recognized by these PRRs with much specificity, ranging from lipids, lipoproteins and proteins to nucleic acids [1]. These ever growing numbers of PRRs include Toll-like receptors (TLRs), RIG-I-like receptors (RLRs), nucleotide oligomerization domain (NOD)-like receptors (NLRs), AIM2 like receptors (ALRs) and other recently identified cytosolic DNA sensors [1,2]. However, nucleic acids are such inherent molecular signatures that are common to both pathogen and host and thus, are at potential risk of self-recognition by innate immune receptors. Self-DNA recognition can lead to deleterious consequences and has been linked to many autoimmune diseases, especially the prototype autoimmune disease systemic lupus erythematosus (SLE), which manifests anti-chromatin and anti-double-stranded (ds) DNA antibodies.

Autoimmune diseases are one of the debilitating diseases that occur when an individual's immune system responds hyperactively against its own cells, tissues, or organs. In light of recent advances it is learnt that self-reactive T and B cells as well as autoantibodies are present in all normal

Biological DNA Sensor.
DOI: http://dx.doi.org/10.1016/B978-0-12-404732-7.00005-8

individuals, but their dysregulation in combination with genetic predisposition and environmental factors can contribute to the development of autoimmune diseases. In other words, autoimmunity is a break in self-immune tolerance, resulting in inflammation and damage. All autoimmune diseases share some common characteristics: the presence of autoantibodies and inflammation. Autoimmune diseases may be systemic or organ specific frequently involving the musculoskeletal system [3]. For a long time, etiopathology of autoimmune diseases has remained enigmatic, as adaptive immunity was known to orchestrate the development and maintenance of autoimmune diseases; however, as current studies show, innate immune responses also contribute strongly to such diseases by various mechanisms ranging from initiation to self-perpetuation of autoantibody production and inflammation. Therefore, growing understanding of the innate immune signaling pathways for host DNA and their priming effects on adaptive immunity has paved the way for dissecting the pathology of autoimmune diseases [4].

This chapter solely emphasizes host DNA sensing mechanisms and their role in autoimmunity. We describe various receptors involved in DNA sensing and their regulatory mechanisms, comprising increased access of host DNA to innate receptors and impaired self-ligand clearance with respect to their relevance in inflammation and autoimmunity.

THE TYPE I IFN SYSTEM AS THE CRUX OF AUTOIMMUNITY

The type I interferon (IFN) production is the key event resulting from recognition of host DNA and consequent activation of signaling pathways by all DNA sensors (with the exception of AIM2). The type I IFN family consists of multiple IFN-α and single IFN-β. The type I IFNs signal through a common heterodimeric receptor, IFNAR, composed of two membrane spanning polypeptide chains, IFNAR1 and IFNAR2, leading to the expression of a large set of IFN-stimulated genes that regulate many biological functions in a cell-specific manner [5]. Almost all types of cells can produce small amounts of type I IFNs, but the major producers are plasmacytoid dendritic cells (pDCs), which produce up to 10^9 IFN-α molecules per cell within 24 h upon activation, in a tightly regulated manner [6]. Type I IFNs influence many critical functions in both innate and adaptive immunity, and several reports show their indispensable role in the development of autoimmune diseases, as evident from the genome-wide association studies that have identified polymorphisms in type I IFN-signaling pathways. As a key feature, reports have also indicated a feed-forward loop for type I IFN production. Usually, type I IFN production is induced by viruses, bacteria, and microbial nucleic acids, but its occurrence in autoimmune diseases has received considerable interest since the findings that there are increased titers of autoantibodies and autoimmune incidences during type I IFN treatment in patients with infectious and malignant diseases without prior history [7,8]. In humans, an increased serum concentration of IFN-α is found in SLE patients [9] and correlated with the SLE disease activity and severity [10,11].

SELF AND NON-SELF-DNA RECOGNITION

Innate immune recognition of DNA is a crucial mechanism to confer protection against invading pathogens. However, this protection comes with a probable cost of self-DNA recognition, if the system goes haywire and the host DNA is recognized, resulting in autoinflammatory or autoimmune diseases. Under physiological conditions, self- and non-self-DNA discrimination is regulated at several levels to ensure reliable innate immune system activation. TLR9 is the most studied DNA sensing receptor, and hence most of the information obtained with regard to self/non-self recognition explained by the regulation of TLR9 activity. However, similar regulatory mechanisms may be implied for cytosolic DNA sensors and will probably be explained in the future with better understanding of cytosolic DNA sensing pathway. The contribution of TLR9 to autoimmune diseases, particularly SLE (the prototype multiorgan disease), has been advanced using mouse models that develop autoimmune diseases spontaneously, especially the MRL strain of mice. Homozygous *lpr* mutant (*lpr/lpr*) mice do not express the functional death receptor FAS and develop an SLE-like disease [12]. The presence of antinuclear antibodies (ANAs) is a major characteristic of SLE and is a sign of a breach in barrier to the nucleic acids. Corresponding to the initial hypothesis, MyD88 deficiency in *lpr/lpr* mice does not lead to ANA production [13]. TLR9 has been implicated in the development of ANAs against DNA, although the mechanism of its involvement is controversial [14,15]. The first study unexpectedly demonstrated that TLR9 deficiency in *lpr/lpr* mice reduced dsDNA and autoantibody production, but glomerulonephritis was still present and some disease parameters were enhanced [14]. Subsequently, another group showed that TLR9 deficiency leads to more severe lupus in the *lpr/lpr* model [16]. Furthermore, the role of TLR9 in severe disease progression has been confirmed in a novel SLE model, *Ali5* mutant mice, which when crossed with TLR9-deficient mice showed increased phospholipase-Cγ2 and exacerbated disease with increased production of ANAs [17]. Similar results were observed in TLR9-deficient *lpr/lpr* B6[lpr] mice and B6.Nba2 mice. Moreover, TLR7 deficiency in these models ameliorated the disease [18]. Recently, it was reported that the receptor for type I IFNs (*Ifnar1*) deficiency prevented the exacerbation of autoimmunity leading to renal disease in the TLR9 deficient-MRL Fas[lpr] mouse model; they also showed that *Ifnar1* had no effect on antinucleosome or anti-sm autoantibody titers, but instead regulated anti-RNA antibody production explaining how loss of IFNAR-I signaling negates many of the disease promoting effects of TLR9 deficiency [19].

The discrepancy in the different murine models is unclear, although it is suggested that cytosolic DNA sensors might redundantly contribute to the production of ANAs. The cytosolic DNA sensors DHX36 and DHX9 have been shown to signal through MyD88, which corresponds to the MyD88 dependency for ANA production in *lpr/lpr* mice [20]. Future studies will surely lead to better understanding of the precise role of TLR 9.

To activate an innate immune response against host DNA, multiple barriers must be breached; firstly, sequence specific recognition by receptors that are unique to pathogens. Until recently, the prevailing paradigm was that TLR9 recognizes hypomethylated 2′-deoxyribo (cytidine-phosphate-guanosine) CpG DNA that is abundantly present in bacterial DNA and is a much more potent activator than mammalian DNA, which is largely methylated [21,22]. However, a number of recent studies showed that TLR9 recognizes the 2′-deoxyribose sugar backbone of natural phosphodiester oligodeoxynucleotides in a sequence-independent manner, while CpG dependent TLR9 activation is indispensable for synthetic phosphorothioate oligodeoxynucleotides [23,24]. The second challenge is the excluded localization of the receptors, which is achieved by subcellular compartmentalization [25]. The endosomal localization of TLR9 reduces the likelihood of host DNA access to the receptor and activation of innate immune response, but this checkpoint can be overcome by several mechanisms that enhance host DNA uptake by pDCs with the help of cytoplasmic proteins or receptor-mediated endocytosis as discussed in the following sections. Finally, there are augmented levels of host DNA released as a result of increased cell death from the processes of apoptosis, necrosis, pyroptosis and NETosis, which can increase the chances of hyperactivation of innate receptors. Apoptosis and necrosis are known to be the major forms of cell death for release of self-DNA, but they differ morphologically and mechanistically. In necrosis, the mitochondria and nucleus swell, the cell membranes rupture, and cellular contents are released. In contrast, during apoptosis, the nucleus condenses and becomes fragmented, the cell membranes convolute, and the cellular contents are packaged into apoptotic bodies/blebs that are engulfed by phagocytes [26]. The DNA within apoptotic bodies retains the ability to activate TLR9 [4]. In addition, apoptotic cells not cleared efficiently can undergo secondary necrosis, thereby releasing self-ligands. Pyroptosis is a form of cell death resulting from inflammasome activation that can release self nucleic acids. Another interesting phenomenon of host DNA release recently studied extensively is NETosis, wherein dying neutrophils release neutrophil extracellular traps (NETs) composed of chromatin and granular contents into the extracellular space. Failure to clear self-DNA released by the abovementioned mechanisms can lead to autoimmune pathology. DNases (DNases I, II and III) present in different compartments of the cells are able to degrade the overload of self-DNA. Moreover, several additional mechanisms work toward efficient self-ligand clearance, as explained in great detail below.

HETEROGENEITY IN INNATE RECEPTORS FOR DNA
DNA Sensing TLR-TLR9

The discovery of TLR9 as a sensor for foreign DNA to trigger innate immune responses was pioneering for the studies encompassing the role of DNA sensors in innate immunity. TLR9 is mainly expressed in pDCs, macrophages and B cells, and is able to stimulate robust type I IFN production, mainly in pDCs, in response to unmethylated CpG DNA. TLR9 was originally identified to recognize only bacterial DNA [27], but was later demonstrated to also recognize viral DNA [28]. Further studies showed that host DNA can also act as an endogenous ligand in the form of immune complexes to activate TLR9 by PDCs [29]. TLR9 is localized in several cellular compartments, including the endoplasmic reticulum (ER), endosomes, and lysosomes, rendering possible the spatiotemporal regulation of its intracellular trafficking, which is a key factor for TLR9 ligand accessibility and its activation. Inactive TLR9 is associated with the ER in unstimulated pDCs. However, CpG DNA internalized via the endocytic pathway goes to the lysosomal compartment to interact with pre-processed active TLR9 that has been trafficked from ER through the Golgi and gained residence in endolysosomes [30,31]. A number of ER-associated proteins, such as GP96 and PRAT4A, have been reported to act as chaperones and mediate the trafficking of TLR9 to various subcellular locations [32,33]. UNC93B1 is an important ER-resident protein controlling the trafficking of TLR9, since a point mutant Unc93b1 protein cannot exit the ER and overexpression of wild-type Unc93b1 increases TLR9 trafficking to endosomes [34,35].

Recent studies have shown that proteolytic cleavage of the major portion of the TLR9 ectodomain by cathepsins is required for its activation in response to CpG DNA [30,31]. This TLR9 processing proceeds in two steps, cleavage of a portion of ectodomain by multiple cathepsins, and trimming of the N-terminal, and the cleaved form of TLR9 is functional and associates with MyD88 [36].

Cytosolic DNA Sensors – Plausible Players in Autoimmunity

The appearance of DNA in the cytoplasm is a sign of danger for the cell as a result of infection or tissue damage. The innate immune system recognizes DNA present in the cytoplasm by a number of sensors, although the direct relationship of autoimmune pathogenesis to cytosolic DNA sensors is a relatively uncharted area.

There has been a voluminous amount of research to identify cytosolic DNA sensor candidates, following the use of TLR9 antagonists and TLR9-deficient mice to demonstrate an additional pathway of DNA recognition that is independent of TLR9. Initial studies showed that intracellular administration of long polymers of a poly(dA-dT).poly(dA-dT) DNA or a synthetic 45-mer random DNA sequence lacking CpG motifs (called immunostimulatory DNA or ISD) triggered type I IFN production in an IRF3-dependent but TLR9-independent manner [37,38]. Consistently, several reports showed that the response to vaccination remained intact in TLR9 deficient mice [39,40].

Observations that mice lacking DNase II die at an embryonic stage owing to autoimmunity and TREX1-deficient mice develop autoimmune myocarditis [41,42] led to proposals of the involvement of cytosolic DNA sensors in autoimmunity and their lack of specificity to discriminate between self- and non-self-DNA. Recognition of cytosolic DNA appears to be more complex than anticipated because it is independent of the degree of methylation or the DNA sequence, hence both foreign DNA and self-DNA may activate the cytosolic pathway. Continuing studies indicate that cytosolic dsDNA is sensed by a broad spectrum of sensors (Figure 5.1).

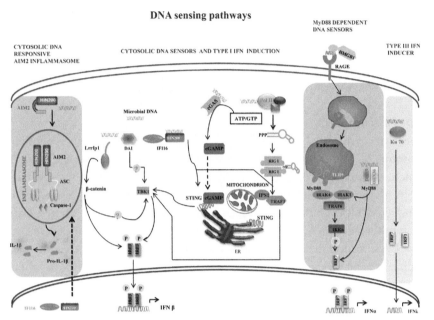

Figure 5.1 *DNA sensing pathways.* A number of DNA sensors have been identified in the cytosol. Five DNA sensors that induce type I IFN production are shown here. LRRFIP1 can bind both dsDNA and dsRNA by activating β-catenin, which migrates to the nucleus to increase IFN-β transcription in an IRF3-dependent and TBK1-independent manner. IFI16 can recognize viral DNA in the nucleus via its HIN200 domains and then migrate to the cytosol to induce IFN-β production mediated by the STING–TBK1 and IRF3 pathways. DAI can recognize both the B-form and left-handed Z-form of DNA and induce IFN-β in a TBK1/IRF3-dependent manner. Cytoplasmic AT-rich dsDNA is recognized by RNA Pol III and transcribed into 5′-triphosphate RNA, which is further recognized by RIG-I to induce IFN-β via IPS-1. The newly identified cyclic GMP-AMP (cGAMP) synthase functions as a cytosolic DNA sensor. In the presence of DNA, it converts cGTP and cATP to cGMP and cAMP, respectively, and the cyclic GMP-AMP interacts with STING and triggers IFN-β production dependent on IRF3. Self-DNA can also gain access to TLR9 with the aid of HMGB1 and RAGE, which can bind to extracellular DNA and facilitate DNA uptake and TLR9 engagement, whereby TLR9 is activated and signals through MyD88 to induce IFN-α production. Another class of DNA sensors dependent on MyD88 are the cytosolic DExDH box helicases DHX9 and DHX36. The AIM2-like receptors AIM2 and IFI16 can form inflammasomes in association with ASC in response to bacterial and viral DNAs, and these inflammasomes mediate caspase-1-dependent cleavage of pro-IL-1β and pro-IL-18 and secretion of active IL-1β and IL-18. Ku-70 induces IRF1/IRF7-dependent IFN-λ production. Please see color plate at the back of the book.

The major portion of cytosolic DNA sensors discovered to date exert their effector role by induction of type I IFNs. However, there is a growing body of evidence suggesting that some DNA sensors can form inflammasomes to activate caspase-1 and can also induce type III IFN. All of these classes of cytosolic DNA sensors are discussed here.

Type I IFN Inducers

The first cytosolic DNA sensor identified was DAI (DNA-dependent activator of IRFs, also known as Z-DNA binding protein-1, ZBP-1), which reportedly recognizes both the B-form and Z-form of cytosolic ds-DNA and its knockdown in L929 cells decreased type I IFN responses to transfected DNA [43]. However, the type I IFN production in response to transfected DNA was comparable in mouse embryonic fibroblasts and DCs from mice lacking DAI, thereby suggesting the existence of additional and/or redundant sensors or a cell type-specific role of these sensors, since many cell types may possess multiple DNA sensors [44,45].

Further studies identified another cytosolic DNA sensor, IFI16 and its mouse homologue p204, a pyrin domain–containing protein belonging to the HIN-200 family (PYHIN). IFI16 is predominantly a nuclear protein, but can come out into the cytosol and recognize DNA to induce IFN-β production. Herpes simplex virus 1 DNA can be detected in the nucleus by IFI16, followed by migration to the cytoplasm to stimulate type I IFN signaling. The IFI16 mediated recognition of DNA is STING-dependent [46]. Autoantibodies against IFI16 have been reported in patients with autoimmune diseases including SLE, Sjogren syndrome and scleroderma [47,48].

An unusual means of cytosolic DNA recognition is attained by DNA-dependent RNA polymerase III, which converts cytosolic AT-rich ds-DNA into 5′-ppp ds-RNA to induce IFN-β production mediated by RIG-I/IPS1 with a partial contribution of STING [49–51].

Recently, other unique DNA sensors have been identified, namely LRRFIP1, Ku70, DHX9, and DHX36. However the relevance of these receptors to autoimmune diseases remains to be identified.

DHX36 and DHX9, members of the RHA subfamily of DExD/H helicases, sense microbial DNA differently, in that DHX36 recognizes CpG-A using the DEAH domain, and DHX9 senses CpG-B using the DUF domain in pDCs. Upon DNA recognition, DHX9 promotes tumor necrosis factor (TNF)-α and interleukin (IL)-6 production via nuclear factor (NF)-κb and DHX36 mediates IFN-α production through IRF7, and both processes are dependent on MyD88 [20].

LRRFIP1 is a cytosolic nucleic acid-binding protein that acts as a sensor for both ds-DNA and ds-RNA by interacting with β-catenin, thereby promoting its activation to facilitate the recruitment of acetyltransferase p300 to the IFN enhanceosome in an IRF3-dependent manner to increase the *IFNB* gene transcription [52].

Ku70 is a component of a heterodimeric Ku protein that is required for non-homologous end-joining DNA repair, VDJ recombination, telomerase maintenance and various other nuclear processes. It also acts as a cytosolic DNA sensor for a variety of DNA forms and promotes the production of type III IFN-λ1 rather than type I IFN mediated by activation of IRF1 and IRF7 [53].

The Chen group recently identified a novel cytosolic DNA sensor, cyclic GMP-AMP (cGAMP) synthase or (cGAS). cGAS is a member of the nucleotidyltransferase family. Overexpression of cGAS activated IRF3 and induced IFN-β in a STING-dependent manner, whereas cGAS knockdown in L929 cells abrogated IRF3 activation and IFN-β induction by DNA virus infection or DNA transfection. Mechanistically, in the presence of DNA, cGAS synthesizes cGAMP, an endogenous second messenger that interacts with STING, leading to IRF3-mediated IFN production [54]. However, whether cGAS has a role in autoimmune diseases is subject to further research.

DNA-Activated Inflammasomes

Among all of the DNA sensing pathways that induce type I IFN production, there are certain members of the ALRs, namely cytosolic murine Aim2, human AIM2 and nuclear IFI16, that stimulate a rather different immune response by activating inflammasomes. Inflammasomes are multiprotein macromolecular structures that cleave pro-IL-1β to its active form, leading to inflammatory responses and induction of cell pyroptosis.

For a long time, NLR family members were known to form inflammasomes in response to various ligands. However, it was then reported that an additional cytosolic DNA-sensing inflammasome exists that is activated in response to 250 bp dsDNA, although the activation occurs in response to poly(dA-dT).poly(dT-dA), but not ISD [4,55]. Subsequently, several groups identified AIM2, a cytoplasmic member of the PYHIN family, as the sensor that activates the inflammasome upon stimulation with cytosolic DNA [56–59]. AIM2 can sense cytoplasmic DNA via its HIN-200 domain. Upon sensing the cytosolic DNA, AIM2 recruits the adapter protein ASC through its PYD domain and ASC then recruits procaspase-1 via a CARD–CARD interaction to form an active AIM2 inflammasome. The activated AIM2 inflammasome activates caspase-1 that cleaves pro-IL-1β and pro-IL-18 to mature forms for secretion. The AIM2 inflammasome activation in macrophages was also reported to induce cell death by pyroptosis [57]. Bone marrow-derived DCs and splenic cells from Aim2-deficient mice were unable to activate the inflammasome in response to DNA. Notably, AIM2 was not required for the production of type I IFN in mice in response to poly(dA-dT). poly(dT-dA) and certain pathogens [60–62]. Since AIM2 recognizes both bacterial and viral DNAs in a sequence-independent manner, it is possible that it can recognize host DNA present in the cytosol in an autoimmune condition. Studies have shown that the Aim2 protein suppresses type I IFN expression and the expression of IFN-regulated genes in mice; additionally lupus-prone NZB and B6.Nba2 female mice express reduced levels of Aim2 mRNA and protein in immune cells, compared with age and sex matched C57Bl/6 mice [63]. The activity of AIM2 inflammasomes is negatively regulated by type I IFN, partly through suppression of pro-IL-1β [64]. Precisely, Aim2 and AIM2 inflammasome activities are negatively regulated by p202 and IFI16 proteins, respectively [59,63,65,66]. Furthermore, p202 was identified as a candidate lupus susceptibility gene in several reports. The AIM2 inflammasome is activated in response to cytosolic DNA to release IL-1β in psoriatic keratinocytes and, furthermore, LL37, which can interact with DNA in psoriatic skin, neutralized cytosolic DNA in keratinocytes and blocked AIM2 inflammasome activation [67]. Although these studies indicate that an imbalance between DNA-responsive inflammasomes and their negative regulators may have a role in the development of autoimmunity, direct evidence of host DNA recognition and activation of DNA-responsive inflammasomes in autoimmune diseases remains an issue to be addressed in further research.

In addition to type I IFN induction, IFI16 has been reported to sense Kaposi sarcoma-associated herpesvirus derived DNA in the nucleus of endothelial cells. Upon sensing viral DNA, IFI16 recruits the adapter molecule ASC and procaspase-1 to form an active inflammasome [68]. However, the question remains as to how host DNA is prevented from being recognized by nuclear IFI16. A role of IFI16 in autoimmunity has been implicated indirectly, such as findings that selective blockade of the inhibitory Fcγ receptor (FcγRIIb) in human monocytes and DCs induces a type I IFN response program that results in transcriptional activation of the IFI16 gene [69]. Furthermore, elevated levels of IFI16 mRNA have been detected in peripheral blood mononuclear cells from SLE patients, compared with healthy subjects [70].

MECHANISMS UNDERLYING HOST DNA INDUCED INFLAMMATION AND AUTOIMMUNITY

As mentioned earlier, multiple levels of checks work to prevent self-DNA recognition, consequent type I IFN production, and perpetuation to autoimmunity. Increasing numbers of studies are unfolding the critical mechanisms involved in the pathogenesis of autoimmune diseases, most of which are centered around augmented cell death, decreased cell debris clearance, and increased internalization and recognition of self-DNA.

Enhanced self-DNA Uptake and Recognition by TLR9

Specific mechanisms have evolved to prevent the engagement of intracellular DNA sensors with self-DNA. However, several mechanisms for increased self-DNA internalization and TLR9 activation have been reported, which are extensively described. These are helpful for understanding the source of exaggerated type I IFN production that leads to autoimmune pathology (Figure 5.2).

LL37 Accompanied Uptake of Self-DNA

LL37, the only cathelicidin with alpha-helical structures, is an endogenous cationic and amphipathic antimicrobial peptide produced by keratinocytes and neutrophils [71]. LL37 is detected in psoriatic skin lesions and as a component of NETs infiltrating the skin and kidneys of patients with SLE and small-vessel vasculitis [72–74]. Fundamentally, LL37 is ascribed as an alarmin that is chemotactic for various leukocytes and can induce the activation of monocytes, macrophages, mast cells, endothelial cells, and keratinocytes. It is predominantly stored in the secondary granules of neutrophils and is released in the course of activation and degranulation of neutrophils as a constituent of NETs [75,76]. Increasing numbers of reports have recently shown that LL37 can convert otherwise inert extracellular self-DNA into a potent ligand for TLR9 and trigger the type I IFN response in pDCs. This is achieved in three distinct steps. First, LL37 binds to self-DNA released by necrotic cells to form aggregated and condensed structures, thereby protecting them from degradation by extracellular nucleases. Second, the LL37/DNA complex is translocated into the endocytic compartment of pDCs through lipid-raft mediated endocytosis. Finally, LL37 retains the DNA complex in the early endocytic compartments of pDCs inducing a specific and sustained IFN response via TLR9/MyD88/IRF7 signaling [72,77]. The role of LL37 in the recognition and translocation of

Different modes of self nucleic acid recognition

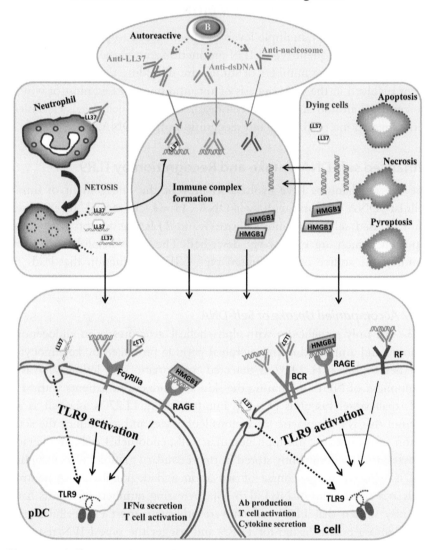

Figure 5.2 *Different modes of self-DNA recognition by TLR9.* Autoreactive B cells can produce ANAs in the form of anti-dsDNA, anti-LL37, and anti-nucleosome antibodies, and these autoantibodies can form immune complexes with subcellular components, nuclear autoantigens (DNA, chromatin, nucleosomes), and proteins such as LL37 and HMGB1 derived from inefficient clearance of dead cells arising from different forms of cell death, namely apoptosis, necroptosis, and pyroptosis. LL37-DNA complexes can also come from neutrophil death due to NETosis, induced by anti-LL37 antibodies binding to LL37 on the neutrophil surface. The various ICs can promote delivery of self-DNA to endosomal compartments of pDCs or B cells to activate TLR9 through

self-DNA is not confined to the activation of TLR9 in pDCs only. This is based on reports showing uncontrolled type I IFN production and SLE-like autoimmunity due to accumulation of self-DNA in DNase II-deficient mice that cannot be reversed in the absence of TLR9 or MyD88 [78]. Self-DNA-containing circulating immune complexes that trigger DC activation via TLR9-independent pathways in SLE [79,80] also suggest that innate sensing of self-DNA may also occur via cytosolic receptors in cells other than DCs. To this end Chamilos et al. recently demonstrated that LL37 efficiently transports self-DNA (dsB-DNA) into monocytes, leading to the production of type I IFNs in a TLR-independent manner via the STING/TBK1 signaling axis, regardless of the DNA sequence, CpG content, or methylation status [81]. Cramp (LL37 counterpart in mice)/self-DNA complex-mediated pDC activation and IFN-α production are implicated in the generation of anti-DNA antibodies in the autoimmunity of atherosclerosis, critically aggravating atherosclerotic lesion formation [82].

HMGB1/RAGE Mediated Internalization of Self Nucleic Acids

Danger associated molecular patterns (DAMPs) are released by dying or necrotic cells and lead to sterile inflammation. HMGB1 is one such DAMP that has distinct roles depending upon its cellular location. It is a small (215 amino acid residues) nonhistone chromatin-associated nuclear protein that binds to DNA in a conformation-dependent but sequence-independent manner and regulates transcription in the nucleus [83,84]. Meanwhile, it can be actively secreted in response to different stimuli such as IFN-λ, type I IFNs, TNF-α, lipopolysaccharide, lipotechoic acid, poly-(I:C) and IL-1β by macrophages [85–87], or passively released by necrotic and apoptotic cells, thereby acting as a prototypic DAMP [83]. HMGB1 is an important modulator of TLR9 activation by CpG. It enhances IL-6 secretion in response to type B CpG oligodeoxynucleotides from bone marrow-derived macrophages and IL-6, IL-12, and TNF-α from conventional DCs in a TLR9-dependent manner [88]. Association of the HMGB1/DNA complex with RAGE (receptor advanced glycation end products), a multiligand receptor, provides another means of access

◀ engagement of a variety of receptors. RAGE binds to HMGB1–DNA complexes, and FcγR binds to the Fc portion of autoantibody complexes with LL37–DNA or DNA on pDCs. On B cells, B cell receptor (BCR) binds to Fc portion of autoantibody complexes with LL37–DNA or dsDNA, RAGE binds to HMGB1–DNA complexes, and RF to anti-dsDNA antibodies. Please see color plate at the back of the book.

to host DNA for TLR9. HMGB1 released from necrotic cells binds to DNA-containing immune complexes in serum, and these HMGB1/DNA complexes bind to RAGE which recruits TLR9 and augments type I IFN production and proliferation of autoreactive B cells, also acting as an amplification loop in autoimmunity and inflammation. HMGB1 does not affect the uptake of CpG-DNA into TLR9-expressing early endosomes, but rather accelerates the recruitment of TLR9 from the ER to endosomes [89]. Furthermore, HMGB1-containing nucleosomes derived from secondary necrotic cells are capable of activating macrophages and DCs to secrete IL-1, IL-6, IL-10, and TNF-α and to express co-stimulatory molecules. In addition, they also induce anti-dsDNA and anti-histone IgG responses in a TLR2-dependent manner in non-autoimmune mice [90]. The autoantibodies formed against dsDNA and nucleosomes are characteristic of SLE, and relevant to this, high expression levels of cytoplasmic and extracellular HMGB1 are observed in skin lesions of patients with cutaneous lupus [91]. Similarly, elevated levels of systemically released HMGB1 are found in serum and plasma from patients with SLE and experimental lupus mouse models. Notably, serum HMGB1 and anti-HMGB1 autoantibody levels are correlated with disease activity [92]. Finally, nephritogenic anti-DNA antibodies synergize with HMGB1 and stimulate proinflammatory gene expression from mesangial cells via RAGE and TLR2/4, which may contribute to antibody-induced kidney damage in SLE [93]. Other HMGB proteins HMGB2 and HMGB3 have also been shown to be sentinels for DNA that activate TLR9 [94].

Receptor Mediated Endocytosis – DNA En Route to TLR9

Another access route of self-DNA to intracellular compartments is receptor mediated endocytosis. Self-DNA complexed with autoantibodies can bind to Fc receptors (FcRs) or B cell receptors (BCRs) and is internalized followed by trafficking to compartments containing TLR9 [4].

In the first mechanism, antigen-presenting cells (APCs) like pDCs express FcRs, which are able to bind to the Fc region of antibody-DNA complexes and facilitate their uptake through FcRγ mediated endocytosis, resulting in large amounts of IFN-α production [75,95,96]. A series of elegant studies showed that serum from SLE patients stimulates IFN-α production from normal pDCs in vitro and that total IgG from SLE serum combined with apoptotic and necrotic cells produces the same effect [96,97]. Further studies in mouse DCs demonstrated that the uptake of immune complexes containing DNA or chromatin is dependent on

FcγIIIR and TLR9 [29,79]. Similarly, DNA containing immune complexes from the serum of SLE patients can bind to FcγRIIa(CD32) on human pDCs and these DNA complexes are further internalized into the endosomal compartment where they are recognized by TLR9 resulting in the production of enormous amounts of IFN-α, which is characteristic of many SLE patients as an 'interferon–alpha signature'[98].

Host DNA as an endogenous ligand can also be recognized and trafficked by a second mechanism dependent on autoreactive BCRs. B cells express TLR9 as well as antigen-specific receptors called BCRs, which can uptake chromatin or related immune complexes by binding to the antigenic epitopes of the macromolecules (BCR specific for DNA/histone) or the Fc moiety of the autoantibodies (BCR with rheumatoid factor (RF) activity) [5]. These materials are transported into endosomal compartments where the DNA interacts with TLR9, and this synergistic engagement of BCR and TLR9 results in the activation and proliferation of autoreactive B cells that are otherwise quiescent. Preliminary evidence supporting this notion came from *in vitro* analysis of mice transgenic for the AM14 BCR. AM14 B cells express an RF-like BCR for autologous IgG2a, a specificity normally found in the B cell repertoire of auto-immune-prone mice. This receptor binds to monomeric IgG2a with a relatively low affinity. Consequently, in non-autoimmune mice, AM14 B cells are not tolerated and develop into normal mature B cells [99]. *In vitro*, AM14 B cells proliferate in response to immune complexes containing IgG2a antibodies bound to DNA or DNA-associated proteins [100], but not to immune complexes bound to proteins or unrelated haptens [101]. Of note, a recent report highlighted that BCR signals are sufficient to cause relocalization of TLR9 to autophagosomes, where it colocalizes with BCR after internalization of immune complexes [102] possibly enhancing the responsiveness of autoreactive B cells to self-DNA ligands.

Altogether, these studies strongly indicate a critical mechanism for breach of host DNA tolerance by internalization of self-DNA-containing immune complexes via BCRs and FcRs, ultimately leading to large amounts of IFN-α production and an increased propensity toward auto-immune diseases.

Amyloid Fibril-Mediated Uptake of DNA

Amyloid precursor proteins can form amyloid fibrils in the presence of nucleic acids [103]. Recently, it was demonstrated that DNA-containing amyloid fibrils can induce high levels of IFN-α/β production by pDCs in response to self-DNA. These self-DNA-containing amyloid fibrils are internalized by pDCs and retained in early endosomes to activate TLR9 and produce large amounts of type I IFN. However, the involvement of specific pDC surface receptors, if any, in mediating this internalization is unknown. *In vivo* as well, DNA-containing amyloid fibrils induced infiltration and activation of pDCs correlating with rapid transcription of type I IFN genes in mice. Furthermore, immunization with DNA-containing amyloid fibrils induced autoantibodies and proteinuria in non-immune mice, thereby establishing an inducible murine model for lupus [104].

Uptake of DNA Mediated by Complex Cationic Lipids

A well-established method for gene delivery in mammalian cells both *in vitro* and *in vivo* is mediated by cationic lipids [75,105,106]. CpG DNA encapsulated with cationic liposomes has been shown to facilitate uptake by B cells, DCs, and macrophages and is protected from serum nucleases [107]. Interestingly, CpG DNA complexed with a cationic lipid N-[1-(2,3-dioleoyl-oxy)propyl]N,N,N-trimethylammonium methylsulfate (DOTAP) in an aggregated form induced potent type I IFN response compared to CpG alone [108]. In addition, DOTAP, which is known to enter cells through an endocytic pathway when complexed with CpG DNA [109], enhances endosomal translocation of vertebrate DNA to strongly activate TLR9 and hence DCs [110]. The assumption that the self-DNA/LL37 complexes implicated in the pathogenesis of psoriasis may have a similar mode of action to that of DNA/DOTAP aggregates [111] might give rise to a tempting idea about their relevance in many biological situations, including autoimmune diseases.

Granulin Mediated Uptake of DNA

The soluble cysteine-rich protein granulin is constitutively secreted into the serum by murine macrophages and DCs. Recently, Park et al. identified granulin as a soluble cofactor that binds to CpG DNA and enhances its delivery to endolysosomal compartments where it can interact with TLR9 [112]. In a recent report, the serum level of progranulin (PGRN), the precursor of granulin, was found to be significantly higher in SLE patients than in healthy controls. Increased PGRN levels were significantly correlated with the systemic disease activity and anti-dsDNA antibody titers and inversely correlated with CH50, C3, and C4 levels. Furthermore, serum PGRN levels were significantly decreased after successful treatment of SLE

[113]. Nonetheless, further studies are needed to reveal the more detailed mechanisms of PGRN and granulin in human autoimmune diseases.

Increased Cell Death

Self-DNA conducing to the pathogenesis of autoinflammatory and autoimmune diseases predominantly arises from dying cells as a consequence of the body's homeostatic processes like apoptosis (or secondary necrosis) to eliminate harmful, useless, or senescent cells. This is accomplished in two steps of cell death and dead cell clearance, both of which are tightly regulated and have been extensively studied in mouse models and humans [114].

Augmented cell death may be a consequence of rapid cell death and a concomitant lack of clearance resulting in increased cell debris and self nucleic acids.

Until recently, apoptotic debris has been thought to be the primary source of nuclear autoantigens, like DNA in the extracellular milieu, where they gain access to the immune system. However, increasing evidence shows that extracellular DNA or chromatin can be released as a result of NETosis, a unique form of cell death characterized by the release of NETs from neutrophils to trap and kill microorganisms. NETs are composed of chromatin fibers with diameters of 15–17 nm, DNA and histones bound to granular and selected cytoplasmic proteins [115]. The major constituents of NETs, host DNA and chromatins, are autoantigens in many systemic autoimmune diseases, and therefore aberrant NET formation may be important in the generation of autoimmune diseases as suggested by various groups [74,116,117]. Recent studies have demonstrated that mature neutrophils from SLE undergo NETosis upon exposure to SLE-derived autoantibodies to release NETs. NETs contain complexes of host DNA, LL37, and HMGB1 that potentially trigger pDC activation via TLR9 to produce high levels of IFN-α to further exacerbate the disease, thus clarifying the long-sought link between SLE and neutrophils. Notably, neutrophils from patients with SLE and various other autoimmune diseases are more prone to NETosis, particularly in response to antibody complexes, and the presence of IFN-α in the serum enhances NETosis by upregulating the surface expression of LL37 [73,118,119].

Finally, efficient and timely clearance of NETs is important to maintain tissue homeostasis and prevent autoimmunity. The serum endonuclease DNase I can degrade NETs *in vitro*. An inherited form of SLE was linked to a mutation in DNase I or to DNase I-like 3 [6,120]. Moreover, a subset of SLE patients had inefficient NET degradation correlating with lupus nephritis owing to either the presence of elevated titers of anti-NET antibodies, which sterically prevented DNase I access to NETs, or the presence of DNase inhibitors [121]. Finally, impaired NET degradation could be caused by complement activation and increased deposition of the complement protein C1q, which directly inhibits DNase I [120].

Defective Self-Ligand Clearance

Several regulatory mechanisms work to eradicate self-DNA originating from apoptotic debris before it engages with the innate receptors, and impaired clearance of self-DNA can promote autoimmunity. This degradation is achieved by DNases present in separate cellular compartments or with the help of several secreted and complement proteins that facilitate and enhance self-DNA clearance (Figure 5.3).

DNases

DNases degrade self nucleic acids under normal conditions in different compartments. To date, three different DNases have been attributed with this function, DNase I, DNase II, and TREX1 (or DNase III).

DNase I is the major serum endonuclease that degrades extracellular dsDNA into trioligonucleotides or tetraoligonucleotides [122]. DNase I-deficient mice manifested classical symptoms of SLE, including presence of ANAs, agglomeration of immune complexes in glomeruli, progression to glomerulonephritis and eventual death [123]. Congruently, in human studies, mutations in *dnase I* are linked to SLE, and a direct correlation was reported between low activity of DNase I and SLE [120]. Decreased DNase I activity is implicated in other systemic and organ-specific autoimmune diseases including Sjogren's syndrome [124], thyroid autoimmunity [125] and severe inflammatory bowel disease [126].

In the course of dead cell clearance, macrophages are not activated by nucleic acids present in the engulfed apoptotic cells because they are equipped with DNase II, an acid DNase located in lysosomes, where it degrades DNA from dead cells into nucleotides as well as nuclear DNA from erythroid precursor cells following their engulfment [41,127] thus serving as a sophisticated mechanism to prevent TLR recognition of self-DNA. Before the engulfment of apoptotic cells, the chromosomal DNA is autonomously cleaved into nucleosomal units (multimers of 180 bp) by another endonuclease, caspase activated DNase (CAD), after which they are engulfed by macrophages and transferred to lysosomes, rendering them susceptible to degradation by DNase II. DNase II is ubiquitously expressed

Impaired clearance of self nucleic acid

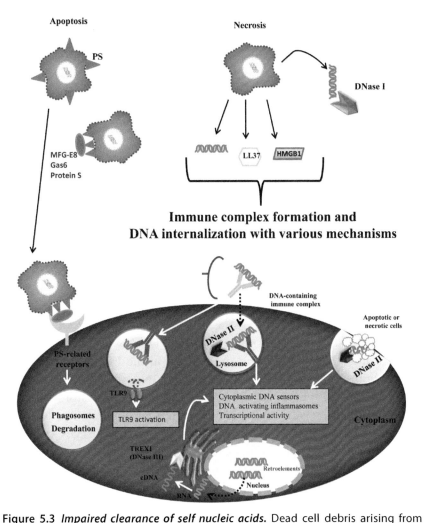

Figure 5.3 *Impaired clearance of self nucleic acids.* Dead cell debris arising from apoptosis and necrosis is the source of self-DNA that can activate various DNA sensors. Several molecules including receptors that bind and clear cell debris and secreted molecules such as MFG-8, GAS6, and Protein S act in concert to effectively clear this self-DNA containing dead cells. DNase I can degrade extracellular DNA and prevent its internalization. DNase II present in lysosomes digests DNA from engulfed apoptotic and necrotic cells or DNA that has escaped from DNase I and is internalized with the help of various receptors (BCR, FcγR, RAGE) and accessory proteins (LL37, HMGB1). DNase III can degrade cell-intrinsic DNA derived from the reverse transcription of endogenous retroelements. Please see color plate at the back of the book.

and has two subtypes, DNase IIα (DNase II) and DNase IIβ (DNase II-like acid DNase (DLAD)). However, lysosomes carry a single DNase i.e. DNase IIα [26]. DNase II deficient mice die late in embryogenesis from severe anemia owing to their inability to degrade expelled nuclear DNA from erythroid precursors during definitive erythropoiesis [41]. High levels of IFN-β are produced by DNase II$^{-/-}$ macrophages carrying undigested DNA, which are responsible for the embryonic lethality of DNase II$^{-/-}$ mice, while the embryonic lethality and severe anemia can be rescued by IFNAR deficiency [78,128]. Although DNase II$^{-/-}$ IFNRI$^{-/-}$ mice are born normal, they develop polyarthritis as they age and inflammatory cytokines such as IL-1β, IL-6, and TNF-α are strongly activated in the affected joints [129]. However, ongoing studies are identifying many candidate cytosolic DNA sensors as described above. Additional deficiencies in TLR9 or TLR3 or their adapters (MyD88 and TRIF) had no effect on the lethality of DNase II deficient mice, suggesting that mammalian DNA activates the innate immune system in a TLR-independent manner [78]. Nonetheless, IRF3 and IRF7 deficiency impaired IFN-β gene expression in the fetal liver of DNase II$^{-/-}$ embryos and rescued the lethality, hinting that a cytosolic sensor might be involved [26].

The third DNase is Trex1 (also called DNase III), which was recently identified as a negative regulator of cytosolic DNA sensors. It is a major mammalian 3′-5′ DNA-exonuclease residing in the ER and cytosol, and acts on both single stranded (ss)DNA and dsDNA [4]. Mutations in the gene encoding TREX1 have been associated with Aicardi–Goutieres syndrome, which is characterized by lethal encephalopathy due to overproduction of type I IFN [130,131]. Trex1-deficient mice develop autoinflammatory myocarditis leading to progressive cardiomyopathy and circulatory failure, and ultimately to death [42]. This myocarditis is rescued by additional deficiency in IRF3, IFNAR, or RAG2, although the IFN-β induction in Trex$^{-/-}$ RAG2$^{-/-}$ double-deficient mice indicates that autoimmunity is emanated by innate immune activation triggered by cytosolic DNA sensors. Moreover, Trex1 is the only 3′-5′ exonuclease known to show IFN-β-inducible expression.

Delineating an unanticipated link between endogenous retroviral elements and autoimmunity, Stetson et al. showed that Trex1 deficiency leads to accumulation of ssDNA reverse-transcribed from retroelements and that Trex1 can metabolize retroviral DNA. Therefore, this study marks a critical negative regulatory mechanism for cytosolic DNA sensors to prevent autoimmunity in a cell-intrinsic manner, in contrast to the other innate immune contributors, TLRs, which function in a non-cell autonomous way [132]. Heterozygous Trex1 mutations cause familial chilblain lupus [133]. Of note, Aicardi–Goutieres syndrome is a heterogeneous disorder and can also be caused by mutations in the three non-allelic components of RNASEH2 (encoded by AGS2, 3, and 4) and SAMHD1 protein (AGS5) [134]. Trex1 deficient mice develop myocarditis, and this was prevented in Trex1$^{-/-}$ STING$^{-/-}$ double knockout mice [135] suggesting that Trex1 is a negative regulator of STING signaling.

Additional Mechanisms for Self-Ligand Clearance

In the process of apoptosis, macrophages can distinguish between healthy cells and apoptotic cells based on the presence of 'eat me' signals such as phosphatidyl serine (PS), ICAM3, annexin I, and cardiolipin on the surface of apoptotic cells, which renders them susceptible to phagocytosis [26]. Soluble innate immune pattern-recognition proteins (sPRPs), MFG-E8, Gas6, and Protein S, bind to PS and facilitate apoptotic cell uptake [136]. MFG-E8 binds to PS via its factor VIII-homology domain and also binds to αvβ3 or αvβ5 integrin, while Gas6 and Protein S bind to Tyro 3, Axl, and Mer (TAM receptors). MFG-E8 plays an essential role in the engulfment of apoptotic cells, as demonstrated by female MFG-E8-null mice that develop SLE type autoimmune diseases with age, and at 40 weeks produce large amounts of anti-dsDNA antibodies and ANAs leading to glomerulonephritis and proteinuria [137]. Mice deficient in all three TAM receptors develop lupus-like autoimmunity with splenomegaly and hyperresponsive APCs [138].

Complement components form the other arm of innate immunity that has been proposed to contribute to the clearance of apoptotic and necrotic cells containing self nucleic acids by facilitation and/or recognition by phagocytes. Specifically, C1q plays a pivotal role in apoptotic cell clearance by the myeloid lineage. C1q can clear apoptotic and necrotic cells in an IgM-dependent manner [139]. In humans, complete deficiency of C1q has been reported in a rare form of SLE-like disease that develops at an early age and tends to be more severe [140]. Similarly, C1q-null mice develop a lupus-like disease characterized by increased apoptotic bodies in the glomeruli *in vivo* and decreased apoptotic cell uptake by C1q$^{-/-}$ macrophages *in vitro* and *in vivo* [75,141,142]. Serum amyloid protein component (SAP), a highly conserved plasma protein, is also involved in avoiding self-ligand recognition. SAP binds to DNA and chromatin contained within the nuclear debris released by necrosis in a calcium-dependent manner and inhibits the formation of pathogenic autoantibodies against chromatin and DNA, probably by regulating their degradation. Moreover, SAP-deficient mice develop ANAs and severe glomerulonephritis [143]. Recently, Mincle and CLEC9A were identified as receptors for necrotic cell clearance [76,144,145], but their role in the context of self-DNA has not been studied. Mice with mutations in the gene encoding flap endonuclease 1 (FEN1) showed accumulation of incompletely digested DNA fragments in apoptotic cells and were predisposed to autoimmunity, chronic inflammation, and cancer [146].

CONCLUSIONS AND FUTURE PERSPECTIVES

Over the years paramount discoveries have been made in the identification of receptors involved in DNA sensing and their downstream signaling. These receptors include the membrane associated TLR9 and several cytosolic receptors that can sense DNA in their respective compartments, thereby leading to the production of type I IFNs and other proinflammatory cytokines. These discoveries have contributed enormously toward the understanding of the etiopathology of various autoimmune diseases, as the friendly protective type I IFNs can turn to foes if they are dysregulated and produced against self-DNA ligands. Type I IFN induction due to aberrant self-DNA recognition appears to be a central pathogenic event in many autoimmune diseases.

The present chapter has elaborated how self-DNA is recognized by multiple innate immune receptors that overcome several mechanisms of surveillance employed for self/non-self discrimination, augmented cell death, and efficient clearance of dead cell debris, and the failure of such surveillance mechanisms can culminate in inflammatory and autoimmune conditions. Along with the present knowledge, further research will be needed to understand more clearly the role of cytosolic DNA sensors in self-DNA recognition, which is comparatively less known, and other unknown mechanisms that are involved in self/non-self discrimination of DNA.

In addition, most autoimmune diseases have protean manifestations and are polygenic, suggesting the involvement of multiple genes and factors in disease progression. Future investigations incorporating genetic, clinical, and experimental approaches are required to illustrate the immune pathways involved in a much more resolved manner. This will provide a new therapeutic arena wherein therapeutics can be designed specifically to inhibit self-DNA recognition.

REFERENCES

[1] Kawai T, Akira S. The role of pattern-recognition receptors in innate immunity: update on Toll-like receptors. Nature Immunol 2010;11(5):373–84.

[2] Sharma S, Fitzgerald KA. Innate immune sensing of DNA. PLoS pathogens 2011;7(4):e1001310.

[3] Doria A, Zen M, Bettio S, Gatto M, Bassi N, Nalotto L, et al. Autoinflammation and autoimmunity: bridging the divide. Autoimmun Rev 2012;12(1):22–30.

[4] Barbalat R, Ewald SE, Mouchess ML, Barton GM. Nucleic acid recognition by the innate immune system. Ann Rev Immunol 2011;29:185–214.

[5] Baccala R, Hoebe K, Kono DH, Beutler B, Theofilopoulos AN. TLR-dependent and TLR-independent pathways of type I interferon induction in systemic autoimmunity. Nature Med 2007;13(5):543–51.

[6] Al-Mayouf SM, Sunker A, Abdwani R, Abrawi SA, Almurshedi F, Alhashmi N, et al. Loss-of-function variant in DNASE1L3 causes a familial form of systemic lupus erythematosus. Nature Genet 2011;43(12):1186–8.

[7] Autoimmune thyroid disease in interferon-treated patients. Lancet 1985;2(8446):100–1.

[8] Ronnblom LE, Alm GV, Oberg KE. Possible induction of systemic lupus erythematosus by interferon-alpha treatment in a patient with a malignant carcinoid tumour. J Intern Med 1990;227(3):207–10.

[9] Hooks JJ, Moutsopoulos HM, Geis SA, Stahl NI, Decker JL, Notkins AL. Immune interferon in the circulation of patients with autoimmune disease. New Engl J Med 1979;301(1):5–8.

[10] Bengtsson AA, Sturfelt G, Truedsson L, Blomberg J, Alm G, Vallin H, et al. Activation of type I interferon system in systemic lupus erythematosus correlates with disease activity but not with antiretroviral antibodies. Lupus 2000;9(9):664–71.

[11] Gota C, Calabrese L. Induction of clinical autoimmune disease by therapeutic interferon-alpha. Autoimmunity 2003;36(8):511–8.

[12] Marshak-Rothstein A. Toll-like receptors in systemic autoimmune disease. Nature Rev Immunol 2006;6(11):823–35.

[13] Lau CM, Broughton C, Tabor AS, Akira S, Flavell RA, Mamula MJ, et al. RNA-associated autoantigens activate B cells by combined B cell antigen receptor/Toll-like receptor 7 engagement. J Exp Med 2005;202(9):1171–7.

[14] Christensen SR, Kashgarian M, Alexopoulou L, Flavell RA, Akira S, Shlomchik MJ. Toll-like receptor 9 controls anti-DNA autoantibody production in murine lupus. J Exp Med 2005;202(2):321–31.

[15] Christensen SR, Shupe J, Nickerson K, Kashgarian M, Flavell RA, Shlomchik MJ. Toll-like receptor 7 and TLR9 dictate autoantibody specificity and have opposing inflammatory and regulatory roles in a murine model of lupus. Immunity 2006;25(3):417–28.

[16] Wu X, Peng SL. Toll-like receptor 9 signaling protects against murine lupus. Arthrit Rheum 2006;54(1):336–42.

[17] Yu P, Wellmann U, Kunder S, Quintanilla-Martinez L, Jennen L, Dear N, et al. Toll-like receptor 9-independent aggravation of glomerulonephritis in a novel model of SLE. Internatl Immunol 2006;18(8):1211–9.

[18] Nickerson KM, Christensen SR, Shupe J, Kashgarian M, Kim D, Elkon K, et al. TLR9 regulates TLR7- and MyD88-dependent autoantibody production and disease in a murine model of lupus. J Immunol 2010;184(4):1840–8.

[19] Nickerson KM, Cullen JL, Kashgarian M, Shlomchik MJ. Exacerbated autoimmunity in the absence of TLR9 in MRL.Faslpr mice depends on Ifnar1. J Immunol 2013;190(8):3889–94.

[20] Kim T, Pazhoor S, Bao M, Zhang Z, Hanabuchi S, Facchinetti V, et al. Aspartate-glutamate-alanine-histidine box motif (DEAH)/RNA helicase A helicases sense

microbial DNA in human plasmacytoid dendritic cells. Proc Natl Acad Sci USA 2010;107(34):15181–6.

[21] Krieg AM, Yi AK, Matson S, Waldschmidt TJ, Bishop GA, Teasdale R, et al. CpG motifs in bacterial DNA trigger direct B-cell activation. Nature 1995;374(6522):546–9.

[22] Krieg AM. CpG motifs in bacterial DNA and their immune effects. Ann Rev Immunol 2002;20:709–60.

[23] Haas T, Metzger J, Schmitz F, Heit A, Muller T, Latz E, et al. The DNA sugar backbone 2′ deoxyribose determines toll-like receptor 9 activation. Immunity 2008;28(3):315–23.

[24] Wagner H. The sweetness of the DNA backbone drives Toll-like receptor 9. Curr Opin Immunol 2008;20(4):396–400.

[25] Barton GM, Kagan JC, Medzhitov R. Intracellular localization of Toll-like receptor 9 prevents recognition of self-DNA but facilitates access to viral DNA. Nature Immunol 2006;7(1):49–56.

[26] Nagata S. Autoimmune diseases caused by defects in clearing dead cells and nuclei expelled from erythroid precursors. Immunol Rev 2007;220:237–50.

[27] Hemmi H, Takeuchi O, Kawai T, Kaisho T, Sato S, Sanjo H, et al. A Toll-like receptor recognizes bacterial DNA. Nature 2000;408(6813):740–5.

[28] Lund J, Sato A, Akira S, Medzhitov R, Iwasaki A. Toll-like receptor 9-mediated recognition of herpes simplex virus-2 by plasmacytoid dendritic cells. J Exp Med 2003;198(3):513–20.

[29] Barrat FJ, Meeker T, Gregorio J, Chan JH, Uematsu S, Akira S, et al. Nucleic acids of mammalian origin can act as endogenous ligands for Toll-like receptors and may promote systemic lupus erythematosus. J Exp Med 2005;202(8):1131–9.

[30] Ewald SE, Lee BL, Lau L, Wickliffe KE, Shi GP, Chapman HA, et al. The ectodomain of Toll-like receptor 9 is cleaved to generate a functional receptor. Nature 2008;456(7222):658–62.

[31] Park B, Brinkmann MM, Spooner E, Lee CC, Kim YM, Ploegh HL. Proteolytic cleavage in an endolysosomal compartment is required for activation of Toll-like receptor 9. Nature Immunol 2008;9(12):1407–14.

[32] Yang Y, Liu B, Dai J, Srivastava PK, Zammit DJ, Lefrancois L, et al. Heat shock protein gp96 is a master chaperone for toll-like receptors and is important in the innate function of macrophages. Immunity 2007;26(2):215–26.

[33] Takahashi K, Shibata T, Akashi-Takamura S, Kiyokawa T, Wakabayashi Y, Tanimura N, et al. A protein associated with Toll-like receptor (TLR) 4 (PRAT4A) is required for TLR-dependent immune responses. J Exp Med 2007;204(12):2963–76.

[34] Tabeta K, Hoebe K, Janssen EM, Du X, Georgel P, Crozat K, et al. The Unc93b1 mutation 3d disrupts exogenous antigen presentation and signaling via Toll-like receptors 3, 7 and 9. Nature Immunol 2006;7(2):156–64.

[35] Kim YM, Brinkmann MM, Paquet ME, Ploegh HL. UNC93B1 delivers nucleotide-sensing toll-like receptors to endolysosomes. Nature 2008;452(7184):234–8.

[36] Ewald SE, Engel A, Lee J, Wang M, Bogyo M, Barton GM. Nucleic acid recognition by Toll-like receptors is coupled to stepwise processing by cathepsins and asparagine endopeptidase. J Exp Med 2011;208(4):643–51.

[37] Ishii KJ, Coban C, Kato H, Takahashi K, Torii Y, Takeshita F, et al. A Toll-like receptor-independent antiviral response induced by double-stranded B-form DNA. Nature Immunol 2006;7(1):40–8.

[38] Stetson DB, Medzhitov R. Recognition of cytosolic DNA activates an IRF3-dependent innate immune response. Immunity 2006;24(1):93–103.

[39] Spies B, Hochrein H, Vabulas M, Huster K, Busch DH, Schmitz F, et al. Vaccination with plasmid DNA activates dendritic cells via Toll-like receptor 9 (TLR9) but functions in TLR9-deficient mice. J Immunol 2003;171(11):5908–12.

[40] Babiuk S, Mookherjee N, Pontarollo R, Griebel P, van Drunen Littel-van den Hurk S, Hecker R, et al. TLR9$^{-/-}$ and TLR9$^{+/+}$ mice display similar immune responses to a DNA vaccine. Immunology 2004;113(1):114–20.

[41] Kawane K, Fukuyama H, Kondoh G, Takeda J, Ohsawa Y, Uchiyama Y, et al. Requirement of DNase II for definitive erythropoiesis in the mouse fetal liver. Science 2001;292(5521):1546–9.

[42] Morita M, Stamp G, Robins P, Dulic A, Rosewell I, Hrivnak G, et al. Gene-targeted mice lacking the Trex1 (DNase III) 3′→5′ DNA exonuclease develop inflammatory myocarditis. Mol Cell Biol 2004;24(15):6719–27.

[43] Takaoka A, Wang Z, Choi MK, Yanai H, Negishi H, Ban T, et al. DAI (DLM-1/ZBP1) is a cytosolic DNA sensor and an activator of innate immune response. Nature 2007;448(7152):501–5.

[44] Ishii KJ, Kawagoe T, Koyama S, Matsui K, Kumar H, Kawai T, et al. TANK-binding kinase-1 delineates innate and adaptive immune responses to DNA vaccines. Nature 2008;451(7179):725–9.

[45] Wang Z, Choi MK, Ban T, Yanai H, Negishi H, Lu Y, et al. Regulation of innate immune responses by DAI (DLM-1/ZBP1) and other DNA-sensing molecules. Proc Natl Acad Sci USA 2008;105(14):5477–82.

[46] Unterholzner L, Keating SE, Baran M, Horan KA, Jensen SB, Sharma S, et al. IFI16 is an innate immune sensor for intracellular DNA. Nature Immunol 2010;11(11):997–1004.

[47] Mondini M, Vidali M, Airo P, De Andrea M, Riboldi P, Meroni PL, et al. Role of the interferon-inducible gene IFI16 in the etiopathogenesis of systemic autoimmune disorders. Ann NY Acad Sci 2007;1110:47–56.

[48] Gugliesi F, De Andrea M, Mondini M, Cappello P, Giovarelli M, Shoenfeld Y, et al. The proapoptotic activity of the interferon-inducible gene IFI16 provides new insights into its etiopathogenetic role in autoimmunity. J Autoimmun 2010;35(2):114–23.

[49] Chiu YH, Macmillan JB, Chen ZJ. RNA polymerase III detects cytosolic DNA and induces type I interferons through the RIG-I pathway. Cell 2009;138(3):576–91.

[50] Ablasser A, Bauernfeind F, Hartmann G, Latz E, Fitzgerald KA, Hornung V. RIG-I-dependent sensing of poly(dA:dT) through the induction of an RNA polymerase III-transcribed RNA intermediate. Nature Immunol 2009;10(10):1065–72.

[51] Barber GN. Innate immune DNA sensing pathways: STING, AIMII and the regulation of interferon production and inflammatory responses. Curr Opin Immunol 2011;23(1):10–20.

[52] Yang P, An H, Liu X, Wen M, Zheng Y, Rui Y, et al. The cytosolic nucleic acid sensor LRRFIP1 mediates the production of type I interferon via a beta-catenin-dependent pathway. Nature Immunol 2010;11(6):487–94.

[53] Zhang X, Brann TW, Zhou M, Yang J, Oguariri RM, Lidie KB, et al. Cutting edge: Ku70 is a novel cytosolic DNA sensor that induces type III rather than type I IFN. J Immunol 2011;186(8):4541–5.

[54] Sun L, Wu J, Du F, Chen X, Chen ZJ. Cyclic GMP-AMP synthase is a cytosolic DNA sensor that activates the type I interferon pathway. Science 2013;339(6121):786–91.

[55] Muruve DA, Petrilli V, Zaiss AK, White LR, Clark SA, Ross PJ, et al. The inflammasome recognizes cytosolic microbial and host DNA and triggers an innate immune response. Nature 2008;452(7183):103–7.

[56] Burckstummer T, Baumann C, Bluml S, Dixit E, Durnberger G, Jahn H, et al. An orthogonal proteomic-genomic screen identifies AIM2 as a cytoplasmic DNA sensor for the inflammasome. Nature Immunol 2009;10(3):266–72.

[57] Fernandes-Alnemri T, Yu JW, Datta P, Wu J, Alnemri ES. AIM2 activates the inflammasome and cell death in response to cytoplasmic DNA. Nature 2009;458(7237):509–13.

[58] Hornung V, Ablasser A, Charrel-Dennis M, Bauernfeind F, Horvath G, Caffrey DR, et al. AIM2 recognizes cytosolic dsDNA and forms a caspase-1-activating inflammasome with ASC. Nature 2009;458(7237):514–8.

[59] Roberts TL, Idris A, Dunn JA, Kelly GM, Burnton CM, Hodgson S, et al. HIN-200 proteins regulate caspase activation in response to foreign cytoplasmic DNA. Science 2009;323(5917):1057–60.

[60] Fernandes-Alnemri T, Yu JW, Juliana C, Solorzano L, Kang S, Wu J, et al. The AIM2 inflammasome is critical for innate immunity to Francisella tularensis. Nature Immunol 2010;11(5):385–93.

[61] Jones JW, Kayagaki N, Broz P, Henry T, Newton K, O'Rourke K, et al. Absent in melanoma 2 is required for innate immune recognition of Francisella tularensis. Proc Natl Acad Sci USA 2010;107(21):9771–6.

[62] Rathinam VA, Jiang Z, Waggoner SN, Sharma S, Cole LE, Waggoner L, et al. The AIM2 inflammasome is essential for host defense against cytosolic bacteria and DNA viruses. Nature Immunol 2010;11(5):395–402.

[63] Panchanathan R, Duan X, Shen H, Rathinam VA, Erickson LD, Fitzgerald KA, et al. Aim2 deficiency stimulates the expression of IFN-inducible Ifi202, a lupus susceptibility murine gene within the Nba2 autoimmune susceptibility locus. J Immunol 2010;185(12):7385–93.

[64] Guarda G, Braun M, Staehli F, Tardivel A, Mattmann C, Forster I, et al. Type I interferon inhibits interleukin-1 production and inflammasome activation. Immunity 2011;34(2):213–23.

[65] Panchanathan R, Shen H, Duan X, Rathinam VA, Erickson LD, Fitzgerald KA, et al. Aim2 deficiency in mice suppresses the expression of the inhibitory Fcgamma receptor (FcgammaRIIB) through the induction of the IFN-inducible p202, a lupus susceptibility protein. J Immunol 2011;186(12):6762–70.

[66] Duan X, Ponomareva L, Veeranki S, Panchanathan R, Dickerson E, Choubey D. Differential roles for the interferon-inducible IFI16 and AIM2 innate immune sensors for cytosolic DNA in cellular senescence of human fibroblasts. Mol Cancer Res: MCR 2011;9(5):589–602.

[67] Dombrowski Y, Peric M, Koglin S, Kammerbauer C, Goss C, Anz D, et al. Cytosolic DNA triggers inflammasome activation in keratinocytes in psoriatic lesions. Sci Transnatl Med 2011;3(82):82ra38.

[68] Kerur N, Veettil MV, Sharma-Walia N, Bottero V, Sadagopan S, Otageri P, et al. IFI16 acts as a nuclear pathogen sensor to induce the inflammasome in response to Kaposi sarcoma-associated herpesvirus infection. Cell Host Microbe 2011;9(5):363–75.

[69] Dhodapkar KM, Banerjee D, Connolly J, Kukreja A, Matayeva E, Veri MC, et al. Selective blockade of the inhibitory Fcgamma receptor (FcgammaRIIB) in human dendritic cells and monocytes induces a type I interferon response program. J Exp Med 2007;204(6):1359–69.

[70] Kimkong I, Avihingsanon Y, Hirankarn N. Expression profile of HIN200 in leukocytes and renal biopsy of SLE patients by real-time RT-PCR. Lupus 2009;18(12):1066–72.

[71] Frohm M, Agerberth B, Ahangari G, Stahle-Backdahl M, Liden S, Wigzell H, et al. The expression of the gene coding for the antibacterial peptide LL-37 is induced in human keratinocytes during inflammatory disorders. J Biol Chem 1997;272(24):15258–63.

[72] Lande R, Gregorio J, Facchinetti V, Chatterjee B, Wang YH, Homey B, et al. Plasmacytoid dendritic cells sense self-DNA coupled with antimicrobial peptide. Nature 2007;449(7162):564–9.

[73] Villanueva E, Yalavarthi S, Berthier CC, Hodgin JB, Khandpur R, Lin AM, et al. Netting neutrophils induce endothelial damage, infiltrate tissues, and expose immunostimulatory molecules in systemic lupus erythematosus. J Immunol 2011;187(1):538–52.

[74] Kessenbrock K, Krumbholz M, Schonermarck U, Back W, Gross WL, Werb Z, et al. Netting neutrophils in autoimmune small-vessel vasculitis. Nature Med 2009;15(6):623–5.

[75] Celhar T, Magalhaes R, Fairhurst AM. TLR7 and TLR9 in SLE: when sensing self goes wrong. Immunol Res 2012;53(1–3):58–77.

[76] Yang D, de la Rosa G, Tewary P, Oppenheim JJ. Alarmins link neutrophils and dendritic cells. Trends Immunol 2009;30(11):531–7.

[77] Gilliet M, Lande R. Antimicrobial peptides and self-DNA in autoimmune skin inflammation. Curr Opin Immunol 2008;20(4):401–7.

[78] Okabe Y, Kawane K, Akira S, Taniguchi T, Nagata S. Toll-like receptor-independent gene induction program activated by mammalian DNA escaped from apoptotic DNA degradation. J Exp Med 2005;202(10):1333–9.

[79] Boule MW, Broughton C, Mackay F, Akira S, Marshak-Rothstein A, Rifkin IR. Toll-like receptor 9-dependent and -independent dendritic cell activation by chromatin-immunoglobulin G complexes. J Exp Med 2004;199(12):1631–40.

[80] Decker P, Singh-Jasuja H, Haager S, Kotter I, Rammensee HG. Nucleosome the main autoantigen in systemic lupus erythematosus, induces direct dendritic cell activation via a MyD88-independent pathway: consequences on inflammation. J Immunol 2005;174(6):3326–34.

[81] Chamilos G, Gregorio J, Meller S, Lande R, Kontoyiannis DP, Modlin RL, et al. Cytosolic sensing of extracellular self-DNA transported into monocytes by the antimicrobial peptide LL37. Blood 2012;120(18):3699–707.

[82] Doring Y, Manthey HD, Drechsler M, Lievens D, Megens RT, Soehnlein O, et al. Auto-antigenic protein-DNA complexes stimulate plasmacytoid dendritic cells to promote atherosclerosis. Circulation 2012;125(13):1673–83.

[83] Sims GP, Rowe DC, Rietdijk ST, Herbst R, Coyle AJ. HMGB1 and RAGE in inflammation and cancer. Ann Rev Immunol 2010;28:367–88.

[84] Thomas JO, Travers AA. HMG1 and 2, and related 'architectural' DNA-binding proteins. Trends Biochem Sci 2001;26(3):167–74.

[85] Rendon-Mitchell B, Ochani M, Li J, Han J, Wang H, Yang H, et al. IFN-gamma induces high mobility group box 1 protein release partly through a TNF-dependent mechanism. J Immunol 2003;170(7):3890–7.

[86] Wang H, Bloom O, Zhang M, Vishnubhakat JM, Ombrellino M, Che J, et al. HMG-1 as a late mediator of endotoxin lethality in mice. Science 1999;285(5425):248–51.

[87] Jiang W, Bell CW, Pisetsky DS. The relationship between apoptosis and high-mobility group protein 1 release from murine macrophages stimulated with lipopolysaccharide or polyinosinic-polycytidylic acid. J Immunol 2007;178(10):6495–503.

[88] Ivanov S, Dragoi AM, Wang X, Dallacosta C, Louten J, Musco G, et al. A novel role for HMGB1 in TLR9-mediated inflammatory responses to CpG-DNA. Blood 2007;110(6):1970–81.

[89] Tian J, Avalos AM, Mao SY, Chen B, Senthil K, Wu H, et al. Toll-like receptor 9-dependent activation by DNA-containing immune complexes is mediated by HMGB1 and RAGE. Nature Immunol 2007;8(5):487–96.

[90] Urbonaviciute V, Furnrohr BG, Meister S, Munoz L, Heyder P, De Marchis F, et al. Induction of inflammatory and immune responses by HMGB1-nucleosome complexes: implications for the pathogenesis of SLE. J Exp Med 2008;205(13):3007–18.

[91] Popovic K, Ek M, Espinosa A, Padyukov L, Harris HE, Wahren-Herlenius M, et al. Increased expression of the novel proinflammatory cytokine high mobility group box chromosomal protein 1 in skin lesions of patients with lupus erythematosus. Arthrit Rheum 2005;52(11):3639–45.

[92] Urbonaviciute V, Voll RE. High-mobility group box 1 represents a potential marker of disease activity and novel therapeutic target in systemic lupus erythematosus. J Intern Med 2011;270(4):309–18.

[93] Qing X, Pitashny M, Thomas DB, Barrat FJ, Hogarth MP, Putterman C. Pathogenic anti-DNA antibodies modulate gene expression in mesangial cells: involvement of HMGB1 in anti-DNA antibody-induced renal injury. Immunol Lett 2008;121(1):61–73.

[94] Yanai H, Ban T, Wang Z, Choi MK, Kawamura T, Negishi H, et al. HMGB proteins function as universal sentinels for nucleic-acid-mediated innate immune responses. Nature 2009;462(7269):99–103.

[95] Ronnblom L, Eloranta ML, Alm GV. The type I interferon system in systemic lupus erythematosus. Arthrit Rheum 2006;54(2):408–20.

[96] Bave U, Magnusson M, Eloranta ML, Perers A, Alm GV, Ronnblom L. Fc gamma RIIa is expressed on natural IFN-alpha-producing cells (plasmacytoid dendritic cells) and is required for the IFN-alpha production induced by apoptotic cells combined with lupus IgG. J Immunol 2003;171(6):3296–302.

[97] Lovgren T, Eloranta ML, Bave U, Alm GV, Ronnblom L. Induction of interferon-alpha production in plasmacytoid dendritic cells by immune complexes containing nucleic acid released by necrotic or late apoptotic cells and lupus IgG. Arthrit Rheum 2004;50(6):1861–72.

[98] Means TK, Latz E, Hayashi F, Murali MR, Golenbock DT. Luster AD. Human lupus autoantibody-DNA complexes activate DCs through cooperation of CD32 and TLR9. J Clin Invest 2005;115(2):407–17.

[99] Hannum LG, Ni D, Haberman AM, Weigert MG, Shlomchik MJ. A disease-related rheumatoid factor autoantibody is not tolerized in a normal mouse: implications for the origins of autoantibodies in autoimmune disease. J Exp Med 1996;184(4):1269–78.

[100] Monestier M, Kotzin BL. Antibodies to histones in systemic lupus erythematosus and drug-induced lupus syndromes. Rheum Dis Clin North Am 1992;18(2):415–36.

[101] Viglianti GA, Lau CM, Hanley TM, Miko BA, Shlomchik MJ, Marshak-Rothstein A. Activation of autoreactive B cells by CpG dsDNA. Immunity 2003;19(6):837–47.

[102] Chaturvedi A, Dorward D, Pierce SK. The B cell receptor governs the subcellular location of Toll-like receptor 9 leading to hyperresponses to DNA-containing antigens. Immunity 2008;28(6):799–809.

[103] Di Domizio J, Zhang R, Stagg LJ, Gagea M, Zhuo M, Ladbury JE, et al. Binding with nucleic acids or glycosaminoglycans converts soluble protein oligomers to amyloid. J Biol Chem 2012;287(1):736–47.

[104] Di Domizio J, Dorta-Estremera S, Gagea M, Ganguly D, Meller S, Li P, et al. Nucleic acid-containing amyloid fibrils potently induce type I interferon and stimulate systemic autoimmunity. Proc Natl Acad Sci USA 2012;109(36):14550–5.

[105] Montier T, Benvegnu T, Jaffres PA, Yaouanc JJ, Lehn P. Progress in cationic lipid-mediated gene transfection: a series of bio-inspired lipids as an example. Current Gene Ther 2008;8(5):296–312.

[106] Audouy S, Hoekstra D. Cationic lipid-mediated transfection in vitro and in vivo (review). Mol Membr Biol 2001;18(2):129–43.

[107] Gursel I, Gursel M, Ishii KJ, Klinman DM. Sterically stabilized cationic liposomes improve the uptake and immunostimulatory activity of CpG oligonucleotides. J Immunol 2001;167(6):3324–8.

[108] Honda K, Ohba Y, Yanai H, Negishi H, Mizutani T, Takaoka A, et al. Spatiotemporal regulation of MyD88–IRF-7 signalling for robust type-I interferon induction. Nature 2005;434(7036):1035–40.

[109] Zabner J, Fasbender AJ, Moninger T, Poellinger KA, Welsh MJ. Cellular and molecular barriers to gene transfer by a cationic lipid. J Biol Chem 1995;270(32):18997–9007.

[110] Yasuda K, Yu P, Kirschning CJ, Schlatter B, Schmitz F, Heit A, et al. Endosomal translocation of vertebrate DNA activates dendritic cells via TLR9-dependent and -independent pathways. J Immunol 2005;174(10):6129–36.

[111] Hou B, Reizis B, DeFranco AL. Toll-like receptors activate innate and adaptive immunity by using dendritic cell-intrinsic and -extrinsic mechanisms. Immunity 2008;29(2):272–82.

[112] Park B, Buti L, Lee S, Matsuwaki T, Spooner E, Brinkmann MM, et al. Granulin is a soluble cofactor for toll-like receptor 9 signaling. Immunity 2011;34(4):505–13.

[113] Tanaka A, Tsukamoto H, Mitoma H, Kiyohara C, Ueda N, Ayano M, et al. Serum progranulin levels are elevated in patients with systemic lupus erythematosus, reflecting disease activity. Arthritis Res Ther 2012;14(6):R244.

[114] Nagata S. Apoptosis and autoimmune diseases. Ann NY Acad Sci 2010;1209:10–16.

[115] Brinkmann V, Reichard U, Goosmann C, Fauler B, Uhlemann Y, Weiss DS, et al. Neutrophil extracellular traps kill bacteria. Science 2004;303(5663):1532–5.

[116] Kaplan MJ, Radic M. Neutrophil extracellular traps: double-edged swords of innate immunity. J Immunol 2012;189(6):2689–95.

[117] Lande R, Ganguly D, Facchinetti V, Frasca L, Conrad C, Gregorio J, et al. Neutrophils activate plasmacytoid dendritic cells by releasing self-DNA-peptide complexes in systemic lupus erythematosus. Science Transl Med 2011;3(73):73ra19.

[118] Garcia-Romo GS, Caielli S, Vega B, Connolly J, Allantaz F, Xu Z, et al. Netting neutrophils are major inducers of type I IFN production in pediatric systemic lupus erythematosus. Science Transl Med 2011;3(73):73ra20.

[119] Martinelli S, Urosevic M, Daryadel A, Oberholzer PA, Baumann C, Fey MF, et al. Induction of genes mediating interferon-dependent extracellular trap formation during neutrophil differentiation. J Biol Chem 2004;279(42):44123–32.

[120] Yasutomo K, Horiuchi T, Kagami S, Tsukamoto H, Hashimura C, Urushihara M, et al. Mutation of DNASE1 in people with systemic lupus erythematosus. Nature Genet 2001;28(4):313–4.

[121] Hakkim A, Furnrohr BG, Amann K, Laube B, Abed UA, Brinkmann V, et al. Impairment of neutrophil extracellular trap degradation is associated with lupus nephritis. Proc Natl Acad Sci USA 2010;107(21):9813–8.

[122] Martinez Valle F, Balada E, Ordi-Ros J, Vilardell-Tarres M. DNase 1 and systemic lupus erythematosus. Autoimmun Rev 2008;7(5):359–63.

[123] Napirei M, Karsunky H, Zevnik B, Stephan H, Mannherz HG, Moroy T. Features of systemic lupus erythematosus in Dnase1-deficient mice. Nature Genet 2000;25(2):177–81.

[124] Belguith-Maalej S, Hadj-Kacem H, Kaddour N, Bahloul Z, Ayadi H. DNase1 exon2 analysis in Tunisian patients with rheumatoid arthritis, systemic lupus erythematosus and Sjogren syndrome and healthy subjects. Rheumatol Internatl 2009;30(1):69–74.

[125] Dittmar M, Bischofs C, Matheis N, Poppe R, Kahaly GJ. A novel mutation in the DNASE1 gene is related with protein instability and decreased enzyme activity in thyroid autoimmunity. J Autoimmun 2009;32(1):7–13.

[126] Malickova K, Duricova D, Bortlik M, Hruskova Z, Svobodova B, Machkova N, et al. Impaired deoxyribonuclease I activity in patients with inflammatory bowel diseases. Autoimmune Dis 2011;2011:945861.

[127] Kawane K, Fukuyama H, Yoshida H, Nagase H, Ohsawa Y, Uchiyama Y, et al. Impaired thymic development in mouse embryos deficient in apoptotic DNA degradation. Nature Immunol 2003;4(2):138–44.

[128] Yoshida H, Okabe Y, Kawane K, Fukuyama H, Nagata S. Lethal anemia caused by interferon-beta produced in mouse embryos carrying undigested DNA. Nature Immunol 2005;6(1):49–56.

[129] Kawane K, Ohtani M, Miwa K, Kizawa T, Kanbara Y, Yoshioka Y, et al. Chronic polyarthritis caused by mammalian DNA that escapes from degradation in macrophages. Nature 2006;443(7114):998–1002.

[130]   Crow YJ, Hayward BE, Parmar R, Robins P, Leitch A, Ali M, et al. Mutations in the gene encoding the 3'-5' DNA exonuclease TREX1 cause Aicardi-Goutieres syndrome at the AGS1 locus. Nature Genet 2006;38(8):917–20.

[131]   Lee-Kirsch MA, Gong M, Chowdhury D, Senenko L, Engel K, Lee YA, et al. Mutations in the gene encoding the 3'-5' DNA exonuclease TREX1 are associated with systemic lupus erythematosus. Nature Genet 2007;39(9):1065–7.

[132]   Stetson DB, Ko JS, Heidmann T, Medzhitov R. Trex1 prevents cell-intrinsic initiation of autoimmunity. Cell 2008;134(4):587–98.

[133]   Rice G, Newman WG, Dean J, Patrick T, Parmar R, Flintoff K, et al. Heterozygous mutations in TREX1 cause familial chilblain lupus and dominant Aicardi-Goutieres syndrome. Am J Hum Genet 2007;80(4):811–5.

[134]   Crow YJ, Rehwinkel J. Aicardi-Goutieres syndrome and related phenotypes: linking nucleic acid metabolism with autoimmunity. Human Mol Genet 2009;18(R2):R130–6.

[135]   Gall A, Treuting P, Elkon KB, Loo YM, Gale Jr. M, Barber GN, et al. Autoimmunity initiates in nonhematopoietic cells and progresses via lymphocytes in an interferon-dependent autoimmune disease. Immunity 2012;36(1):120–31.

[136]   Fadok VA, Bratton DL, Rose DM, Pearson A, Ezekewitz RA, Henson PM. A receptor for phosphatidylserine-specific clearance of apoptotic cells. Nature 2000;405(6782):85–90.

[137]   Hanayama R, Tanaka M, Miyasaka K, Aozasa K, Koike M, Uchiyama Y, et al. Autoimmune disease and impaired uptake of apoptotic cells in MFG-E8-deficient mice. Science 2004;304(5674):1147–50.

[138]   Rothlin CV, Ghosh S, Zuniga EI, Oldstone MB, Lemke G. TAM receptors are pleiotropic inhibitors of the innate immune response. Cell 2007;131(6):1124–36.

[139]   Ogden CA, Kowalewski R, Peng Y, Montenegro V, Elkon KB. IGM is required for efficient complement mediated phagocytosis of apoptotic cells in vivo. Autoimmunity 2005;38(4):259–64.

[140]   Arkwright PD, Abinun M, Cant AJ. Autoimmunity in human primary immunodeficiency diseases. Blood 2002;99(8):2694–702.

[141]   Taylor PR, Carugati A, Fadok VA, Cook HT, Andrews M, Carroll MC, et al. A hierarchical role for classical pathway complement proteins in the clearance of apoptotic cells in vivo. J Exp Med 2000;192(3):359–66.

[142]   Botto M, Dell'Agnola C, Bygrave AE, Thompson EM, Cook HT, Petry F, et al. Homozygous C1q deficiency causes glomerulonephritis associated with multiple apoptotic bodies. Nature Genet 1998;19(1):56–9.

[143]   Bickerstaff MC, Botto M, Hutchinson WL, Herbert J, Tennent GA, Bybee A, et al. Serum amyloid P component controls chromatin degradation and prevents antinuclear autoimmunity. Nature Med 1999;5(6):694–7.

[144]   Sancho D, Joffre OP, Keller AM, Rogers NC, Martinez D, Hernanz-Falcon P, et al. Identification of a dendritic cell receptor that couples sensing of necrosis to immunity. Nature 2009;458(7240):899–903.

[145]   Yamasaki S, Ishikawa E, Sakuma M, Hara H, Ogata K, Saito T. Mincle is an ITAM-coupled activating receptor that senses damaged cells. Nature Immunol 2008;9(10):1179–88.

[146]   Zheng L, Dai H, Zhou M, Li M, Singh P, Qiu J, et al. Fen1 mutations result in autoimmunity, chronic inflammation and cancers. Nature Med 2007;13(7):812–9.

CHAPTER 6

Toll-Like Receptor 9 and Toll-Like Receptor 7 in the Development and Regulation of Systemic Autoimmune Disease

Ann Marshak-Rothstein[1] and Michael P. Cancro[2]

[1]Department of Medicine/Rheumatology, University of Massachusetts Medical School, Worcester, MA, USA and
[2]Department of Pathology and Laboratory Medicine, Perelman School of Medicine at the University of Pennsylvania, Philadelphia, PA, USA

INTRODUCTION

Pattern recognition receptors (PRRs), expressed by most innate immune effector cells, were initially characterized by their capacity to recognize conserved molecular patterns common to various microbes. The early detection of pathogens afforded by PRRs leads to the prompt activation of dendritic cells, macrophages, neutrophils, and other innate immune populations required to contain the initial infection. These cells subsequently express an appropriate array of cytokines, chemokines, and co-stimulatory molecules that in turn activate and fine-tune the adaptive immune response, leading to full pathogen clearance.

In addition to these well-established interactions with microbial ligands, many PRRs also recognize endogenous ligands that become exposed in the context of infection, tissue damage, and/or inflammation. The detection of these self-determinants provides a second tier of danger signals. Unfortunately, failure to appropriately contain PRR-mediated self-recognition can eventually lead to sustained autoimmune activity. Examples include diseases such as systemic lupus erythematosis (SLE), scleroderma, Sjogren's disease, and even psoriasis, all of which have been linked to a specialized subset of PRRs, the nucleic acid-sensing members of the Toll-like receptor (TLR) gene family, which are predominantly located in endolysosomal compartments.

Biological DNA Sensor.
DOI: http://dx.doi.org/10.1016/B978-0-12-404732-7.00006-X

NUCLEIC ACID SENSING TLRs

To date, five TLRs have been linked to the detection of microbial nucleic acids. TLR3 is generally described as a receptor for dsRNA, TLR7 and TLR8 as receptors for ssRNA, TLR9 as a receptor for DNA, and TLR13 as a receptor for bacterial ribosomal RNA [1–5]. These molecules function as dimers made up of solenoid-shaped ectodomains, a transmembrane domain, and a cytosolic Toll IL-1 receptor (TIR) domain that serves as the signaling platform. Each of these receptors plays a remarkably specific role in microbial immunity [6], reflecting both pathogen compartmentalization and cell-type specific expression patterns. Importantly, TLR3, TLR7, TLR8, and TLR9 also have all been shown to recognize endogenous ligands and thereby contribute to the development of systemic autoimmune disease.

In order to function properly, TLR3, 7, 8, and 9 must traffic to acidic endosomal/lysosomal compartments, and agents that interfere with endosome acidification, such as concanamycin B and bafilomycin A, effectively block TLR activation and thus responses to endogenous ligands [7,8]. These TLRs are chaperoned from the endoplasmic reticulum to appropriate endolysosomal compartments by Unc93B1 [9]. Within these compartments, the TLR ectodomains are cleaved through a cathepsin-dependent process to produce signaling competent receptors [10–12]. Ligand binding can then induce a conformational change that facilitates interaction of the TLR TIR domains with the TIR domains of either TRIF (TLR3) or MyD88 (TLR7, 8, and 9) [13]. It is then generally accepted that the MyD88 death domain (DD) interacts with the death domain of IL-1R-associated kinase 4 (IRAK4), which in turn phosphorylates IRAK1, and activates the E3 ubiquitin ligase TRAF6 [14,15]. Assembly of the IRAK/TRAF signaling complex results in activation of both the NFκB and MAPK signaling cascades. The TLR and cell type specific features of the TLR7 and TLR9 signaling cascades remain poorly defined, but it is clear that aberrant TLR activation by endogenous nucleic acids promotes the activation of autoreactive B cells, dendritic cells, neutrophils, and various other effector populations involved in SLE and related conditions.

IMMUNE COMPLEX ACTIVATION OF pDCs AND TYPE I IFN

Plasmacytoid dendritic cells (pDCs) are among the most notorious TLR-activated effectors in the context of SLE. As first appreciated by Lars Ronnblom and colleagues, immune complexes (ICs) isolated from the sera of SLE patients are unexpectedly potent inducers of pDC type I interferon (IFN) [16]. Although the onset of systemic autoimmune disease is frequently associated with viral infection, and therefore potential exposure to viral nucleic acids, this IFN-inducing activity can be mimicked by combining purified SLE IgG with sterile apoptotic or necrotic cell debris. Activation by IgG-bound cell debris depends on engagement of the ICs by stimulatory FcγRs and their subsequent delivery to a TLR-associated stimulatory compartment [17]. Importantly, IC activation of pDCs is sensitive to both DNAses and RNAses, thereby linking endogenous nucleic acids to pDC activation and IFN production.

Type I IFN production has also been tied to SLE pathogenesis through correlative and functional studies on human patients. The first indication of an IFN/SLE link came from individuals given type I IFNs for the treatment of malignancies or viral infections; these individuals occasionally developed high serum autoantibody titers and other symptoms of autoimmune disease that cleared when the IFN therapy was terminated [18–20]. Moreover, gene expression profiling of PBMCs obtained from SLE patients and control populations revealed a dramatic upregulation of many IFN-inducible genes in SLE patients, referred to as an IFN-signature [21,22]. The mechanistic basis for this correlation has been explored by Banchereau and colleagues, and other investigators who identified pleiotrophic effects of IFN on dendritic cells, CD8 T cells, B cells, and neutrophils, rendering them hyperresponsive to immunostimulatory factors present in SLE serum samples [23–25]. Not only does IFN lower the BCR signaling threshold, induce the expression of co-stimulatory molecules, and drive isotype switching, it can promote pDCs and macrophages to make cytokines such as IL-6 and IP-10, which further promote B cell differentiation to plasmablasts [26–28].

The importance of IFN in SLE pathogenesis has been further tested in murine models. All type I IFNs signal through the same receptor, so the role of IFN can be directly tested in autoimmune prone mice that lack the type I IFN receptor (IFNaR). For the NZM 2328 and NZB strains, IFNaR deficient cohorts had much lower autoantibody titers, reduced proteinuria, and dramatically improved survival [29]. IFNaR deficiency also improved the clinical prognosis of several other autoimmune prone strains, such as NBA/2 and B6 FcγR2b Yaa mice, to varying degrees [30–35]. Unexpectedly, IFNaR deficiency was reported to result in more severe disease in MRL/lpr and MRL/+ backcross mice [36], although this result has been challenged by subsequent studies [34].

IFNaR blocking antibodies can also attenuate disease in NZB and BxSB mice [37]. Whether IFN blockade will be an effective therapeutic strategy for human SLE remains to be determined.

ACTIVATION OF AUTOREACTIVE B CELLS DEPENDS ON BCR/TLR CO-ENGAGEMENT

The expression of PRRs is not limited to innate immune cells; B lymphocytes also express nucleic-acid sensing PRRs, including TLRs 9, 7, 8, and 3. As with DCs, nucleic acid-associated ligands must be delivered to these endolysosomal receptors by a cell surface receptor, and the relevant receptor is the B cell antigen receptor (BCR). The significance of this BCR/TLR activation scheme was first demonstrated in vitro using BCR transgenic mice. SLE-prone mice often make high titers of rheumatoid factor (RF) [38], autoantibodies that bind autologous IgG. The AM14 transgenic line, developed by Weigert and colleagues [39], expresses a prototypic RF+ antibody, derived from Fas-deficient MRL/lpr mice. B cells expressing the AM14 receptor have proven extremely useful for evaluating the role of TLRs in autoreactive B cell activation. AM14 B cells bind IgG2a with sufficiently low affinity that they survive both central and peripheral negative selection checkpoints and persist in normal mice as resting naïve B cells, even in the presence of (monomeric) serum IgG2a. However, AM14 B cells are spontaneously activated by ICs in autoimmune prone mice, and the resulting plasmablasts accumulate in extrafollicular foci, where they undergo isotype switching and somatic hypermutation [40].

In vitro, AM14 B cells can be activated by sera from autoimmune mice or by IgG2a monoclonal autoantibodies that either bind DNA or RNA *per se*, or bind DNA- or RNA-associated autoantigens. In primary B cell cultures, it is not necessary to provide a further source of autoantigen, as these autoantibodies bind autoantigens released from dead or dying cells. Activation by these 'spontaneous' autoantibody/autoantigen ICs can be blocked by the addition of DNAse or RNAase to the cultures, or by chloroquine, bafilomycin, or [41] other inhibitors of endosome/lysosome acidification [42]. AM14 B cells can also be activated by 'defined' ICs such as biotinylated dsDNA bound by IgG2a anti-biotin antibodies, and the use of defined mammalian dsDNA fragments has revealed that TLR9 maintains its preference for CG-rich DNA among potential endogenous sequences [43]. ICs that do not incorporate nucleic acids cannot induce the AM14 B cells to proliferate. Following the identification of TLR9 and

TLR7 as the endosomal/lysosomal receptors for DNA and RNA [2–4], and the production of the corresponding knockout MyD88$^{-/-}$, TLR9$^{-/-}$, and TLR7$^{-/-}$ mice by the Akira laboratory, it was possible to use TLR-deficient AM14 B cells to establish a critical role for TLR9 and TLR7 in autoreactive B cell activation [7,8]. Importantly, AM14 B cells recapitulate the activation requirements of autoreactive B cells that directly recognize DNA, RNA, or DNA- or RNA-associated autoantigens, as evidenced by subsequent *in vivo* studies (see below). In addition to providing proof that TLR9 and TLR7 can detect and respond to endogenous, as well as microbial ligands, RF+ B cells provide a unique experimental tool for parsing the unique features of BCR/TLR7 and BCR/TLR9 signaling cascades by using IgG2a monoclonal autoantibodies reactive with either DNA-associated or RNA-associated autoantibodies as ligands.

AUTOANTIGENS AND TLR ACTIVATION

TLR9 was initially characterized by its ability to distinguish hypomethylated CpG motifs present in bacterial and viral DNA from mammalian DNA; cytosine methylation and the paucity of CG dinucleotides rendered mammalian DNA significantly less stimulatory [44]. Analogously, TLR7 preferentially recognizes U-rich RNAs that incorporate unmodified nucleosides, which are again more common in microbial RNA [45]. How then can ICs derived from SLE patient sera or autoimmune-prone mice – which contain mammalian nucleic acids – drive TLR9-dependent responses? One possibility is that proinflammatory cytokines, such as the type I IFNs, upregulate TLR expression levels [22,46], or amplify downstream intracellular signal cascades, so otherwise weak agonists signal effectively. In addition, mammalian DNA or RNA may include certain segments that are themselves inherently more stimulatory. For example, CpG islands, frequently present in promoter regions of mammalian genes, are hypomethylated during active gene transcription, and dsDNA fragments derived from CpG islands are stronger TLR ligands than dsDNA fragments that lack CG dinucleotides. Mitochondria are endosymbionts derived from bacteria, and mitochondrial DNA is not methylated; therefore it is not surprising that mitochondrial DNA also effectively activates TLR9 [47]. In the case of mammalian RNA, many of the recurring SLE-related autoantigens are proteins associated with small non-coding RNAs or microRNAs that are commonly U-rich and relatively unmodified [3,48,49]. Finally, it is also possible that oxidation or other forms of nucleic acid damage enhance TLR immunogenicity [41].

Since DNA- and RNA-associated macromolecules are intracellular, a second question is how these self-components become available to the immune system. A likely explanation is that they are released from dead and dying cells, making the mode of cell death and clearance a key etiologic factor. Cells that die by necrosis or pyroptosis release inflammatory mediators such as the HMGB proteins or the antimicrobial peptide LL37. HMGB1 can enhance the uptake of nucleic acids through antibody-independent mechanisms, and then direct these molecules to TLR-associated endosomal/lysosomal compartments [50]. The cationic antimicrobial peptide LL37 can also facilitate the uptake of endogenous nucleic acids. LL37 binds to both DNA and RNA to form complexes that are protected from enzymatic degradation. These LL37-bound complexes then bind to anionic proteoglycans located in lipid rafts, thus allowing transport across the membrane into endocytic compartments [51,52]. The pathogenic consequences of LL37-mediated DNA and RNA delivery have been best described in models of psoriasis, where LL37-bound complexes have been found to trigger pDCs to make type I IFN. LL37-complexes have also been found to activate myeloid DCs, and keratinocytes are most likely to contribute to inflammatory processes beyond psoriasis [52,53].

Cells dying by apoptosis have generally been considered non-inflammatory because apoptotic cells and apoptotic bodies are rapidly detected and silently removed by an arsenal of well-positioned phagocytes. Apoptotic cells and debris display surface phosphotidylserine (PS) that, when detected by scavenger receptors on phagocytes, enables uptake of the apoptotic debris through a process designated efferocytosis [54]. However, when these scavenger systems are saturated or impaired, persistence of apoptotic cells may lead to secondary necrosis, along with the subsequent release of proinflammatory mediators. Accordingly, the failure to appropriately clear nucleic acid-associated debris through non-inflammatory mechanisms may have dire consequences, as it has the potential to activate both innate and adaptive immune cells, with the subsequent risk of autoimmune disease.

ABERRANT SELF-ANTIGEN CLEARANCE AND DETECTION

In fact, macrophages from some SLE patients have a compromised capacity to phagocytose apoptotic debris [55,56]. Similarly, mice deficient in the expression of scavenger receptors such as TIM-4 or CD36, which recognize PS directly, or mer, which recognizes PS through bridging molecules such as Protein S or GAS6, develop features of SLE including autoantibodies, splenomegaly, and glomerulonephritis [57,58]. The complement subunit C1q, an exceedingly strong risk factor for SLE, also contributes to the clearance of extracellular DNA [59], but in addition may regulate other aspects of antigen processing that preclude access to TLR-associated compartments [60]. The recruitment of C1q to the surface of apoptotic cells can be facilitated by natural IgM antibodies that bind to phosphorylcholine or modified lipids that mark the outer apoptotic cell membrane [61]. Other serum proteins, such as milk fat globule EGF factor 8 (MFG-E8), also enhance clearance of cell debris. MFG-E8 binds PS and is then detected by phagocytic cells through specific αv integrin receptors. As might be expected in mice with clearance defects, MFG-8-deficient mice develop symptoms of SLE [62]. However, it has recently been shown that MFG-E8 does more than simply promote the uptake of apoptotic debris by phagocytes. Rather, when apoptotic cells coated with MFG-E8 are taken up by bone marrow-derived dendritic cells (BMDC), the internalized apoptotic cells transit to an acidic lysosomal compartment, where they are rapidly degraded and thus unavailable for antigen processing and cross-presentation. In the absence of MFG-8, BMDC appear to take up smaller fragments of cellular debris, which are then retained in a higher pH compartment where antigen processing and presentation can take place [63]. These data point out the potential pathogenic consequences of both cell debris accumulation and trafficking of the cell debris to an incorrect compartment.

An additional example whereby failure to degrade DNA might yield increased availability of endogenous TLR9 agonists is the recently described neutrophil extracellular traps (NETS), whose formation results from the extrusion of chromatin strands during a specialized form of neutrophil cell death. This type of death, termed NETosis, is normally triggered by a variety of microbial ligands. The sticky intertwining chromatin strands released during NETosis are coated with an assortment of antimicrobial peptides, including LL37, and serve as an important mechanism for immobilizing and killing extracellular bacteria and fungi [64]. However, IFN-primed neutrophils can also be induced to form NETs by the ICs present in SLE sera through a TLR9-dependent mechanism and SLE patients often have increased numbers of immature neutrophils [22], pointing to an increased turnover rate. Therefore, it has been proposed that neutrophil NETS may provide an important source of DNA-associated autoantigen [25,65]. NETS are eventually degraded by circulating DNAse. Intriguingly, sera obtained from certain SLE patients with more severe clinical disease have been reported to interfere with

NET degradation by extracellular DNAse, further implicating NETs in SLE pathogenesis [66]. However, NETs do not appear to be required for autoantibody production or disease pathogenesis, as SLE-prone mice lacking NADPH oxidase activity, and therefore the ability to form NETS, still develop severe SLE [67].

PARADOXICAL ROLES OF TLR9 AND TLR7 IN MURINE MODELS OF SLE

The *in vitro* analyses linking TLR9 and TLR7 to the immune activation of DCs and B cells predicted that both TLR9 and TLR7 would contribute to immune activation *in vivo* in the context of SLE. This premise has now been tested in a wide range of autoimmune models. Consistent with the *in vitro* data, autoimmune prone mice that lack molecules required for both TLR7 and TLR9 signal transduction such as MyD88 or IRF5, or proper TLR7/9 endosome trafficking such as Unc93b, or fail to express both functional TLR7 and TLR9, invariably have dramatically reduced autoantibody titers and clinical disease [7,35,68–73]. Also as predicted by the *in vitro* studies, TLR7-deficient autoimmune prone mice fail to make antibodies reactive with RNA-associated autoantigens [68,74–76], and TLR9-deficient autoimmune prone mice fail to make autoantibodies that are DNA or chromatin reactive, as determined by a homogeneous nuclear immunofluorescent staining pattern of HEp2 cells or a chromatin-specific ELISA [72,77,78].

Furthermore, TLR7-deficient autoimmune-prone strains all have markedly reduced renal disease and much improved survival rates [68,75,76], while overexpression of TLR7 has the opposite outcome. Examples include the Yaa mutation, essentially a duplication and Y chromosome translocation of a genetic interval, derived from the X chromosome, that incorporates the gene for TLR7 (and TLR8) [79,80]. Originally described as the locus responsible for autoimmune disease in BXSB mice, Yaa serves as an autoimmune accelerator locus in male mice that inherit this mutation, as Yaa male mice essentially express twice the normal level of TLR7 [81]. The effect of the Yaa mutation disappears in Yaa TLR7-deficient males where the functional copy number for TLR7 is reduced back to 1 [82]. Expression of several copies of a TLR7 transgene recapitulates many of the immunoregulatory defects of the Yaa mutation [82], although other genes encoded within the translocated interval contribute to certain features of the Yaa phenotype [31]. Overall these data demonstrate a clear role for TLR7 in promoting autoantibody production and other clinical features of SLE.

By contrast, the role of TLR9 in murine models of autoimmunity appears more complex. Despite the fact that TLR9-deficiency severely limits the production of anti-DNA and anti-chromatin antibodies, TLR9-deficient autoimmune-prone mice invariably develop more severe autoimmune renal disease and a shortened lifespan [72,77,78,83,84]. As mentioned above, the exacerbated disease triggered by TLR9-deficiency remains TLR7 dependent, since double deficient mice develop only limited disease profiles. Together, these observations suggest that TLR9 negatively regulates TLR7-driven responses of self-ligands. Nevertheless, TLR9-engagement can clearly lead to the production of DNA/chromatin-reactive autoantibodies.

Apart from B cells, TLR9 detection of LL37-transported DNA has been implicated in the LL37-dependent activation of plasmacytoid dendritic cells that leads to psoriatic skin inflammation [51,85]. However, LL37 has recently been reported to also bind RNA and thereby enhance RNA endocytosis [53]. Therefore the relative impact of TLR9 and TLR7 on the development of psoriasis remains unresolved and warrants further investigation.

The exact role of TLR9 in vivo has been further confounded by experimental manipulation of the cellular localization of TLR9. The nucleic acid sensing receptors are essentially restricted to intracellular compartments where they cannot be activated by extracellular sources of DNA or RNA. However, substitution of the TLR9 transmembrane and cytosolic domains with those of TLR4, or mutations in only the TLR9 transmembrane domain, can apparently result in the persistent cell surface expression of an uncleaved, yet functional, form of TLR9 [86,87]. As a cell surface receptor, TLR9 acquires the capacity to detect DNA from extracellular sources, presumably present in the serum or associated with cell debris. Radiation chimeras reconstituted with stem cells transduced with the transmembrane-mutant TLR9 develop systemic inflammation at an early age, through a lymphocyte independent mechanism. They develop profound anemia and drastically reduced numbers of B cells. Inflammation presumably reflects a cytokine storm associated with elevated serum levels of TNF, RANTES, and IL-18. Although this response depends on TLR9-expressing CD11c+ dendritic cells, most likely pDCs, it does not depend on type I IFN [87]. Overall this phenotype appears to be more of an example of sterile inflammation than of an autoimmune condition. It will be interesting to determine how overexpression of an unmutated form of TLR9, expressed in its appropriate cellular context, influences its protective or pathogenic activity.

CRITICAL ROLE FOR B CELLS IN TLR-MEDIATED REGULATION OF AUTOIMMUNE DISEASE AND TLR9-DEPENDENT SUPPRESSION OF AUTOIMMUNE DISEASE

B cells can promote systemic autoimmune diseases through both antibody dependent and independent mechanisms. It is now clear that autoantibodies and/or autoantibody associated ICs are directly involved in the inflammatory processes that contribute to joint inflammation, glomerulonephritis, and vasculitis. Moreover, ICs further activate a variety of cell types, through TLR-dependent mechanisms, to further drive the production of type I IFN and other proinflammatory cytokines, to upregulate co-stimulatory molecules that further contribute to the activation of autoreactive T cells, and even to induce the production of profibrotic mediators [88]. B cells can also produce cytokines and very effectively present peptides derived from their cognate antigen (or autoantigen), and therefore promote immune activation and autoimmune disease even when they cannot secrete antibody [89,90].

A major question that has emerged from the analysis of TLR9-deficient autoimmune prone mice is how TLR9 functions to negatively regulate systemic autoimmune disease. The simplest explanation is that TLR9 is preferentially expressed by a particularly effective suppressor population that negatively regulates immune responses to endogenous TLR ligands. Alternatively, TLR9 and TLR7 may serve unique functions within the same cell type. Quite remarkably, while a number of cell types express TLR9, and could be responsible for the unanticipated *in vivo* suppressive role of TLR9, considerable data now point to B cells as the pivotal regulators, although the actual mechanism is still controversial.

It is generally accepted that 'natural' IgM antibodies contribute to the clearance of apoptotic cells and debris [61,91,92]. Ehlers and colleagues reconstituted lethally irradiated mice with a mixture of 80% $JH^{-/-}$ (B cell deficient) and 20% TLR9-deficient autoimmune prone stem cells to make chimeric mice in which all the B cells, and only a small percentage of the non-B cells, were TLR9-deficient. When compared to a comparable group of mice reconstituted with 80% $JH^{-/-}$ (B cell deficient) and 20% $TLR9^+$ autoimmune prone stem cells, they found that the $JH^{-/-}+TLR9^{-/-}$ chimeras developed a more rapid onset systemic autoimmune disease, consistent with the premise that disease severity resulted from a B cell-intrinsic loss of TLR9 function [84]. They were further able to show that TLR9-deficient autoimmune mice could be protected by the transfer of protective antibodies or by the transfer of TLR9-sufficient

B1b B cells. Based on these studies, they proposed that production of protective antibodies depends on the development and activation of B1b B cells responding to endogenous TLR9 ligands [84]. However, other groups have found that B1 cell numbers are normal in MyD88 deficient cells and that B1 cells respond to both TLR9 and TLR7 ligands [93], so the overall impact of B1b cells will need to be confirmed in additional autoimmune prone strains.

Other studies have pointed to a direct effect of TLR9 on inherent B cell activation. The 3H9 heavy chain transgene can pair with endogenous lambda light chains to form autoantibodies that recognize dsDNA. Remarkably, Nickerson et al. found that TLR9-deficient autoimmune-prone B cells expressing a 3H9λ BCR had a significantly longer lifespan in vivo than TLR9-sufficient autoimmune-prone B cells expressing the same BCR [94]. Moreover, at a time when the TLR9-sufficient mice accumulated at the T/B border, a higher percentage of TLR9-deficient 3H9λ cells entered the B cell follicle, even though they were unable to respond to DNA ligands. By contrast, at a later stage in the disease process, only TLR9-sufficient 3H9 B cells convert to full scale antibody producing cells. Overall, these data indicate that inherent TLR9 regulation of autoreactive B cell development occurs in two phases; TLR9 initially limits the activation of 3H9λ B cells by both compromising the survival and follicular entry of these cells, while at a later stage in the disease process it is required for the activation and differentiation of these cells.

CELL INTRINSIC REGULATION OF TLR9 AND TLR7 IN B CELLS

In order to integrate the data documenting the paradoxical effects of TLR7 and TLR9 on murine models of SLE with the TLR9 B cell intrinsic regulation of 3H9 activation and differentiation described above, it is reasonable to assume that TLR7 and TLR9 activate distinct functional programs in B cells. Moreover under conditions of normal homeostasis, TLR9 engagement, and not TLR7 engagement, most likely serves to limit the activation and expansion of a major proportion of the autoreactive B pool, not just 3H9λ cells. According to the model outlined in Figure 6.1, the autoreactive B cell repertoire presumably includes B cells specific for molecular complexes that incorporate: (a) only DNA associated ligands (e.g. chromatin); (b) only RNA associated ligands (e.g. Sm/RNP); or (c) both DNA and RNA ligands. Engagement of TLR9, either alone or simultaneously with TLR7, provides a dominant negative regulatory signal that constrains the initial response to

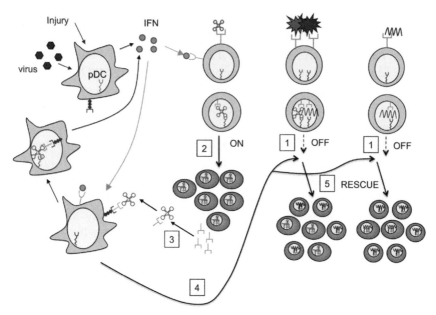

Figure 6.1 *Biphasic role of TLR9 in the etiology of SLE.* [1] TLR9 negatively regulates resting B cells that are stimulated either by BCR/TLR9 engagement, or by a combination of BCR/TLR9 and BCR/TLR7 engagement. [2] TLR7 promotes the activation of resting B cells stimulated by BCR/TLR7 engagement, especially in the presence of type I IFN, and these cells divide and differentiate into antibody producing cells. [3] Autoantibodies produced by BCR/TLR7 activated cells form ICs that stimulate dendritic cells to produce cytokines or survival factors or upregulate costimulatory factors. [4] Cytokines/survival factors produced by DCs or activated T cells reprogram resting B cells so that they can no longer be negatively regulated by BCR/TLR9 engagement. [5] Reprogrammed B cells are now activated through a BCR/TLR9 dependent mechanism. Please see color plate at the back of the book.

autoantigen, eventually shutting down the early response. However, engagement of only TLR7 has a different outcome, eventually leading to the production of autoantibodies that can form ICs if sufficient RNA-associated autoantigen is available. These RNA-associated ICs then activate FcγR+ accessory cells leading to a reprogramming event that in some way reverses the negative regulatory activity of TLR9. Reprogramming could be mediated through soluble factors secreted by activated accessory cells, or alternatively, by accessory cell driven activation of autoreactive T cells that in turn provide soluble or cell-bound 'rescue' signals. Once the cells are reprogrammed, TLR9 is required for the autoreactive B cells to respond positively to DNA or DNA-associated autoantigens. Thus TLR9 initially limits, but then eventually promotes, the activation of DNA-only or DNA/RNA reactive B cells.

COMPARTMENTAL RESTRICTION OF TLR RESPONSES

A major implication of the model outlined above is that BCR/TLR9 and BCR/TLR7 activation of B cells have different functional outcomes, despite the fact that both TLR7 and TLR9 depend on UNC93B1 for transport to the endolysosomal compartments where they are cleaved, detect ligand, and then signal through a MyD88 → IRAK4/IRAK1/ TRAF6 pathway. These distinct functional outcomes could reflect the capacity of the TLR7 and TLR9 TIR domains to bind unique sets of adaptor proteins that activate functionally distinct downstream pathways. Alternatively, unique sets of adaptor proteins could associate with distinct cellular compartments and would be preferentially accessed by TLR9 and TLR7 if they traffic or accumulate in one or another of these compartments.

A growing body of literature suggests the transport adaptor Unc93B1 acts as a rheostat to determine the relative participation of TLR9, TLR7, and other nucleic acid binding receptors during the activation process. As mentioned previously, Unc93B1 regulates trafficking of the endosomal TLRs from the endoplasmic reticulum to the Golgi and then on to various endosomal/lysosomal compartments. Recent studies have indicated that more than one type of compartment may allow for TLR signaling and, for dendritic cells, different compartments may be associated with type I IFN versus proinflammatory cytokine production [95–97]. Thus, the activity of Unc93B1 likely reflects differential triage of these receptors to particular intracellular signaling microenvironments.

Intriguingly, it has recently been shown that the trafficking routes for TLR7 and TLR9 are not the same. Distinct sites on Unc93B1 interact with TLR7 and TLR9, but both molecules cannot be bound efficiently at the same time. Importantly, under normal conditions, Unc93B1 preferentially binds to TLR9, as might be expected for a negative regulator. However an N-terminal D34A mutation in Unc93B1 results in an inability to bind TLR9; instead the D34A mutant shows enhanced binding for TLR7 and cells expressing the D34A mutant are hyperresponsive to TLR7 ligands [98,99]. As in the case of other mice that overexpress TLR7, gene targeted mice that inherit the Unc93B1 D34A mutation develop a B cell-dependent systemic autoimmune disease, in this case associated with autoantibody production, splenomegaly, increased numbers of Th17 and Th1 effector cells, granulocytosis, and thrombocytopenia, and decreased survival due at least in part to hepatic injury and glomerulonephritis [100].

Proper TLR9 trafficking can also be disrupted by truncations or mutations of the Unc93B1 C-terminus that prevent Unc93B1 interaction with the adaptor protein AP-2 [99]. AP-2 complexes are thought to specifically deliver clathrin-coated vesicles from the cell surface to the endocytic compartment, suggesting that TLR9 is transported from to the cell surface by Unc93B1 and then back to the endocytic compartment as part of an AP-2 complex. By contrast, TLR7 associates with Unc93B1 and AP4, an adaptor known to direct transport vesicles from the Golgi directly to endosomes, and therefore most likely follows a different route than TLR9 [99]. Whether AP2-directed TLR9 and AP4-directed TLR7 eventually accumulate in the same or different endolysosomal compartments is less clear. AP3 is known to direct TLR9 and TLR7 to a specialized lysosome related organelle required for type I IFN induction [95,96] and the AP2 and AP4 targeted compartments may also engage distinct signaling cascades. Although currently available data demonstrate clear differences in TLR7 and TLR9 cell trafficking behavior, studies so far are largely limited to HEK and macrophage cell lines, and will need be confirmed for specific primary cell types. Together, these data suggest that Unc93B1 preferentially binds TLR9 in unperturbed B cells and functions as a negative regulator. In the absence of TLR9, not only is the negative feedback loop lost, but Unc93B1 is now more available to TLR7, and TLR7 driven responses are more likely to drive a pathogenic program.

SUMMARY AND UNANSWERED QUESTIONS

Both TLR7 and TLR9 play major roles in the etiology of systemic autoimmune disease. In mice, TLR7 is expressed by DCs, macrophages, neutrophils, and B cells and functions, through cell type specific mechanisms, to promote the production of type I IFNs, proinflammatory cytokines, and autoantibodies. The role of TLR9 is more complicated. Certainly TLR9 is both a negative and positive regulator of B cell function, and B cells play a pivotal role in SLE pathogenesis. Whether TLR9 plays similarly paradoxical roles in other cell types remains to be determined. Remarkably little is known about the distinct features of the TLR7 and TLR9 downstream signaling cascades in cell lines, and even less is understood regarding the primary cell types involved in SLE or other systemic autoimmune diseases. The unique functions of TLR7 and TLR9 are most likely tied to the paths they travel within the cell and the compartments that these paths lead to. Therefore, it will be necessary to gain a better understanding of the mechanisms governing TLR7 and TLR9 trafficking in B cells, and how these basal trafficking patterns are perturbed by inflammation or other danger signals. Such factors may disrupt the preferential binding of Unc93 to TLR9, perhaps by phosphorylation or dephosphorylation of critical binding motifs. One other limitation of our current view of TLRs and SLE has been the focus on murine models. Going forward, it will be important to determine whether the negative regulatory role of TLR9, so apparent in mice, also applies to human B cell activation.

REFERENCES

[1] Alexopoulou L, Holt AC, Medzhitov R, Flavell RA. Recognition of double-stranded RNA and activation of NF-kappaB by Toll-like receptor 3. Nature 2001;413(6857):732–8.

[2] Diebold SS, Kaisho T, Hemmi H, Akira S, Reis E, Sousa C. Innate antiviral responses by means of TLR7-mediated recognition of single-stranded RNA. Science 2004;303(5663):1529–31.

[3] Heil F, Hemmi H, Hochrein H, Ampenberger F, Kirschning C, Akira S, et al. Species-specific recognition of single-stranded RNA via toll-like receptor 7 and 8. Science 2004;303(5663):1526–9.

[4] Hemmi H, Takeuchi O, Kawai T, Kaisho T, Sato S, Sanjo H, et al. A Toll-like receptor recognizes bacterial DNA. Nature 2000;408(6813):740–5.

[5] Oldenburg M, Kruger A, Ferstl R, Kaufmann A, Nees G, Sigmund A, et al. TLR13 recognizes bacterial 23S rRNA devoid of erythromycin resistance-forming modification. Science 2012;337(6098):1111–5.

[6] Desmet CJ, Ishii KJ. Nucleic acid sensing at the interface between innate and adaptive immunity in vaccination. Nat Rev Immunol 2012;12(7):479–91.

[7] Lau CM, Broughton C, Tabor AS, Akira S, Flavell RA, Mamula MJ, et al. RNA-associated autoantigens activate B cells by combined B cell antigen receptor/Toll-like receptor 7 engagement. J Exp Med 2005;202(9):1171–7.

[8] Leadbetter EA, Rifkin IR, Hohlbaum AM, Beaudette BC, Shlomchik MJ, Marshak-Rothstein A. Chromatin-IgG complexes activate B cells by dual engagement of IgM and Toll-like receptors. Nature 2002;416(6881):603–7.

[9] Tabeta K, Hoebe K, Janssen EM, Du X, Georgel P, Crozat K, et al. The Unc93b1 mutation 3d disrupts exogenous antigen presentation and signaling via Toll-like receptors 3, 7 and 9. Nat Immunol 2006;7(2):156–64.

[10] Ewald SE, Lee BL, Lau L, Wickliffe KE, Shi GP, Chapman HA, et al. The ectodomain of Toll-like receptor 9 is cleaved to generate a functional receptor. Nature 2008;56(7222):658–62.

[11] Park B, Brinkmann MM, Spooner E, Lee CC, Kim YM, Ploegh HL. Proteolytic cleavage in an endolysosomal compartment is required for activation of Toll-like receptor 9. Nat Immunol 2008;9(12):1407–14.

[12] Ewald SE, Engel A, Lee J, Wang M, Bogyo M, Barton GM. Nucleic acid recognition by Toll-like receptors is coupled to stepwise processing by cathepsins and asparagine endopeptidase. J Exp Med 2011;208(4):643–51.

[13] Latz E, Verma A, Visintin A, Gong M, Sirois CM, Klein DC, et al. Ligand-induced conformational changes allosterically activate Toll-like receptor 9. Nat Immunol 2007;8(7):772–9.

[14] Suzuki N, Suzuki S, Duncan GS, Millar DG, Wada T, Mirtsos C, et al. Severe impairment of interleukin-1 and Toll-like receptor signalling in mice lacking IRAK-4. Nature 2002;416(6882):750–6.

[15] Kawagoe T, Sato S, Matsushita K, Kato H, Matsui K, Kumagai Y, et al. Sequential control of Toll-like receptor-dependent responses by IRAK1 and IRAK2. Nat Immunol 2008;9(6):684–91.

[16] Ronnblom L, Alm GV. A pivotal role for the natural interferon alpha-producing cells (plasmacytoid dendritic cells) in the pathogenesis of lupus. J Exp Med 2001;194(12):F59–63.

[17] Bave U, Magnusson M, Eloranta ML, Perers A, Alm GV, Ronnblom L. Fc gamma RIIa is expressed on natural IFN-alpha-producing cells (plasmacytoid dendritic cells) and is required for the IFN-alpha production induced by apoptotic cells combined with lupus IgG. J Immunol 2003;171(6):3296–302.

[18] Ronnblom LE, Alm GV, Oberg KE. Possible induction of systemic lupus erythematosus by interferon-alpha treatment in a patient with a malignant carcinoid tumour. J Intern Med 1990;227(3):207–10.

[19] Gota C, Calabrese L. Induction of clinical autoimmune disease by therapeutic interferon-alpha. Autoimmunity 2003;36(8):511–8.

[20] Wilson LE, Widman D, Dikman SH, Gorevic PD. Autoimmune disease complicating antiviral therapy for hepatitis C virus infection. Semin Arthritis Rheum 2002;32(3):163–73.

[21] Baechler EC, Batliwalla FM, Karypis G, Gaffney PM, Ortmann WA, Espe KJ, et al. Interferon-inducible gene expression signature in peripheral blood cells of patients with severe lupus. Proc Natl Acad Sci USA 2003;100(5):2610–5.

[22] Bennett L, Palucka AK, Arce E, Cantrell V, Borvak J, Banchereau J, et al. Interferon and granulopoiesis signatures in systemic lupus erythematosus blood. J Exp Med 2003;197(6):711–23.

[23] Blanco P, Palucka AK, Gill M, Pascual V, Banchereau J. Induction of dendritic cell differentiation by IFN-alpha in systemic lupus erythematosus. Science 2001;294(5546):1540–3.

[24] Jego G, Palucka AK, Blanck JP, Chalouni C, Pascual V, Banchereau J. Plasmacytoid dendritic cells induce plasma cell differentiation through type I interferon and interleukin 6. Immunity 2003;19(2):225–34.

[25] Garcia-Romo GS, Caielli S, Vega B, Connolly J, Allantaz F, Xu Z, et al. Netting neutrophils are major inducers of type I IFN production in pediatric systemic lupus erythematosus. Sci Transl Med 2011;3(73):73ra20.

[26] Braun D, Caramalho I, Demengeot J. IFN-alpha/beta enhances BCR-dependent B cell responses. Int Immunol 2002;14(4):411–9.

[27] Mathian A, Gallegos M, Pascual V, Banchereau J, Koutouzov S. Interferon-alpha induces unabated production of short-lived plasma cells in pre-autoimmune lupus-prone (NZB × NZW)F1 mice but not in BALB/c mice. Eur J Immunol 2011;41(3):863–72.

[28] Xu W, Joo H, Clayton S, Dullaers M, Herve MC, Blankenship D, et al. Macrophages induce differentiation of plasma cells through CXCL10/IP-10. J Exp Med 2012;209(10):1813–23, S1–2.

[29] Agrawal H, Jacob N, Carreras E, Bajana S, Putterman C, Turner S, et al. Deficiency of type I IFN receptor in lupus-prone New Zealand mixed 2328 mice decreases dendritic cell numbers and activation and protects from disease. J Immunol 2009;183(9):6021–9.

[30] Nacionales DC, Kelly-Scumpia KM, Lee PY, Weinstein JS, Lyons R, Sobel E, et al. Deficiency of the type I interferon receptor protects mice from experimental lupus. Arthritis Rheum 2007;56(11):3770–83.

[31] Santiago-Raber ML, Kikuchi S, Borel P, Uematsu S, Akira S, Kotzin BL, et al. Evidence for genes in addition to Tlr7 in the Yaa translocation linked with acceleration of systemic lupus erythematosus. J Immunol 2008;181(2):1556–62.

[32] Jorgensen TN, Roper E, Thurman JM, Marrack P, Kotzin BL. Type I interferon signaling is involved in the spontaneous development of lupus-like disease in B6.Nba2 and (B6.Nba2 × NZW)F(1) mice. Genes Immun 2007;8(8):653–62.

[33] Braun D, Geraldes P, Demengeot J, Type I. Interferon controls the onset and severity of autoimmune manifestations in lpr mice. J Autoimmun 2003;20(1):15–25.

[34] Nickerson KM, Cullen JL, Kashgarian M, Shlomchik MJ. Exacerbated autoimmunity in the absence of TLR9 in MRL.Faslpr mice depends on Ifnar1. J Immunol 2013;190(8):3889–94.

[35] Richez C, Yasuda K, Bonegio RG, Watkins AA, Aprahamian T, Busto P, et al. IFN regulatory factor 5 is required for disease development in the FcgammaRIIB −/−Yaa and FcgammaRIIB−/− mouse models of systemic lupus erythematosus. J Immunol 2010;184(2):796–806.

[36] Hron JD, Peng SL. Type I IFN protects against murine lupus. J Immunol 2004;173(3):2134–42.

[37] Baccala R, Gonzalez-Quintial R, Schreiber RD, Lawson BR, Kono DH, Theofilopoulos AN. Anti-IFN-alpha/beta receptor antibody treatment ameliorates disease in lupus-predisposed mice. J Immunol 2012;189(12):5976–84.

[38] Wolfowicz CB, Sakorafas P, Rothstein TL, Marshak-Rothstein A. Oligoclonality of rheumatoid factors arising spontaneously in lpr/lpr mice. Clin Immunol Immunopathol 1988;46(3):382–95.

[39] Shlomchik MJ, Zharhary D, Saunders T, Camper SA, Weigert MG. A rheumatoid factor transgenic mouse model of autoantibody regulation. Int Immunol 1993;5(10):1329–41.

[40] William J, Euler C, Shlomchik MJ. Short-lived plasmablasts dominate the early spontaneous rheumatoid factor response: differentiation pathways, hypermutating cell types, and affinity maturation outside the germinal center. J Immunol 2005;174(11):6879–87.

[41] Cooke MS, Mistry N, Wood C, Herbert KE, Lunec J. Immunogenicity of DNA damaged by reactive oxygen species–implications for anti-DNA antibodies in lupus. Free Radic Biol Med 1997;22(1–2):151–9.

[42] Rifkin IR, Leadbetter EA, Beaudette BC, Kiani C, Monestier M, Shlomchik MJ, et al. Immune complexes present in the sera of autoimmune mice activate rheumatoid factor B cells. J Immunol 2000;165(3):1626–33.

[43] Uccellini MB, Busconi L, Green NM, Busto P, Christensen SR, Shlomchik MJ, et al. Autoreactive B cells discriminate CpG-rich and CpG-poor DNA and this response is modulated by IFN-alpha. J Immunol 2008;181(9):5875–84.

[44] Krieg AM, Yi AK, Matson S, Waldschmidt TJ, Bishop GA, Teasdale R, et al. CpG motifs in bacterial DNA trigger direct B-cell activation. Nature 1995;374(6522):546–9.

[45] Kariko K, Buckstein M, Ni H, Weissman D. Suppression of RNA recognition by Toll-like receptors: the impact of nucleoside modification and the evolutionary origin of RNA. Immunity 2005;23(2):165–75.

[46] Green NM, Laws A, Kiefer K, Busconi L, Kim YM, Brinkmann MM, et al. Murine B cell response to TLR7 ligands depends on an IFN-beta feedback loop. J Immunol 2009;183(3):1569–76.

[47] Zhang Q, Raoof M, Chen Y, Sumi Y, Sursal T, Junger W, et al. Circulating mitochondrial DAMPs cause inflammatory responses to injury. Nature 2010;464(7285):104–7.

[48] Jakymiw A, Ikeda K, Fritzler MJ, Reeves WH, Satoh M, Chan EK. Autoimmune targeting of key components of RNA interference. Arthritis Res Ther 2006;8(4):R87.

[49] Green NM, Moody KS, Debatis M, Marshak-Rothstein A. Activation of auto-reactive B cells by endogenous TLR7 and TLR3 RNA ligands. J Biol Chem 2012;287(47):39789–99.

[50] Tian J, Avalos AM, Mao SY, Chen B, Senthil K, Wu H, et al. Toll-like receptor 9-dependent activation by DNA-containing immune complexes is mediated by HMGB1 and RAGE. Nat Immunol 2007;8(5):487–96.

[51] Lande R, Gregorio J, Facchinetti V, Chatterjee B, Wang YH, Homey B, et al. Plasmacytoid dendritic cells sense self-DNA coupled with antimicrobial peptide. Nature 2007;449(7162):564–9.

[52] Ganguly D, Chamilos G, Lande R, Gregorio J, Meller S, Facchinetti V, et al. Self-RNA-antimicrobial peptide complexes activate human dendritic cells through TLR7 and TLR8. J Exp Med 2009;206(9):1983–94.

[53] Morizane S, Yamasaki K, Muhleisen B, Kotol PF, Murakami M, Aoyama Y, et al. Cathelicidin antimicrobial peptide LL-37 in psoriasis enables keratinocyte reactivity against TLR9 ligands. J Invest Dermatol 2012;132(1):135–43.

[54] Erwig LP, Henson PM. Clearance of apoptotic cells by phagocytes. Cell Death Differ 2008;15(2):243–50.

[55] Herrmann M, Voll RE, Zoller OM, Hagenhofer M, Ponner BB, Kalden JR. Impaired phagocytosis of apoptotic cell material by monocyte-derived macrophages from patients with systemic lupus erythematosus. Arthritis Rheum 1998;41(7):1241–50.

[56] Kruse K, Janko C, Urbonaviciute V, Mierke CT, Winkler TH, Voll RE, et al. Inefficient clearance of dying cells in patients with SLE: anti-dsDNA autoantibodies, MFG-E8, HMGB-1 and other players. Apoptosis 2010;15(9):1098–113.

[57] Rodriguez-Manzanet R, Sanjuan MA, Wu HY, Quintana FJ, Xiao S, Anderson AC, et al. T and B cell hyperactivity and autoimmunity associated with niche-specific defects in apoptotic body clearance in TIM-4-deficient mice. Proc Natl Acad Sci USA 2010;107(19):8706–11.

[58] Viorritto IC, Nikolov NP, Siegel RM. Autoimmunity versus tolerance: can dying cells tip the balance? Clin Immunol 2007;122(2):125–34.

[59] Manderson AP, Botto M, Walport MJ. The role of complement in the development of systemic lupus erythematosus. Annu Rev Immunol 2004;22:431–56.

[60] Santer DM, Wiedeman AE, Teal TH, Ghosh P, Elkon KB. Plasmacytoid dendritic cells and C1q differentially regulate inflammatory gene induction by lupus immune complexes. J Immunol 2012;188(2):902–15.

[61] Elkon KB, Silverman GJ. Naturally occurring autoantibodies to apoptotic cells. Adv Exp Med Biol 2012;750:14–26.

[62] Hanayama R, Tanaka M, Miyasaka K, Aozasa K, Koike M, Uchiyama Y, et al. Autoimmune disease and impaired uptake of apoptotic cells in MFG-E8-deficient mice. Science 2004;304(5674):1147–50.

[63] Peng Y, Elkon KB. Autoimmunity in MFG-E8-deficient mice is associated with altered trafficking and enhanced cross-presentation of apoptotic cell antigens. J Clin Invest 2011;121(6):2221–41.

[64] Brinkmann V, Zychlinsky A. Beneficial suicide: why neutrophils die to make NETs. Nat Rev Microbiol 2007;5(8):577–82.

[65] Lande R, Ganguly D, Facchinetti V, Frasca L, Conrad C, Gregorio J, et al. Neutrophils activate plasmacytoid dendritic cells by releasing self-DNA-peptide complexes in systemic lupus erythematosus. Sci Transl Med 2011;3(73):73ra19.

[66] Hakkim A, Furnrohr BG, Amann K, Laube B, Abed UA, Brinkmann V, et al. Impairment of neutrophil extracellular trap degradation is associated with lupus nephritis. Proc Natl Acad Sci USA 2010;107(21):9813–8.

[67] Campbell AM, Kashgarian M, Shlomchik MJ. NADPH oxidase inhibits the patho-genesis of systemic lupus erythematosus. Sci Transl Med 2012;4(157):157ra41.

[68] Nickerson KM, Christensen SR, Shupe J, Kashgarian M, Kim D, Elkon K, et al. TLR9 regulates TLR7- and MyD88-dependent autoantibody production and disease in a murine model of lupus. J Immunol 2010;184(4):1840–8.

[69] Groom JR, Fletcher CA, Walters SN, Grey ST, Watt SV, Sweet MJ, et al. BAFF and MyD88 signals promote a lupuslike disease independent of T cells. J Exp Med 2007;204(8):1959–71.

[70] Silver KL, Crockford TL, Bouriez-Jones T, Milling S, Lambe T, Cornall RJ. MyD88-dependent autoimmune disease in Lyn-deficient mice. Eur J Immunol 2007;37(10):2734–43.

[71] Kono DH, Haraldsson MK, Lawson BR, Pollard KM, Koh YT, Du X, et al. Endosomal TLR signaling is required for anti-nucleic acid and rheumatoid factor autoantibodies in lupus. Proc Natl Acad Sci USA 2009;106(29):12061–6.

[72] Santiago-Raber ML, Dunand-Sauthier I, Wu T, Li QZ, Uematsu S, Akira S, et al. Critical role of TLR7 in the acceleration of systemic lupus erythematosus in TLR9-deficient mice. J Autoimmun 2010;34(4):339–48.

[73] Becker-Herman S, Meyer-Bahlburg A, Schwartz MA, Jackson SW, Hudkins KL, Liu C, et al. WASp-deficient B cells play a critical, cell-intrinsic role in triggering autoimmunity. J Exp Med 2011;208(10):2033–42.

[74] Berland R, Fernandez L, Kari E, Han JH, Lomakin I, Akira S, et al. Toll-like receptor 7-dependent loss of B cell tolerance in pathogenic autoantibody knockin mice. Immunity 2006;25(3):429–40.

[75] Savarese E, Steinberg C, Pawar RD, Reindl W, Akira S, Anders HJ, et al. Requirement of Toll-like receptor 7 for pristane-induced production of autoantibodies and development of murine lupus nephritis. Arthritis Rheum 2008;58(4):1107–15.

[76] Lee PY, Kumagai Y, Li Y, Takeuchi O, Yoshida H, Weinstein J, et al. TLR7-dependent and FcgammaR-independent production of type I interferon in experimental mouse lupus. J Exp Med 2008;205(13):2995–3006.

[77] Christensen SR, Shupe J, Nickerson K, Kashgarian M, Flavell RA, Shlomchik MJ. Toll-like receptor 7 and TLR9 dictate autoantibody specificity and have opposing inflammatory and regulatory roles in a murine model of lupus. Immunity 2006;25(3):417–28.

[78] Lartigue A, Courville P, Auquit I, Francois A, Arnoult C, Tron F, et al. Role of TLR9 in anti-nucleosome and anti-DNA antibody production in lpr mutation-induced murine lupus. J Immunol 2006;177(2):1349–54.

[79] Pisitkun P, Deane JA, Difilippantonio MJ, Tarasenko T, Satterthwaite AB, Bolland S. Autoreactive B cell responses to RNA-related antigens due to TLR7 gene duplication. Science 2006;312(5780):1669–72.

[80] Subramanian S, Tus K, Li QZ, Wang A, Tian XH, Zhou J, et al. A Tlr7 translocation accelerates systemic autoimmunity in murine lupus. Proc Natl Acad Sci USA 2006;103(26):9970–5.

[81] Bolland S, Yim YS, Tus K, Wakeland EK, Ravetch JV. Genetic modifiers of systemic lupus erythematosus in FcgammaRIIB(−/−) mice. J Exp Med 2002;195(9):1167–74.

[82] Deane JA, Pisitkun P, Barrett RS, Feigenbaum L, Town T, Ward JM, et al. Control of toll-like receptor 7 expression is essential to restrict autoimmunity and dendritic cell proliferation. Immunity 2007;27(5):801–10.

[83] Yu P, Wellmann U, Kunder S, Quintanilla-Martinez L, Jennen L, Dear N, et al. Toll-like receptor 9-independent aggravation of glomerulonephritis in a novel model of SLE. Int Immunol 2006;18(8):1211–9.

[84] Stoehr AD, Schoen CT, Mertes MM, Eiglmeier S, Holecska V, Lorenz AK, et al. TLR9 in peritoneal B-1b cells is essential for production of protective self-reactive IgM to control Th17 cells and severe autoimmunity. J Immunol 2011;187(6):2953–65.

[85] Gilliet M, Lande R. Antimicrobial peptides and self-DNA in autoimmune skin inflammation. Curr Opin Immunol 2008;20(4):401–7.

[86] Barton GM, Kagan JC, Medzhitov R. Intracellular localization of Toll-like receptor 9 prevents recognition of self DNA but facilitates access to viral DNA. Nat Immunol 2006;7(1):49–56.

[87] Mouchess ML, Arpaia N, Souza G, Barbalat R, Ewald SE, Lau L, et al. Transmembrane mutations in Toll-like receptor 9 bypass the requirement for ectodomain proteolysis and induce fatal inflammation. Immunity 2011;35(5):721–32.

[88] Alvarez D, Briassouli P, Clancy RM, Zavadil J, Reed JH, Abellar RG, et al. A novel role of endothelin-1 in linking Toll-like receptor 7-mediated inflammation to fibrosis in congenital heart block. J Biol Chem 2011;286(35):30444–54.

[89] Lund FE, Randall TD. Effector and regulatory B cells: modulators of CD4(+) T cell immunity. Nat Rev Immunol 2010;10(4):236–47.

[90] Shlomchik MJ, Madaio MP. The role of antibodies and B cells in the pathogenesis of lupus nephritis. Springer Semin Immunopathol 2003;24(4):363–75.

[91] Boes M, Schmidt T, Linkemann K, Beaudette BC, Marshak-Rothstein A, Chen J. Accelerated development of IgG autoantibodies and autoimmune disease in the absence of secreted IgM. Proc Natl Acad Sci USA 2000;97(3):1184–9.

[92] Werwitzke S, Trick D, Kamino K, Matthias T, Kniesch K, Schlegelberger B, et al. Inhibition of lupus disease by anti-double-stranded DNA antibodies of the IgM isotype in the (NZB × NZW)F1 mouse. Arthritis Rheum 2005;52(11):3629–38.

[93] Meyer-Bahlburg A, Rawlings DJ. Differential impact of Toll-like receptor signaling on distinct B cell subpopulations. Front Biosci 2012;17:1499–516.

[94] Nickerson KM, Christensen SR, Cullen JL, Meng W, Luning Prak ET, Shlomchik MJ. TLR9 promotes tolerance by restricting survival of anergic anti-DNA B cells, yet is also required for their activation. J Immunol 2013;190(4):1447–56.

[95] Sasai M, Linehan MM, Iwasaki A. Bifurcation of Toll-like receptor 9 signaling by adaptor protein 3. Science 2010;329(5998):1530–4.

[96] Blasius AL, Arnold CN, Georgel P, Rutschmann S, Xia Y, Lin P, et al. Slc15a4, AP-3, and Hermansky-Pudlak syndrome proteins are required for Toll-like receptor signaling in plasmacytoid dendritic cells. Proc Natl Acad Sci USA 2010;107(46):19973–8.

[97] Henault J, Martinez J, Riggs JM, Tian J, Mehta P, Clarke L, et al. Noncanonical autophagy is required for type I interferon secretion in response to DNA-immune complexes. Immunity 2012;37(6):986–97.

[98] Fukui R, Saitoh S, Matsumoto F, Kozuka-Hata H, Oyama M, Tabeta K, et al. Unc93B1 biases Toll-like receptor responses to nucleic acid in dendritic cells toward DNA- but against RNA-sensing. J Exp Med 2009;206(6):1339–50.

[99] Lee BL, Moon JE, Shu JH, Yuan L, Newman ZR, Schekman R, et al. UNC93B1 mediates differential trafficking of endosomal TLRs. Elife 2013;2:e00291.

[100] Fukui R, Saitoh S, Kanno A, Onji M, Shibata T, Ito A, et al. Unc93B1 restricts systemic lethal inflammation by orchestrating Toll-like receptor 7 and 9 trafficking. Immunity 2011;35(1):69–81.

CHAPTER 7

Bacterial Infections and the DNA Sensing Pathway

Jan Naujoks and Bastian Opitz

Department of Internal Medicine/Infectious Diseases and Pulmonary Medicine, Charité Universitätsmedizin Berlin, Augustenburger Platz 1, 13353 Berlin, Germany

INTRODUCTION

The innate immune system recognizes the presence of pathogens or tissue damage by employing pattern recognition receptors (PRRs) [1]. These PRRs sense several foreign molecules shared by a large group of microbes. For example, many Toll-like receptors (TLRs), NOD-like receptors (NLRs) or RIG-I-like receptors (RLRs) recognize lipopolysaccharide, peptidoglycan or RNA motifs that are expressed by different bacteria or viruses but not by the host [2–4]. Other receptors, however, are activated by 'common' molecules expressed by microbes and/or the host if these molecules are localized to unphysiological cellular compartments. For instance, ATP, the central molecule of intracellular energy transfer, triggers an inflammatory signal if released into the extracellular space [5]. Similarly, bacterial and host DNA are known to activate innate immune responses [6–9]. DNA is found in the nucleus and the mitochondria but is absent from the cytosol or the extracellular space of host cells. However, microbial DNA can be translocated into the host cell cytosol or released into the extracellular space during bacterial infections [10,11]. Moreover, tissue damage might release endogenous DNA that subsequently triggers inflammatory responses [12].

The innate immune system relies on several sensor molecules localized in the cytosol and endosomal compartments that detect exogenous and endogenous DNA molecules. Among those sensors are TLR9 on endosomal membranes [13], and cyclic GMP-AMP synthase [14], RNA polymerase III [15,16], DAI [17], DDX41 [18], IFI16 [19], and AIM2 [20–23] in the cytosol. These receptors signal through specific adapter molecules including MyD88, STING, MAVS, and ASC to stimulate production of proinflammatory cytokines or type I IFNs, or to activate inflammasomes

Biological DNA Sensor.
DOI: http://dx.doi.org/10.1016/B978-0-12-404732-7.00007-1

(Figure 7.1). Whereas proinflammatory cytokine production and inflammasome activation are induced by several PRRs during most bacterial infections, the type I IFN response is often triggered by bacterial nucleic acid sensing. This chapter summarizes the current knowledge about the function of DNA sensors in bacterial infection.

THE ROLE OF TLR9 IN BACTERIAL INFECTIONS

It has long been known that bacterial DNA and synthetic DNA oligonucleotides bearing unmethylated CpG base pairs are strong inducers of proinflammatory mediators such as TNF-α [6–9]. Subsequent studies identified TLR9 as the receptor for unmethylated CpG-rich DNA as well as DNAs with a phosphodiester (PD) 2′ deoxyribose sugar backbone, and MyD88 as its central signal transduction adaptor molecule [13,24,25]. Similar to other TLRs, TLR9 consists of leucin–rich repeats (LRRs), a transmembrane domain and a cytosolic Toll/IL-1 receptor homology (TIR) domain [26]. It is expressed in different host cells including

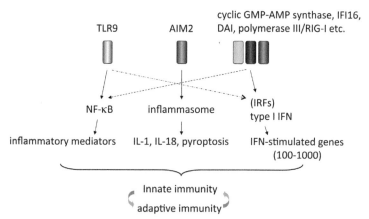

Figure 7.1 *Major pathways activated by DNA sensors of the innate immune system.* After recognition of DNA, TLR9 activates production of NF-κB-dependent proinflammatory genes. In pDCs, TLR9 additionally stimulates type I IFN responses through IRF7. AIM2 forms an inflammasome that regulates IL-1 family cytokines at a posttranslational level and activates an inflammatory cell death called pyroptosis. Many other cytosolic DNA sensors including cyclic GMP-AMP synthase, IFI16, polymerase III/RIG-I and DAI mainly stimulate type I IFN responses via IRF3/7. Most receptors are also capable of activating NF-κB-dependent gene expression. In addition, some of these receptors might stimulate autophagy (not depicted).

plasmocytoid dendridic cells (pDCs), classical dendritic cells (cDCs), macrophages, B cells, as well as intestinal and lung epithelial cells. TLR9 is localized in the ER and migrates into DNA-containing endosomal structures [27]. Here, the LRR of TLR9 is on the inside to bind DNA motifs. Activated TLR9 triggers MyD88-dependent signal transduction via its TIR domain localized at the cytosolic side. In most cell types, TLR9–MyD88 signaling results in NF-κB activation and the production of pro-inflammatory mediators. In pDCs, however, TLR9–MyD88 signaling is additionally linked to the transcription factor IRF7, which controls the expression of type I IFNs and IFN-inducible genes [28–30]. Interestingly, previous studies have indicated the presence of TLR9-independent, MyD88-dependent DNA sensors in pDCs, and one recent study suggested that the helicases DHX36 and DHX9 played an important role in these responses (see also below) [31].

In *Mycobacterium tuberculosis* infection, TLR9 contributes to innate as well as adaptive immune responses [32]. Infected TLR9$^{-/-}$ mice displayed impaired mycobacteria-induced IL-12p40 and IFNγ responses. Furthermore, TLR2/9$^{-/-}$ mice exhibited markedly enhanced susceptibility to infection as compared to TLR2$^{-/-}$ or TLR9$^{-/-}$ mice [32]. Similarly, TLR9 activation played a role in the defense against *Streptococcus pneumoniae* and *Neisseria meningitidis* during pneumonia or sepsis, respectively [33,34]. It was suggested that TLR9 stimulated the phagocytic activity of resident macrophages leading to early bacterial clearance in the lower respiratory tract [33]. Accordingly, another study found an association between single nucleotide polymorphisms in TLR9 and higher susceptibility to bacterial meningitis in children that is typically caused by *S. pneumoniae* and *N. meningitidis* [35]. In pneumonia caused by *Klebsiella pneumoniae* or *Legionella pneumophila*, expression of TLR9 in cDCs and macrophages was required for effective innate immune responses, and TLR9$^{-/-}$ mice showed enhanced susceptibility to these pathogens [36,37]. Moreover, synergistic stimulation of TLR2 and TLR9 with synthetic agonists protected mice against pneumonia caused by *S. pneumoniae* and *P. aeruginosa* [38]. In contrast to those infections, TLR9$^{-/-}$ mice were more resistant to intranasal *Staphylococcus aureus* infection [39].

TLR9 also plays an important role in the intestinal homeostasis through sensing gut flora or probiotic DNA. It was shown that TLR9 signaling mediated the anti-inflammatory effects of probiotics in a murine experimental colitis model [40], and administration of TLR9 agonists was shown to trigger an anti-inflammatory type I IFN response [41]. These studies together suggested that the type I IFNs were involved in the TLR9-dependent protective effects of probiotics. Moreover TLR9 signaling regulated the optimal ratio of CD4(+)Foxp3(+) regulatory T and IL-17- and IFNγ-producing effector T cells in the intestine required to induce inflammatory responses to oral infections. It was shown that triggering of TLR9 by gut commensal DNA decreased frequencies of regulatory T cells and enhanced constitutive effector T cells in the intestine [42]. On the other hand, TLR9 strongly contributed to the development of immunopathology and the regulation of gut microbiota dynamics in a murine model of intestinal graft-versus-host disease [43].

TLR9 can also play a critical role in dysregulated immune responses and death associated with experimental peritonitis and sepsis models in mice [44]. TLR9 in DCs appeared to mediate the detrimental effects for the host. Interestingly, administration of an inhibitory CpG sequence that blocked TLR9 signaling protected WT mice from sepsis-related mortality [44].

Taken together, TLR9 plays a critical role in pathogen recognition and host resistance to bacterial infections, as well as homeostasis in the intestinal tract. On the other hand, TLR9 can also be involved in the development of immunopathologies during sepsis.

THE AIM2 INFLAMMASOME AS A SENSOR OF BACTERIAL DNA

AIM2 (absent in melanoma 2) is the founding member of the AIM2-like receptors consisting of 13 genes in C57BL/6 mice and 4 in humans. AIM2 also belongs to the PYHIN or HIN200 families of proteins based on its possession of an N-terminal pyrin and C-terminal HIN domain [45–47]. AIM2 binds to bacterial, viral and endogenous dsDNA and activates inflammasomes. Inflammasomes are large multiprotein complexes consisting of a receptor (e.g. AIM2 or members of the NOD-like receptor family), the adapter molecule ASC, and the protease caspase-1 [48]. These inflammasomes regulate the activation of caspase-1 which cleaves pro-IL-1β and pro-IL-18 and enables the release of the active mature cytokines. IL-1β and IL-18 are both key cytokines involved in the early innate immune response as well as the priming and regulation of the adaptive immunity.

AIM2 recognition of non-sequence-specific DNA is accomplished through electrostatic attraction between the positively charged HIN domain residues and the dsDNA sugar–phosphate backbone [49]. This

DNA binding releases the autoinhibitory intramolecular complex of the HIN and pyrin domains and might facilitate the homophilic interaction between the pyrin domains of AIM2 and the inflammasome adapter molecule ASC.

Other ALR/HIN200 proteins related to AIM2 in mice are p202, MNDA and MNDAL [23,47]. p202 binds to dsDNA through its two HIN domains but it lacks a PYD domain. Therefore, p202 is not able to recruit the adaptor molecule ASC, and studies have shown that p202 negatively regulates the AIM2 inflammasome [23]. In contrast, MNDA and MNDAL have been shown to activate inflammasomes when overexpressed in HEK293 cells [47]. However, whether these regulatory mechanisms operate during bacterial infections needs to be determined.

AIM2 is mainly expressed in the cytosol of hematopoetic cells. AIM2 is activated by several bacterial species e.g. *Listeria monocytogenes*, *Francisella tularensis*, *S. pneumoniae* and *M. tuberculosis* [50–57]. *L. monocytogenes* is phagocytosed but subsequently escapes from the primary phagosome into the cytosol through the activity of the cytolysin listeriolysin [58]. Some bacteria, however, appear to be degraded intracellularly, which releases free bacterial DNA into the cytosol and activates the AIM2 inflammasome. The AIM2 inflammasome together with the NLRP3 and NLRC4 inflammasomes regulate IL-1β and IL-18 production as well as the caspase-1-dependent cell death called pyroptosis [52,55,56,59].

Similarly to *L. monocytogenes*, *F. tularensis* escapes from phagosomes into the macrophage cytosol where some bacteria lyse, which results in release of bacterial DNA and its binding to AIM2 [52–54,60]. Interestingly, AIM2 is a type I IFN-induced protein, and bacterial DNA also triggers type I IFNs through an AIM2-independent but STING-dependent mechanism (see below) that leads to AIM2 upregulation. Accordingly, IFNβ stimulation of macrophages enhanced inflammasome activation in response to *F. tularensis* [54].

The AIM2 inflammasome appears to be the main inflammasome activated by *Francisella* infection. AIM2-deficient mice were susceptible to *F. tularensis* infection, exhibiting higher mortality and bacterial burden than wild-type mice [53]. In agreement with the central role of AIM2 in innate immunity responses to *F. tularensis* on the one hand, and with the fact that this bacterium is nonetheless a successful pathogen on the other, *F. tularensis* expresses the protein mviN to limit AIM2 inflammasome activation *in vitro* and enhance virulence *in vivo* [60].

Furthermore, AIM2$^{-/-}$ macrophages produced less IL-1β and IL-18 after infection with *M. tuberculosis*. AIM2-deficient mice were highly susceptible to intratracheal infection with *M. tuberculosis* and this was associated with defective IL-1β and IL-18 production together with impaired Th1 responses [57].

In addition to its role in the response against intracellular pathogens, AIM2 also plays an important role in extracellular bacterial infections. After *S. pneumoniae* infection, macrophages lacking AIM2 expression showed markedly reduced caspase-1 activation and IL-1β production [50,51]. The current model proposes that phagocytic uptake of *S. pneumoniae* by macrophages is followed by the proteolytic bacterial degradation, phagolysosomal membrane destruction by the pore-forming toxin pneumolysin, and the leakage of bacterial DNA into the cytosol where it binds to AIM2 [50,51].

Taken together, AIM2 is an essential cytosolic sensor of DNA from different bacterial species that forms inflammasomes together with ASC and caspase-1. The AIM2 inflammasome regulates IL-1β as well as IL-18 production at the posttranslational level, mediates pyroptosis, and is required for host defense against bacteria including *F. tularensis* and *M. tuberculosis*.

STING-DEPENDENT CYTOSOLIC DNA SENSORS AND TYPE I IFN RESPONSES DURING BACTERIAL INFECTIONS

It has long been known that intracellular delivery of mammalian or bacterial dsDNA triggers proinflammatory responses in the host independently of TLR9 [61], and that microbial and host DNA delivered into the cytosol stimulate type I IFNs [62,63]. Moreover, infections with bacteria that replicate in the host cell cytosol induced production of type I IFNs [64–66]. Subsequent studies showed that bacteria expressing secretion systems capable of injecting microbial molecules into the host cell, or expressing pore-forming toxins that destroy the phagolysosomal membrane after bacterial phagocytosis also stimulated type I IFN responses [63,67–71]. The IFNα/β responses to cytosolic DNA stimulation or bacterial infections induced a similar pattern of gene expression, and were both dependent on the kinase TBK1 and the transcription factor IRF3, but independent of the TLRs [62,63,68,70,71,73]. These results together with others led to the conclusion that sensing of bacterial DNA by cytosolic PRRs was responsible for activating type I IFN responses during infections with various bacteria. However, certain bacteria appear to stimulate type I IFN production in some cell types through different mechanisms activated by, for example, the bacterial second messengers cyclic-di-GMP and cyclic-di-AMP or by bacterial RNA [72,74–76].

Importantly, type I IFN responses induced by cytosolic microbial DNA, c-di-GMP and c-di-AMP require the molecule STING (also known as MPYS and MITA) [77–80]. STING appears to be both an adapter molecule, which relays signals from most cytosolic DNA sensors to TBK1 and IRF3, and a direct sensor (or co-sensor) of cyclic-di-GMP and cyclic-di-AMP [78,81]. STING is a transmembrane protein component of the ER. In the presence of intracellular DNA, STING rapidly traffics from the ER region through to the Golgi to reside in a distinct perinuclear endosomes, and activates the transcription factors IRF3/7 as well as NF-κB to stimulate type I IFN production [77,82]. Moreover, STING has been shown to activate STAT6-dependent chemokine production and induce selected autophagy (see below) [83,84].

STING-deficient macrophages, cDCs or MEFs did not trigger the production of type I IFN in response to *L. monocytogenes, F. tularensis, M. tuberculosis* or several dsDNA species [54,77,80,85]. Similarly, knockdown of STING inhibited type I IFN production induced by *Streptococcus pyogenes, S. pneumoniae, Chlamydia muridarum,* and *L. pneumophila* [50,68,86,87]. Interestingly, putative loss-of-function mutations in STING have been identified in approximately 3% of Americans, suggesting that individuals carrying these mutations exhibit altered susceptibilities to bacterial infections [88].

Several putative cytosolic DNA sensors have been identified during the last couple of years (Figure 7.2). The first cytosolic host protein indicated to mediate type I IFN responses to bacterial DNA was DAI (DNA-dependent activator of interferon regulatory factor; also called Z-DNA binding protein-1) [17]. However, subsequent studies showed that DAI was not required for IFNα/β responses to cytosolic DNA stimulation or bacterial infections in most cell types [89,90].

Other studies indicated that AT-rich DNA can indirectly stimulate type I IFN responses through RIG-I [15,16]. The authors of these studies showed that synthetic or bacterial DNA could be transcribed into RNA and that this was RNA polymerase III-dependent. In addition, they showed that the transcribed RNA activated RIG-I. RIG-I signals via the adapter molecule MAVS leading to IRF3-dependent induction of type I IFN genes [15,16]. However, this pathway does not operate in all cell types or might be redundant in some cells. For example, the type

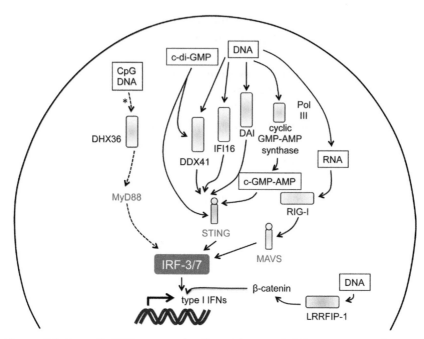

Figure 7.2 *Cytosolic DNA sensors implicated in type I IFN responses to bacterial infections.* Shown are cytosolic proteins that were implicated in sensing bacterial DNA and inducing type I IFN responses, as described in the main text. *This pathway might only operate in pDCs.

I IFN response to poly(dA:dT) was almost completely abolished only in macrophages deficient in both STING and MAVS, and not in macrophages lacking STING alone or MAVS alone [47]. Moreover, many DNA sequences do not activate the RNA polymerase III–RIG-I pathway, and evidence for a role of this pathway in bacterial infections is lacking.

The third identified intracellular DNA sensor able to stimulate type I IFN responses was the human protein IFI16 (gamma–interferon-inducible protein 16) [19]. IFI16 is closely related to AIM2 and member of the AIM2-like receptor and PYHIN families. It was shown that IFI16 binds to viral and synthetic DNA and recruits the adapter STING to stimulate a TBK1–IRF3 pathway leading to type I IFN production. Knockdown of murine p204, considered a structural counterpart of human IFI16, led to a reduction in IFNβ production in response to DNA and *M. tuberculosis* infection in mouse cells [19,85]. Moreover, IFI16 was shown to act as a nuclear pathogen sensor to activate the inflammasome in response to Kaposi sarcoma–associated herpes virus [91]. However, the role of IFI16 as a sensor of bacterial DNA and in the activation of type I IFN responses and/or inflammasomes during bacterial infection is still unknown. The steady state levels of IFI16 in the nucleus might point towards a critical role of IFI16 in infections with DNA viruses rather than bacteria.

Recently, additional molecules including LRRFIP1 [92], DHX9, DHX36 [31], Ku70 [93], LSm14A [94] and DDX41 [18] have been suggested to function as potential cytosolic DNA sensors. For example, LRRFIP1 has been indicated to bind to poly(dA:dT) and poly(dG:dC) dsDNA, and to enhance IFNβ transcription through activation of β-catenin in response to dsDNA as well as *L. monocytogenes* [92]. DHX36 and DHX9 are members of the DExD/H box helicases that have been shown to bind to CpG DNA in the cytosol of pDCs and activate MyD88-dependent signal transduction [31]. Whereas DHX36 was associated with IRF7 activation and subsequent type I IFN production, DHX9 mainly mediated NF-κB-dependent IL-6 and TNFα responses. Moreover, the related DExD/H box helicase DDX41 has been implicated in sensing intracellular synthetic or viral DNA as well as *L. monocytogenes* and in activating STING-dependent type I IFN responses in mDCs and THP1 cells [18]. However, since most of these studies rely on knockdown data only, the final evaluation of the role of these molecules in DNA sensing awaits the generation of respective knockout mice. Moreover, the function of most of these molecules in bacterial infections, especially *in vivo*, is still unknown.

The latest identified putative DNA sensor in the cytosol is the cyclic GMP-AMP synthase [14,95]. In two companion papers the authors demonstrate the role of this enzyme in converting the signal stimulated by cytosolic DNA or DNA virus infection into the endogenous second messenger cyclic GMP-AMP. The authors further demonstrated that this second messenger was then sensed by STING as shown in previous studies of the bacterial molecule cyclic-di-GMP [14,78,95]. Taken together, cyclic GMP-AMP synthase might be the long-sought universal cytosolic DNA sensor. Further studies should evaluate this concept and characterize the function of cyclic GMP-AMP synthase in sensing bacterial infections.

Function of the STING-Dependent DNA Sensing Pathway in Bacterial Infection

Although the upstream receptor molecule involved has not been found yet for most bacterial pathogens and the function of some of the above-mentioned DNA sensors needs to be examined, several studies clearly demonstrate the impact of the DNA sensing and type I IFN inducing pathway in bacterial infection.

L. pneumophila, for example, stimulates type I IFN production in macrophages depending on the bacterial type IV secretion system (T4SS) encoded by the dot/icm genes [63,67]. This response was largely dependent on STING, TBK1 and IRF3 but independent of the TLRs and DAI [63,68,89]. Although sensing of *Legionella* RNA might contribute to this response to some extent [69], and detection of *Legionella* second messengers like cyclic-di-GMP might also play a role, sensing of bacterial DNA appears to be the main mechanism that triggers type I IFN production during *Legionella* infection. First, *Legionella* DNA may be translocated via the T4SS into the host cell cytosol during infection, as has been previously proposed [96,97]. Second, delivery of *Legionella* DNA into the host cell cytosol activated type I IFN production and enzymatic digestion of DNA from *Legionella* extracts inhibited their capabilities to trigger this response [63,68]. Third, type I IFN responses to the bacterium required the critical adapter molecule of the cytosolic DNA (and cyclic-di-GMP) sensing pathways STING [68], as mentioned above. Interestingly, we and others showed that the subsequently produced type I IFNs acted in an autocrine manner to stimulate a cell-autonomous resistance pathway that helps to control *L. pneumophila* in macrophages [67,68,98,99]. Type I IFNs also contribute together with type II IFN to a protective host response in mice intranasally infected with *L. pneumophila* [68].

M. tuberculosis is also recognized by a STING-dependent cytosolic DNA sensing pathway in macrophages. The current concept states that the bacterial ESX-1 secretion system perforates the *Mycobacteria*-containing phagosome membrane and allows leakage of bacterial DNA into the cytosol [70,85]. This mycobacterial DNA recognition activates (i) an IRF3/type I IFN-dependent transcriptional program with detrimental consequences for the host; and (ii) a host protective autophagy pathway. Type I IFN production stimulated by mycobacterial DNA depends on STING and IRF3, and is negatively regulated by the cytosolic DNAse Trex1 [85,100]. Importantly, IRF3$^{-/-}$ and IFNAR$^{-/-}$ mice were more resistant against the bacteria suggesting that the STING/IRF3-mediated cytosolic DNA recognition plays a key role in *M. tuberculosis* pathogenesis [70,85]. The autophagy pathway depends on a STING-mediated recognition of mycobacterial DNA and a subsequent colocalization of ubiquitin as well as the autophagy adapters NDP52 and p62. Importantly, this pathway was independent of IRF3 and type I IFNs, and contributed to mycobacterial restriction in macrophages and mice [84]. These results indicate a new role of the cytosolic DNA sensing pathway in activating autophagy.

Other bacteria that are recognized by the STING-mediated cytosolic DNA sensing pathway in macrophages are streptococci including group A and B streptococci and *S. pneumoniae*. This sensing is dependent on bacterial uptake and phagosomal degradation [50,71,86]. Moreover, phagosomal membrane destruction by bacterial pore-forming toxins is required to allow translocation of bacterial DNA into the host cell cytosol. Although the upstream PRR is still unknown, the streptococcal DNA subsequently activates a type I IFN response via STING, TBK1 and IRF3 [50,71,86]. However, other cell types might have additional mechanisms to sense the bacteria and induce type I IFNs [86].

CONCLUSION

Sensing of microbial nucleic acids is a key mechanism of the host to detect pathogens and to induce immune responses. Bacterial DNA is recognized by TLR9 in endosomes and by AIM2-like receptors, cyclic GMP-AMP synthase and polymerase III/RIG-I in the cytosol. Depending on the receptor involved, bacterial DNA stimulates different host reactions with inflammasome-dependent cytokine production and pyroptosis as well as type I IFN responses possibly being most characteristic. In addition, recent studies indicated a function of the cytosolic DNA sensing pathway in stimulating autophagy. Whereas many inflammasomes are known as important mediators of the host defence in bacterial infections, the role of the AIM2 inflammasome in particular needs further investigation. Moreover, the function of the cytosolic DNA sensors and their type I IFN and autophagy responses during infections with bacteria have been incompletely understood. Given that possibly most bacteria are recognized by those cytosolic DNA sensors, and that type I IFNs as well as autophagy are important regulators of many immunological processes, further research is required.

ACKNOWLEDGMENTS

The authors work is supported by the Deutsche Forschungsgemeinschaft (grants SFB/TR84/A1 and OP 86/7-2).

REFERENCES

[1] Medzhitov R, Janeway Jr CA. Innate immunity: the virtues of a nonclonal system of recognition. Cell 1997;91(3):295–8.

[2] Sorbara MT, Philpott DJ. Peptidoglycan: a critical activator of the mammalian immune system during infection and homeostasis. Immunol Rev 2011;243(1):40–60.

[3] Kato H, Takahasi K, Fujita T. RIG-I-like receptors: cytoplasmic sensors for non-self RNA. Immunol Rev 2011;243(1):91–8.

[4] Kawai T, Akira S. The role of pattern-recognition receptors in innate immunity: update on Toll-like receptors. Nat Immunol 2010;11(5):373–84.

[5] Eltzschig HK, Sitkovsky MV, Robson SC. Purinergic signaling during inflammation. N Engl J Med 2012;367(24):2322–33.

[6] Tokunaga T, Yamamoto H, Shimada S, Abe H, Fukuda T, Fujisawa Y, et al. Antitumor activity of deoxyribonucleic acid fraction from *Mycobacterium bovis* BCG. I. Isolation, physicochemical characterization, and antitumor activity. J Natl Cancer Inst 1984;72(4):955–62.

[7] Sparwasser T, Miethke T, Lipford G, Borschert K, Hacker H, Heeg K, et al. Bacterial DNA causes septic shock. Nature 1997;386(6623):336–7.

[8] Yamamoto S, Yamamoto T, Kataoka T, Kuramoto E, Yano O, Tokunaga T. Unique palindromic sequences in synthetic oligonucleotides are required to induce IFN [correction of INF] and augment IFN-mediated [correction of INF] natural killer activity. J Immunol 1992;148(12):4072–6.

[9] Krieg AM, Yi AK, Matson S, Waldschmidt TJ, Bishop GA, Teasdale R, et al. CpG motifs in bacterial DNA trigger direct B-cell activation. Nature 1995;374(6522):546–9.

[10] Desmet CJ, Ishii KJ. Nucleic acid sensing at the interface between innate and adaptive immunity in vaccination. Nat Rev Immunol 2012;12(7):479–91.

[11] Hornung V, Latz E. Intracellular DNA recognition. Nat Rev Immunol. 2010;10(2):123–30.

[12] Zhang Q, Raoof M, Chen Y, Sumi Y, Sursal T, Junger W, et al. Circulating mitochondrial DAMPs cause inflammatory responses to injury. Nature 2010;464(7285):104–7.

[13] Hemmi H, Takeuchi O, Kawai T, Kaisho T, Sato S, Sanjo H, et al. A Toll-like receptor recognizes bacterial DNA. Nature 2000;408(6813):740–5.

[14] Sun L, Wu J, Du F, Chen X, Chen ZJ. Cyclic GMP-AMP Synthase is a cytosolic DNA sensor that activates the type I interferon pathway. Science 2013;339(6121):786–91.

[15] Ablasser A, Bauernfeind F, Hartmann G, Latz E, Fitzgerald KA, Hornung V. RIG-I-dependent sensing of poly(dA:dT) through the induction of an RNA polymerase III-transcribed RNA intermediate. Nat Immunol 2009;10(10):1065–72.

[16] Chiu YH, Macmillan JB, Chen ZJ. RNA polymerase III detects cytosolic DNA and induces type I interferons through the RIG-I pathway. Cell 2009;138(3):576–91.

[17] Takaoka A, Wang Z, Choi MK, Yanai H, Negishi H, Ban T, et al. DAI (DLM-1/ZBP1) is a cytosolic DNA sensor and an activator of innate immune response. Nature 2007;448(7152):501–5.

[18] Zhang Z, Yuan B, Bao M, Lu N, Kim T, Liu YJ. The helicase DDX41 senses intracellular DNA mediated by the adaptor STING in dendritic cells. Nat Immunol 2011;12(10):959–65.

[19] Unterholzner L, Keating SE, Baran M, Horan KA, Jensen SB, Sharma S, et al. IFI16 is an innate immune sensor for intracellular DNA. Nat Immunol 2010;11(11):997–1004.

[20] Burckstummer T, Baumann C, Bluml S, Dixit E, Durnberger G, Jahn H, et al. An orthogonal proteomic-genomic screen identifies AIM2 as a cytoplasmic DNA sensor for the inflammasome. Nat Immunol 2009;10(3):266–72.

[21] Fernandes-Alnemri T, Yu JW, Datta P, Wu J, Alnemri ES. AIM2 activates the inflammasome and cell death in response to cytoplasmic DNA. Nature 2009;458(7237):509–13.

[22] Hornung V, Ablasser A, Charrel-Dennis M, Bauernfeind F, Horvath G, Caffrey DR, et al. AIM2 recognizes cytosolic dsDNA and forms a caspase-1-activating inflammasome with ASC. Nature 2009;458(7237):514–8.

[23] Roberts TL, Idris A, Dunn JA, Kelly GM, Burnton CM, Hodgson S, et al. HIN-200 proteins regulate caspase activation in response to foreign cytoplasmic DNA. Science 2009;323(5917):1057–60.

[24] Hacker H, Vabulas RM, Takeuchi O, Hoshino K, Akira S, Wagner H. Immune cell activation by bacterial CpG-DNA through myeloid differentiation marker 88 and tumor necrosis factor receptor-associated factor (TRAF)6. J Exp Med 2000;192(4):595–600.

[25] Haas T, Metzger J, Schmitz F, Heit A, Muller T, Latz E, et al. The DNA sugar backbone 2' deoxyribose determines toll-like receptor 9 activation. Immunity 2008;28(3):315–23.

[26] Bauer S, Pigisch S, Hangel D, Kaufmann A, Hamm S. Recognition of nucleic acid and nucleic acid analogs by Toll-like receptors 7, 8 and 9. Immunobiology 2008;213(3-4):315–28.

[27] Latz E, Schoenemeyer A, Visintin A, Fitzgerald KA, Monks BG, Knetter CF, et al. TLR9 signals after translocating from the ER to CpG DNA in the lysosome. Nat Immunol 2004;5(2):190–8.

[28] Lund J, Sato A, Akira S, Medzhitov R, Iwasaki A. Toll-like receptor 9-mediated recognition of herpes simplex virus-2 by plasmacytoid dendritic cells. J Exp Med 2003;198(3):513–20.

[29] Krug A, Luker GD, Barchet W, Leib DA, Akira S, Colonna M. Herpes simplex virus type 1 activates murine natural interferon-producing cells through toll-like receptor 9. Blood 2004;103(4):1433–7.

[30] Hemmi H, Kaisho T, Takeda K, Akira S. The roles of Toll-like receptor 9, MyD88, and DNA-dependent protein kinase catalytic subunit in the effects of two distinct CpG DNAs on dendritic cell subsets. J Immunol 2003;170(6):3059–64.

[31] Kim T, Pazhoor S, Bao M, Zhang Z, Hanabuchi S, Facchinetti V, et al. Aspartate-glutamate-alanine-histidine box motif (DEAH)/RNA helicase A helicases sense microbial DNA in human plasmacytoid dendritic cells. Proc Natl Acad Sci USA 2010;107(34):15181–6.

[32] Bafica A, Scanga CA, Feng CG, Leifer C, Cheever A, Sher A. TLR9 regulates Th1 responses and cooperates with TLR2 in mediating optimal resistance to *Mycobacterium tuberculosis*. J Exp Med 2005;202(12):1715–24.

[33] Albiger B, Dahlberg S, Sandgren A, Wartha F, Beiter K, Katsuragi H, et al. Toll-like receptor 9 acts at an early stage in host defence against pneumococcal infection. Cell Microbiol 2007;9(3):633–44.

[34] Sjolinder H, Mogensen TH, Kilian M, Jonsson AB, Paludan SR. Important role for Toll-like receptor 9 in host defense against meningococcal sepsis. Infect Immun 2008;76(11):5421–8.

[35] Sanders MS, van Well GT, Ouburg S, Lundberg PS, van Furth AM, Morre SA. Single nucleotide polymorphisms in TLR9 are highly associated with susceptibility to bacterial meningitis in children. Clin Infect Dis 2011;52(4):475–80.

[36] Bhan U, Lukacs NW, Osterholzer JJ, Newstead MW, Zeng X, Moore TA, et al. TLR9 is required for protective innate immunity in Gram-negative bacterial pneumonia: role of dendritic cells. J Immunol 2007;179(6):3937–46.

[37] Bhan U, Trujillo G, Lyn-Kew K, Newstead MW, Zeng X, Hogaboam CM, et al. Toll-like receptor 9 regulates the lung macrophage phenotype and host immunity in murine pneumonia caused by *Legionella pneumophila*. Infect Immun 2008;76(7):2895–904.

[38] Duggan JM, You D, Cleaver JO, Larson DT, Garza RJ, Guzman Pruneda FA, et al. Synergistic interactions of TLR2/6 and TLR9 induce a high level of resistance to lung infection in mice. J Immunol 2011;186(10):5916–26.

[39] Parker D, Prince A. *Staphylococcus aureus* induces type I IFN signaling in dendritic cells via TLR9. J Immunol 2012;189(8):4040–6.

[40] Rachmilewitz D, Katakura K, Karmeli F, Hayashi T, Reinus C, Rudensky B, et al. Toll-like receptor 9 signaling mediates the anti-inflammatory effects of probiotics in murine experimental colitis. Gastroenterology 2004;126(2):520–8.

[41] Katakura K, Lee J, Rachmilewitz D, Li G, Eckmann L, Raz E. Toll-like receptor 9-induced type I IFN protects mice from experimental colitis. J Clin Invest 2005;115(3):695–702.

[42] Hall JA, Bouladoux N, Sun CM, Wohlfert EA, Blank RB, Zhu Q, et al. Commensal DNA limits regulatory T cell conversion and is a natural adjuvant of intestinal immune responses. Immunity 2008;29(4):637–49.

[43] Heimesaat MM, Nogai A, Bereswill S, Plickert R, Fischer A, Loddenkemper C, et al. MyD88/TLR9 mediated immunopathology and gut microbiota dynamics in a novel murine model of intestinal graft-versus-host disease. Gut 2010;59(8):1079–87.

[44] Plitas G, Burt BM, Nguyen HM, Bamboat ZM, DeMatteo RP. Toll-like receptor 9 inhibition reduces mortality in polymicrobial sepsis. J Exp Med 2008;205(6):1277–83.

[45] Schattgen SA, Fitzgerald KA. The PYHIN protein family as mediators of host defenses. Immunol Rev 2011;243(1):109–18.

[46] Keating SE, Baran M, Bowie AG. Cytosolic DNA sensors regulating type I interferon induction. Trends Immunol 2011;32(12):574–81.

[47] Brunette RL, Young JM, Whitley DG, Brodsky IE, Malik HS, Stetson DB. Extensive evolutionary and functional diversity among mammalian AIM2-like receptors. J Exp Med 2012;209(11):1969–83.

[48] Schroder K, Tschopp J. The inflammasomes. Cell 2010;140(6):821–32.

[49] Jin T, Perry A, Jiang J, Smith P, Curry JA, Unterholzner L, et al. Structures of the HIN domain:DNA complexes reveal ligand binding and activation mechanisms of the AIM2 inflammasome and IFI16 receptor. Immunity 2012;36(4):561–71.

[50] Koppe U, Hogner K, Doehn JM, Muller HC, Witzenrath M, Gutbier B, et al. *Streptococcus pneumoniae* stimulates a STING- and IFN regulatory factor 3-dependent type I IFN production in macrophages, which regulates RANTES production in macrophages, cocultured alveolar epithelial cells, and mouse lungs. J Immunol 2012;188(2):811–7.

[51] Fang R, Tsuchiya K, Kawamura I, Shen Y, Hara H, Sakai S, et al. Critical roles of ASC inflammasomes in caspase-1 activation and host innate resistance to *Streptococcus pneumoniae* infection. J Immunol 2011;187(9):4890–9.

[52] Rathinam VA, Jiang Z, Waggoner SN, Sharma S, Cole LE, Waggoner L, et al. The AIM2 inflammasome is essential for host defense against cytosolic bacteria and DNA viruses. Nat Immunol 2010;11(5):395–402.

[53] Fernandes-Alnemri T, Yu JW, Juliana C, Solorzano L, Kang S, Wu J, et al. The AIM2 inflammasome is critical for innate immunity to *Francisella tularensis*. Nat Immunol 2010;11(5):385–93.

[54] Jones JW, Kayagaki N, Broz P, Henry T, Newton K, O'Rourke K, et al. Absent in melanoma 2 is required for innate immune recognition of *Francisella tularensis*. Proc Natl Acad Sci USA 2010;107(21):9771–6.

[55] Kim S, Bauernfeind F, Ablasser A, Hartmann G, Fitzgerald KA, Latz E, et al. *Listeria monocytogenes* is sensed by the NLRP3 and AIM2 inflammasome. Eur J Immunol 2010;40(6):1545–51.

[56] Sauer JD, Witte CE, Zemansky J, Hanson B, Lauer P, Portnoy DA. *Listeria monocytogenes* triggers AIM2-mediated pyroptosis upon infrequent bacteriolysis in the macrophage cytosol. Cell Host Microbe 2010;7(5):412–9.

[57] Saiga H, Kitada S, Shimada Y, Kamiyama N, Okuyama M, Makino M, et al. Critical role of AIM2 in *Mycobacterium tuberculosis* infection. Int Immunol 2012;24(10):637–44.

[58] Eitel J, Suttorp N, Opitz B. Innate immune recognition and inflammasome activation in *Listeria monocytogenes* infection. Front Microbiol 2010;1:149.

[59] Tsuchiya K, Hara H, Kawamura I, Nomura T, Yamamoto T, Daim S, et al. Involvement of absent in melanoma 2 in inflammasome activation in macrophages infected with *Listeria monocytogenes*. J Immunol 2010;185(2):1186–95.

[60] Ulland TK, Buchan BW, Ketterer MR, Fernandes-Alnemri T, Meyerholz DK, Apicella MA, et al. Cutting edge: mutation of *Francisella tularensis* mviN leads to increased macrophage absent in melanoma 2 inflammasome activation and a loss of virulence. J Immunol 2010;185(5):2670–4.

[61] Yasuda K, Ogawa Y, Yamane I, Nishikawa M, Takakura Y. Macrophage activation by a DNA/cationic liposome complex requires endosomal acidification and TLR9-dependent and -independent pathways. J Leukoc Biol 2005;77(1):71–9.

[62] Ishii KJ, Coban C, Kato H, Takahashi K, Torii Y, Takeshita F, et al. A Toll-like receptor-independent antiviral response induced by double-stranded B-form DNA. Nat Immunol 2006;7(1):40–8.

[63] Stetson DB, Medzhitov R. Recognition of cytosolic DNA activates an IRF3-dependent innate immune response. Immunity 2006;24(1):93–103.

[64] O'Riordan M, Yi CH, Gonzales R, Lee KD, Portnoy DA. Innate recognition of bacteria by a macrophage cytosolic surveillance pathway. Proc Natl Acad Sci USA 2002;99(21):13861–6.

[65] Stockinger S, Materna T, Stoiber D, Bayr L, Steinborn R, Kolbe T, et al. Production of type I IFN sensitizes macrophages to cell death induced by *Listeria monocytogenes*. J Immunol 2002;169(11):6522–9.

[66] Hess CB, Niesel DW, Cho YJ, Klimpel GR. Bacterial invasion of fibroblasts induces interferon production. J Immunol 1987;138(11):3949–53.

[67] Opitz B, Vinzing M, van Laak V, Schmeck B, Heine G, Gunther S, et al. *Legionella pneumophila* induces IFNbeta in lung epithelial cells via IPS-1 and IRF3, which also control bacterial replication. J Biol Chem 2006;281(47):36173–9.

[68] Lippmann J, Muller HC, Naujoks J, Tabeling C, Shin S, Witzenrath M, et al. Dissection of a type I interferon pathway in controlling bacterial intracellular infection in mice. Cell Microbiol 2011;13(11):1668–82.

[69] Monroe KM, McWhirter SM, Vance RE. Identification of host cytosolic sensors and bacterial factors regulating the type I interferon response to *Legionella pneumophila*. PLoS Pathog 2009;5(11):e1000665.

[70] Stanley SA, Johndrow JE, Manzanillo P, Cox JS. The Type I IFN response to infection with *Mycobacterium tuberculosis* requires ESX-1-mediated secretion and contributes to pathogenesis. J Immunol 2007;178(5):3143–52.

[71] Charrel-Dennis M, Latz E, Halmen KA, Trieu-Cuot P, Fitzgerald KA, Kasper DL, et al. TLR-independent type I interferon induction in response to an extracellular bacterial pathogen via intracellular recognition of its DNA. Cell Host Microbe 2008;4(6):543–54.

[72] Mancuso G, Gambuzza M, Midiri A, Biondo C, Papasergi S, Akira S, et al. Bacterial recognition by TLR7 in the lysosomes of conventional dendritic cells. Nat Immunol 2009;10(6):587–94.

[73] Stockinger S, Reutterer B, Schaljo B, Schellack C, Brunner S, Materna T, et al. IFN regulatory factor 3-dependent induction of type I IFNs by intracellular bacteria is mediated by a TLR- and Nod2-independent mechanism. J Immunol 2004;173(12):7416–25.

[74] Woodward JJ, Iavarone AT, Portnoy DA. c-di-AMP secreted by intracellular *Listeria monocytogenes* activates a host type I interferon response. Science 2010;328(5986):1703–5.

[75] Sander LE, Davis MJ, Boekschoten MV, Amsen D, Dascher CC, Ryffel B, et al. Detection of prokaryotic mRNA signifies microbial viability and promotes immunity. Nature 2011;474(7351):385–9.

[76] McWhirter SM, Barbalat R, Monroe KM, Fontana MF, Hyodo M, Joncker NT, et al. A host type I interferon response is induced by cytosolic sensing of the bacterial second messenger cyclic-di-GMP. J Exp Med 2009;206(9):1899–911.

[77] Ishikawa H, Ma Z, Barber GN. STING regulates intracellular DNA-mediated, type I interferon-dependent innate immunity. Nature 2009;461(7265):788–92.

[78] Burdette DL, Monroe KM, Sotelo-Troha K, Iwig JS, Eckert B, Hyodo M, et al. STING is a direct innate immune sensor of cyclic di-GMP. Nature 2011;478(7370):515–8.

[79] Sauer JD, Sotelo-Troha K, von Moltke J, Monroe KM, Rae CS, Brubaker SW, et al. The N-ethyl-N-nitrosourea-induced Goldenticket mouse mutant reveals an essential function of Sting in the in vivo interferon response to *Listeria monocytogenes* and cyclic dinucleotides. Infect Immun 2011;79(2):688–94.

[80] Jin L, Hill KK, Filak H, Mogan J, Knowles H, Zhang B, et al. MPYS is required for IFN response factor 3 activation and type I IFN production in the response of cultured phagocytes to bacterial second messengers cyclic-di-AMP and cyclic-di-GMP. J Immunol 2011;187(5):2595–601.

[81] Parvatiyar K, Zhang Z, Teles RM, Ouyang S, Jiang Y, Iyer SS, et al. The helicase DDX41 recognizes the bacterial secondary messengers cyclic di-GMP and cyclic di-AMP to activate a type I interferon immune response. Nat Immunol 2012;13(12):1155–61.

[82] Ishikawa H, Barber GN. STING is an endoplasmic reticulum adaptor that facilitates innate immune signalling. Nature 2008;455(7213):674–8.

[83] Chen H, Sun H, You F, Sun W, Zhou X, Chen L, et al. Activation of STAT6 by STING is critical for antiviral innate immunity. Cell 2011;147(2):436–46.

[84] Watson RO, Manzanillo PS, Cox JS. Extracellular *M. tuberculosis* DNA targets bacteria for autophagy by activating the host DNA-sensing pathway. Cell 2012;150(4):803–15.

[85] Manzanillo PS, Shiloh MU, Portnoy DA, Cox JS. *Mycobacterium tuberculosis* activates the DNA-dependent cytosolic surveillance pathway within macrophages. Cell Host Microbe 2012;11(5):469–80.

[86] Gratz N, Hartweger H, Matt U, Kratochvill F, Janos M, Sigel S, et al. Type I interferon production induced by *Streptococcus pyogenes*-derived nucleic acids is required for host protection. PLoS Pathog 2011;7(5):e1001345.

[87] Prantner D, Darville T, Nagarajan UM. Stimulator of IFN gene is critical for induction of IFN-beta during *Chlamydia muridarum* infection. J Immunol 2010;184(5):2551–60.

[88] Jin L, Xu LG, Yang IV, Davidson EJ, Schwartz DA, Wurfel MM, et al. Identification and characterization of a loss-of-function human MPYS variant. Genes Immun 2011;12(4):263–9.

[89] Lippmann J, Rothenburg S, Deigendesch N, Eitel J, Meixenberger K, van Laak V, et al. IFNbeta responses induced by intracellular bacteria or cytosolic DNA in different human cells do not require ZBP1 (DLM-1/DAI). Cell Microbiol 2008;10(12):2579–88.

[90] Ishii KJ, Kawagoe T, Koyama S, Matsui K, Kumar H, Kawai T, et al. TANK-binding kinase-1 delineates innate and adaptive immune responses to DNA vaccines. Nature 2008;451(7179):725–9.

[91] Kerur N, Veettil MV, Sharma-Walia N, Bottero V, Sadagopan S, Otageri P, et al. IFI16 acts as a nuclear pathogen sensor to induce the inflammasome in response to Kaposi sarcoma-associated herpesvirus infection. Cell Host Microbe 2011;9(5):363–75.

[92] Yang P, An H, Liu X, Wen M, Zheng Y, Rui Y, et al. The cytosolic nucleic acid sensor LRRFIP1 mediates the production of type I interferon via a beta-catenin-dependent pathway. Nat Immunol 2010;11(6):487–94.

[93] Zhang X, Brann TW, Zhou M, Yang J, Oguariri RM, Lidie KB, et al. Cutting edge: Ku70 is a novel cytosolic DNA sensor that induces type III rather than type I IFN. J Immunol 2011;186(8):4541–5.

[94] Li Y, Chen R, Zhou Q, Xu Z, Li C, Wang S, et al. LSm14A is a processing body-associated sensor of viral nucleic acids that initiates cellular antiviral response in the early phase of viral infection. Proc Natl Acad Sci USA 2012;109(29):11770–5.

[95] Wu J, Sun L, Chen X, Du F, Shi H, Chen C, et al. Cyclic GMP-AMP is an endogenous second messenger in innate immune signaling by cytosolic DNA. Science 2013;339(6121):826–30.

[96] Segal G, Purcell M, Shuman HA. Host cell killing and bacterial conjugation require overlapping sets of genes within a 22-kb region of the *Legionella pneumophila* genome. Proc Natl Acad Sci USA 1998;95(4):1669–74.

[97] Vogel JP, Andrews HL, Wong SK, Isberg RR. Conjugative transfer by the virulence system of *Legionella pneumophila*. Science 1998;279(5352):873–6.

[98] Coers J, Vance RE, Fontana MF, Dietrich WF. Restriction of *Legionella pneumophila* growth in macrophages requires the concerted action of cytokine and Naip5/Ipaf signalling pathways. Cell Microbiol 2007;9(10):2344–57.

[99] Schiavoni G, Mauri C, Carlei D, Belardelli F, Pastoris MC, Proietti E, et al. IFN protects permissive macrophages from *Legionella pneumophila* infection through an IFN-gamma-independent pathway. J Immunol 2004;173(2):1266–75.

[100] Stetson DB, Ko JS, Heidmann T, Medzhitov R. Trex1 prevents cell-intrinsic initiation of autoimmunity. Cell 2008;134(4):587–98.

CHAPTER 8

Viral Infections and the DNA Sensing Pathway: Lessons from Herpesviruses and Beyond

Søren R. Paludan[1,2] and Andrew G. Bowie[3]
[1]Department of Biomedicine
[2]Aarhus Research Center for Innate Immunology, University of Aarhus, The Bartholin Building, Aarhus C, Denmark
[3]School of Biochemistry and Immunology, Trinity Biomedical Sciences Institute, Trinity College Dublin, Ireland

INTRODUCTION

The immune system serves the purpose of protecting the organism against infections, and is also intimately involved in the subsequent healing process including clearance of dead cells and cellular debris [1]. The first line of defense is mediated by the innate immune system, which is characterized by the ability to rapidly respond to infection and tissue damage, but with limited specificity and no in-built memory [1]. The innate immune system utilizes a limited number of germline-encoded receptors, called pattern recognition receptors (PRRs), to recognize microbial and host molecules and initiate intracellular signaling, eventually inducing expression of genes responsible for the function of the given PRR. The first identified – and best characterized – class of PRRs is the Toll-like receptors (TLR)s [2–4], which are expressed on the cell surface and in endosomal compartments. More recently, intracellular PRRs have been discovered [5–7], which sense viruses and intracellular bacteria and together constitute what has been termed the 'cytosolic surveillance pathway'.

In recent years it has emerged that DNA is a potent stimulator of innate immune responses, and is involved in both early activation of defense against infections [8,9], and subsequent bridging to activation of adaptive immune responses [10,11]. Figure 8.1 summarizes the current knowledge on DNA-activated responses related to the innate immune system [8,9,12–19]. Besides infections, the immuno-stimulatory activity of DNA is involved in the pathogenesis of some autoinflammatory diseases,

Biological DNA Sensor.
DOI: http://dx.doi.org/10.1016/B978-0-12-404732-7.00008-3

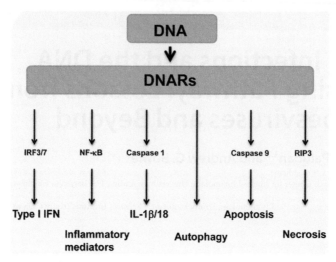

Figure 8.1 *Cellular functions stimulated by DNA.* Intracellular DNA is recognized by DNA sensors leading to activation of multiple pathways. The best characterized DNA-stimulated pathway is the one leading to activation of IRFs and induction of IFNs. Other well-characterized pathways activated by DNA recognition are the inflammatory NF-κB and inflammasome pathways, which stimulate expression of inflammatory genes and cleavage of pro-IL-1β/IL-18, respectively. More recently, it has emerged that intracellular DNA also stimulates autophagy and different types of cell death.

most notably systemic lupus erythematosus (SLE) and Aicardi–Goutières syndrome [20,21]. DNA is sensed by a heterogeneous group of PRRs, and the current paradigm is that these sensors normally distinguish self-DNA from nonself-DNA based on cytosolic versus nuclear localization respectively, although this view has been challenged by recent findings [22–24]. Since the identification of the first endosomal and cytoplasmic DNA sensors this field has experienced an immense expansion [25,26], but although many proteins are now implicated as innate DNA sensors, our current knowledge still falls short of explaining key phenomena, which are essential for understanding protective *versus* pathological immune responses during infection. Much of the current knowledge on innate DNA recognition is based on work with herpesviruses, which are DNA viruses that replicate in the nucleus of host cells and have the unique ability to establish latent infections in specific cell types. In fact, almost every DNA sensor proposed to date has been shown to mediate herpesvirus-induced cytokine and interferon responses. The family of herpesviruses can be subdivided into α-, β-, and γ-herpesviruses and includes several human pathogenic viruses including herpes simplex virus (HSV, α), cytomegalovirus (CMV, β), and Epstein–Barr virus (EBV, γ). Herpesviruses have large double-stranded DNA genomes (120–220 kb), which are protected by a protein capsid and a lipid bilayer in the intact virus particle, and are released into the nucleus in productively infected cells [27]. The replication cycle occurs in three separate steps called immediate early, early and late, each with their specific set of viral genes being produced [28].

At this stage more than 10 DNA sensors have been proposed, and the field is at a stage where we need to sort out the functions and modes of actions of DNA sensors in defense and disease. For instance: Are all proposed DNA receptors (DNARs) *bona fide* PRRs? Is there a division-of-labor between the DNARs, for example do they function in different cell types, or respond to different types of DNA? Is there interaction between the different DNA sensing systems? Do all DNARs detect DNA in the same way? Where in the cell do the DNARs detect DNA and how is downstream signaling transduced? What is the role of DNARs in protective immune responses? In this chapter we review the literature on DNA sensing and viral evasion during herpesvirus infections, and try to identify outstanding questions that need to be addressed. Finally, we extend the discussion beyond herpesviruses to intracellular bacteria, parasites, and autoinflammatory diseases.

INNATE IMMUNOLOGICAL RECOGNITION OF HERPESVIRUS DNA

Today there is evidence for innate immunological recognition of DNA in endosomes, the cytoplasm, and the nucleus [8,9,22,29]. However, there is only limited knowledge on how microbial DNA actually gets exposed to DNA sensors. Herpesviruses enter into cells via fusion of viral and cellular membranes on the cell surface. From here, the DNA-containing capsid is delivered to nuclear pores along microtubules, and DNA is eventually released into the nucleus to initiate productive infection. Concerning how herpesvirus DNA gets exposed to TLR9 in endosomes, it has been proposed that uptake of viruses by phagocytosis delivers the particle into the degrading environment of late endosomes and lysosomes, which may break down the capsid and expose the DNA [30]. There are also data to suggest that cytosolic HSV and CMV DNA induce autophagy [15,16], and that the autophagy pathway delivers cytosolic cargo to endosomes for recognition by TLRs [31]. There is much less knowledge on how the cell may sense viral DNA as foreign in the nucleus. Since the nucleus is the location of host DNA, a machinery to distinguish self from non-self DNA needs to be more sophisticated in this organelle than in other locations where DNA is not present under normal circumstances. Possible signals of danger in the nucleus could be free DNA ends, non-histonylated DNA, the presence of DNA in areas of the nucleus normally not occupied by DNA, and/or, finally, extensive RNA transcription from viral DNA by RNA polymerase III as discussed below (Figure 8.2). The vast majority of proposed DNA sensors localize to the cytoplasm, but it was only recently described that herpesvirus capsids get ubiquitinated in the cytoplasm, and subsequently degraded by the proteasome, which releases the genomic DNA to the cytoplasm for innate recognition [32]. With this knowledge it will be interesting to learn if the cell is in possession of sensors targeting viral capsids for degradation and also whether cytosolic DNA recognition occurs in random locations in the cytoplasm or preferentially at recognition hotspots, a candidate for which could be around the proteasome. Collectively, depending on the subcellular localization of herpesviruses, cells use different mechanisms to present the foreign DNA to PRRs.

Most work on DNA recognition during herpesvirus infection has relied on the assumption that the pathogen-associated molecular pattern (PAMP) driving these responses is exclusively B-form DNA. Herpesviruses have their genome as linear dsDNA that becomes circular in the nucleus. Thus, although it seems likely that dsDNA is the main herpesvirus DNA PAMP, there are reasons to believe that several DNA PAMPs are present at different stages of infection (Figure 8.2). For example, the cytosolic 5′ exonuclease Trex1 could digest 5′ ends of B-form DNA to now contain both regions of ssDNA and dsDNA. This could lead to recognition of one piece of DNA by more than one type of DNA sensor. There are data in the literature demonstrating that dsDNA is indeed targeted by Trex1 [33]. Moreover, the linear genomic DNA leads to exposure of free DNA ends, which are not normally present in the cytoplasm – and a signature of damage in the nucleus. Finally, recent work on the malaria parasite *Plasmodium* has shown that although its genomic DNA is double-stranded, it stimulates innate immune responses at least partly via AT-rich stem–loop structures through an as yet unidentified DNA sensor [34]. Thus, although herpesviruses are dsDNA viruses and B-form DNA is likely to represent the main herpes DNA PAMP, we still do not have a clear picture of what DNA forms actually mediate the stimulation of innate immune responses to herpesvirus infections.

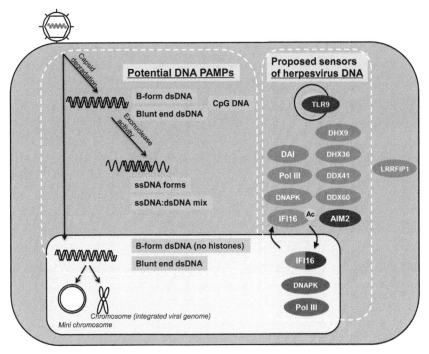

Figure 8.2 *Herpesvirus DNA PAMPs and PRRs.* The herpesvirus genome is in the form of dsDNA, which is believed to be the prime herpesvirus PAMP in the cytoplasm, and potentially also in the nucleus. The incoming genomic DNA is linear, and free DNA ends can potentially also be sensed by host PRRs. Finally, genomic DNA partially digested by exonucleases, and hence containing stretches of ssDNA, could also contribute to stimulation of innate responses during herpesvirus infection. It is currently not known if nuclear herpesvirus DNA in the form of episomes or integrated into the host genome is detected by the innate immune system. DNA sensors are well-established to detect DNA in the cytoplasm, and there is also evidence for innate DNA recognition in the nucleus. All proposed DNA sensors – except LRRFIP1 – have been suggested to recognize herpesvirus DNA. Purple, DNA sensors stimulating IFN expression upon DNA recognition in endosomes. Red, DNA sensors stimulating IFN expression upon DNA recognition in the cytoplasm. Green, DNA sensors potentially stimulating IFN expression upon DNA recognition in the nucleus. Blue, DNA sensors stimulating inflammasome activation. Please see color plate at the back of the book.

DNA SENSORS INVOLVED IN RECOGNITION OF HERPESVIRUS DNA

At this stage more than 10 proteins have been proposed to be intracellular DNA sensors. All of these, except one – LRRFIP1 [35] – have been demonstrated to have impact on innate immune response to herpesvirus infections in different systems (Table 8.1).

Table 8.1 Proposed DNA sensors and their involvement in recognition of herpesviruses

DNA Sensor	Virus	Cell Type	Response	Evidence for DNA Binding	References
TLR9	HSV-1/2, VZV, M/ HCMV, EBV, MHV68	pDCs*	Type I IFN	Interaction between TLR9-Fc fusion protein and CpG DNA in α-screen	[29,40–48]
DAI/ ZBP1	HSV-1, HCMV	Fibroblasts, astrocytes, microglia	IFN-β; TNF-α, IL-6	FRET between B-DNA and DAI; Pull-down of DAI +/− B-DNA competition	[26,57,59]
AIM2	MCMV	Macrophages	IL-1β, IL-18	Affinity purification of Myc-tagged AIM2 with dsDNA-coupled beads; Interaction between rAIM2 and dsDNA in α-screen	[65,66,69]
IFI16/p204	HSV-1, HCMV	Macrophages, endothelial cells	IFN-β, IL-1β	Co-precipitation of cytosolic IFI16 with dsDNA-couples beads; Interaction between rIFI16 and dsDNA in α-screen	[32,58]
Pol III	HSV-1, EBV	EBV+ B cell, macrophage cell line	IFN-β	Production of IFN-inducing RNA transcripts sensitive to Pol III inhibition; Purified core RNA Pol III complex produced IFN-inducing RNAs	[60,61]
DNA-PK	HSV-1, HSV-2	HEK293T, MEFs	IFN-λ1, IFN-β, IL-6	Co-precipitation of cytosolic Ku70, Ku80, and DNA-PKsc with dsDNA-couples beads	[77,78]
LRRFIP1	No herpesviruses reported (*L. monocytogenes*)	Mouse primary peritoneal macrophages	IFN-β	Co-precipitation of FLAG-tagged LRRFIP1 with synthetic dsDNA	[35]
DHX9	HSV-1	pDCs	TNF-α	Co-precipitation of DHX9 with biotin-CpG-B	[81]
DHX36	HSV-1	pDCs	IFN-α	Co-precipitation of DHX36 with biotin-CpG-A	[81]
DDX41	HSV-1	DCs	IFN-α, β	Co-precipitation of B- and Z-form DNA with HA-tagged DDX41	[73]
DDX60	HSV-1	HeLa cells	IFN-β, CXCL10	dsDNA-dependent shift of purified His-tagged DDX60 migration in gel shift assay	[84]

*In humans TLR9 is expressed mainly in pDCs. Much of the information on TLR9 is based on work with murine macrophages, in which TLR9 is expressed.

TLR9. The first-described cellular sensor of DNA was TLR9 [25], which is expressed preferentially in plasmacytoid dendritic cells (DC)s [36]. Active TLR9 is localized in endosomes, is activated by a pathway involving proteolytic cleavage of the ectodomain, and recognizes mainly unmethylated CpG-rich DNA [25,37–39]. Some herpesviruses have CG genome content as high as 70%; TLR9 has been demonstrated to sense infections with all classes of herpesviruses [29,40–48], and is the only DNA sensor that has been studied extensively in *in vivo* models of herpesvirus infections. In models for α-herpesvirus infections, TLR9 has generally been found to play only minor roles [42]. In the murine ocular HSV-1 infection model, TLR9 has, however, been reported to play essential roles in both antiviral responses and immunopathology [49,50]. Finally, one report provides data on an essential role for synergy between TLR2 and TLR9 in activation of antiviral defense in the brain during HSV-1 and -2 infections [51]. This finding suggests that the concept of joint action of PRRs in activation of antiviral defense is important and needs more attention from the research community.

Concerning β-herpesviruses, TLR9$^{-/-}$ mice exhibit reduced type I interferon (IFN) expression during MCMV infection [52,53]. The TLR9 defect leads to impaired antiviral response due to reduced natural killer cell-derived IFN-γ production [52], whereas the reduced type I IFN expression seems to be compensated for by other mechanisms [54]. Interestingly, the phenotype of TLR9$^{-/-}$ mice is amplified by additional deficiency of TLR7, again emphasizing the role of synergy between PRRs in activation of innate responses *in vivo* [55].

For γ-herpesviruses, the MHV68 model has demonstrated an essential role for TLR9 in MHV68 infection. Interestingly, the TLR9$^{-/-}$ mice exhibited elevated splenic viral load during both acute and latent infection, hence providing direct evidence for a role for TLR9 in both acute and chronic herpesvirus infection [48]. All together, the data available suggest a clear but minor role for TLR9 alone in defense against herpesvirus infections. This is supported by data from humans, where individuals unresponsive to TLR9 stimulation due to defective expression of the TLR9 signaling molecule IRAK4 do not experience elevated susceptibility to herpesvirus infections [56].

DAI. DAI/ZBP1 was reported as the first innate sensor of cytosolic DNA [26]. It was shown that the murine fibroblast cell line L929 induces IFN-β in response to HSV-1 infection in a manner partially dependent on DAI/ZBP1. Subsequent work has confirmed a role for DAI/ZBP1 in induction of the type I IFN response during HCMV infection in human foreskin fibroblasts through a pathway dependent on DEAD box protein 3 (DDX3) and stimulator of IFN genes (STING) [57]. In contrast to this, it has been difficult to find essential roles for DAI/ZBP1 in DNA sensing in leukocytes [10,58]. Using siRNA knockdown, it has been reported that HSV-1-induced expression of IL-6 and TNF-α is dependent on DAI in the neuroglial cells microglia and astrocytes [59]. Therefore, based on the published work on DAI/ZBP1 it seems likely that this DNA sensor is mainly involved in DNA sensing in cells of the fibroblast lineage and it seems plausible that essential roles for DAI/ZBP1 can be revealed in model systems where these cell types are central to prevention or development of disease. Importantly, it was recently reported that DAI/ZBP1 stimulates necrosis in fibroblasts during MCMV infection through a RIP3-dependent pathway and this activates an antiviral response [18,19].

Pol III. RNA polymerase III has been reported to use herpesvirus DNA as template to produce 5′-triphosphate RNAs which induce type I IFN through RIG-I [60,61]. Subsequent studies in primary human macrophages have not been able to confirm a role for the Pol III system in sensing of HSV-1 [62], and we therefore still do not have a clear picture of the role of Pol III in DNA sensing, IFN induction, and host restriction of herpesviruses. Despite this, it is interesting that Pol III is well described to play central roles in viral gene expression in both the lytic cycle and latent stages of herpesvirus infection [63,64]. With this in mind, it is therefore tempting to speculate that Pol III does not act as a prime sensor of incoming DNA in the infecting virion, but rather as a cellular sensor of the virus at different stages of the viral replication cycle. In such a scenario, Pol III is a sensor of herpesvirus DNA mainly in permissive cells. Another issue we still remain to address is the exact subcellular site of DNA sensing by Pol III. Is Pol III a nuclear or cytoplasmic innate sensor of DNA?

Aim2. In addition to inducing IFN expression, cytosolic DNA can also trigger inflammasome activation and maturation of pro-IL-1β and pro-IL-18 [14]. AIM2 is a PYHIN protein that localizes to the cytoplasm and activates an ASC-caspase 1 inflammasome in response to dsDNA and poly(dA:dT) [65–68]. Macrophages from AIM2$^{-/-}$ mice exhibit impaired production of IL-1β after MCMV infection, whereas the IL-1β response to HSV-1 infection is unaffected [69]. Likewise, another α-herpesvirus, varicella–zoster virus, also stimulates inflammasome activation, independent of AIM2 in THP-1 monocytes but dependent on NLRP3 [70]. Thus, although both α- and β-herpesviruses induce type I

IFN expression via intracellular DNA sensors, only β-herpesviruses activate the AIM2 inflammasome. $AIM2^{-/-}$ mice challenged with MCMV responded with reduced serum IL-18 levels, impaired splenic IFN-γ production and elevated viral load in the spleen [69]. In the future, it will be interesting to learn how α-herpesviruses evade activation of the AIM2 inflammasome and how the different DNA-sensing systems interact to achieve specificity in the innate antiviral responses *in vivo*. For instance, does AIM2 have preferences for other DNA species than the IFN-inducing DNA sensors? Do IFN-inducing and inflammasome-activating DNA sensors recognize DNA in the same cytoplasmic compartments or does the subcellular site of DNA accumulation have an impact on the nature of the DNA-stimulated innate response?

IFI16. Work by Unterholzner *et al.* demonstrated that human IFI16, which like AIM2 is a PYHIN family protein, binds dsDNA and induces STING-dependent IFN-β responses [58]. Unlike many proposed DNA sensors, IFI16 does have distinct protein domains for ligand binding (HIN domain) and potentially for signal transduction (PYRIN domain). Furthermore, the structure of the HIN domains directly bound to immune stimulatory DNA has been solved [71]. In a murine macrophage cell line, p204, which has a similar domain structure to human IFI16, was found to be essential for HSV-1-induced IFN-β expression [58]. A direct colocalization between HSV-1 and CMV DNA and IFI16 has subsequently been found in the cytoplasm of human macrophages in which the type I IFN response to HSV-1 and HCMV infection was dependent on IFI16 [32]. Interestingly, IFI16 is predominantly localized in the nucleus, and it was speculated early on that IFI16 might also act as a PRR in the nucleus, although how it distinguishes between self and non-self DNA in the nucleus was unclear [72]. A report by Kerur et al. has demonstrated that IFI16 acts as a sensor for Kaposi's sarcoma-associated herpesvirus (KSHV, HHV8) DNA in the nucleus of endothelial cells leading to activation of an IFI16-ASC-Caspase 1 inflammasome [22]. This was followed by a report showing that IFI16 senses HSV-1 DNA in the nucleus of permissive cells [23,24]. Overexpression of wild-type IFI16 in HEK293 cells allowed HSV-1 to induce IFN-β mRNA by about 5-fold, although production of IFN protein was not reported [23]. Kerur et al. demonstrated that DNA sensing was dependent on the ability of the virus to replicate, hence suggesting that the potential function of IFI16 as a sensor of nuclear herpesvirus DNA can be activated only in cell types permissive for infection. Since myeloid cells represent the main activators of inflammasomes, it will be interesting to learn if productive herpesvirus infection in myeloid cells (e.g. human herpesvirus (HHV)-6 infection in macrophages) leads to potent activation of the IFI16 inflammasome through nuclear DNA recognition. An important question raised by the work on nuclear DNA sensing by IFI16 is what regulates the transduction of signaling from the nucleus to the cytoplasm after nuclear DNA sensing. Cristea and associates identified acetylation of the nuclear localization signals in IFI16 as a mechanism to promote cytoplasmic localization [23], and it will be interesting to learn more about how cells and microbes control IFI16 localization. Also the issue of how nuclear DNA sensing transduces signals to the cytoplasm, which is essential for IFN responses after nuclear DNA sensing by IFI16 [24], needs further attention.

In contrast to the modest, but potentially physiologically relevant, stimulation of IL-1β production and IFN-β expression after nuclear recognition of herpesvirus DNA, the strong IFI16-dependent IFN-β response to HSV-1 and CMV infections in macrophages is strictly dependent on DNA recognition in the cytoplasm [32]. This raises the question as to what extent the site of IFI16-mediated DNA sensing itself (cytoplasm vs. nucleus) determines the quality of the response downstream of IFI16. This issue needs further clarification. Another important point from the work by Horan et al. is that IFI16 is a primary sensor of foreign DNA in cell types constitutively expressing this protein, such as macrophages [32]. In other cell types, such as monocytes, IFI16 is IFN-inducible, and more likely to play a role in amplifying the DNA-dependent innate response at later time points [73].

It is now well established that IFI16 is a *bona fide* innate sensor of intracellular DNA driving IFN-β responses to herpesvirus infections – at least in macrophages [32,58]. This DNA sensor has also been proposed to be involved in DNA-driven stimulation of adaptive immunity [11]. However, many key questions remain unanswered. For instance, what is the role of IFI16 in control of infections and disease pathogenesis in living organisms? How does IFI16 distinguish between self and non-self DNA in the nucleus, given that DNA recognition by IFI16 is base sequence-independent [71]? What controls the acetylation of IFI16 which determines the subcellular localization of IFI16 [23] and what determines whether IFI16 acts as a PRR or participates in other cellular activities? Data by Carr and associates on RNAi-mediated p204 knockdown in cornea epithelial cells *in vivo* suggest a role for p204 in control of ocular HSV-1 infection [74]. The generation of a p204-deficient mouse strain will allow further clarification of the role of p204 in antiviral defense *in vivo*. In addition, the

PYHIN family has four members in humans and 13 in mice, and recent work has demonstrated the ability of many PYHIN proteins to enhance IL-1β production and IFN-β expression when overexpressed [75]. It will be interesting to gain more knowledge on the role of the PYHIN family proteins in DNA-driven innate immune responses, and to determine which other family members, apart from IFI16, p204 and AIM2, may also act as 'AIM2-like receptors' for sensing DNA.

DNA-PK. The DNA-dependent protein kinase (DNA-PK) is a heterotrimeric protein complex consisting of three subunits, DNA-PKcs, Ku70, and Ku80. The latter two proteins form a heterodimeric complex and are involved in detection of dsDNA breaks during DNA damage, leading to rapid activation of the catalytic DNA-PKcs subunit and initiation of nonhomologous end joining in the DNA damage response (DDR) [76]. Two publications now demonstrate a role for DNA-PK in DNA sensing in HEK293T cells and murine fibroblasts [77,78]. Imamichi and associates first demonstrated that knockdown of Ku70 suppressed the IFN-λ1 response to linear plasmid DNA in HEK293 cells, with no effect on the type I IFN response [77]. It was also reported that HSV-2 infection stimulated expression of IFN-λ1, but this response was not formally demonstrated to be dependent on Ku70 [77]. Recently a second study reported that DNA-PK detects DNA in the cytosol of fibroblasts and induces expression of IFN-β, ISGs and also the inflammatory gene IL-6 [78]. These authors also demonstrated reduced expression of IL-6 expression in DNA-PKcs-deficient mice after intradermal HSV-1 infection as compared to wild-type mice [78]. In further support of a role for DNA-PK in control of herpesvirus infection is the finding that HSV-1 replicates to high titers in murine Ku70$^{-/-}$ fibroblasts [79].

Since Ku70/Ku80 detects DNA ends in the DDR, it appears likely that this proposed PRR senses the same motif in the innate immunological response. The published data showed that long DNA induced more IFN-λ1 than short DNA and that linear DNA also induced to a higher extent than circular DNA [77]. Such data may suggest that DNA-PK detects free DNA ends and acts in concert with other DNA sensors recognizing dsDNA to mount IFN responses. This issue needs to be clarified. Other outstanding questions include, what are the downstream IRFs activated by DNA-PK and does this DNA sensor activate non-IRF pathways? Does nuclear DNA-PK also act as a nuclear PRR similar to IFI16, and is the DDR capable of activating signaling to IFN expression, hence providing a potential mechanism for IFN expression after nuclear DNA sensing?

DDX41. Liu and associates reported that the DExD/H-box helicase DDX41 can interact with synthetic dsDNA and induce type I IFN expression in myeloid DCs [73]. The signaling downstream of DDX41 was dependent on STING and TBK1, as also reported for DAI and IFI16 [57,58,73]. How exactly this helicase would engage with downstream signaling pathways is unclear, although the paper reported that the DEAD domain was essential for both DNA binding and recruitment of STING. It was found that the type I IFN response to infection with HSV-1, and also adenovirus and the intracellular bacterium *Listeria monocytogenes*, was largely abolished when DDX41 expression was reduced by RNA interference. Interestingly, the authors found that in cell types with limited basal IFI16 expression, DDX41 seems to be a primary sensor of cytoplasmic DNA, inducing IFN and subsequent IFI16 expression, which then served as an amplifier of the innate response [73]. Thus, data from studies involving IFI16 and DDX41 suggest that the pattern of DNAR expression in cells may define which sensor mediates the innate response to intracellular DNA. Thus it will be interesting to determine the level of redundancy that exists between different DNA sensors in particular cell types and *in vivo*, or whether there are qualitative differences in the nature of the responses evoked by different STING-dependent IFN-inducing DNA sensors. The findings that DAI can stimulate programmed necrosis and that IFI16 can sense nuclear DNA to activate the inflammasome, while both sensors can induce type I IFN, indicate that the different DNARs do have both shared and unique functions [19,22]. Finally, it was recently reported that DDX41, in addition to sensing DNA, is also a sensor for cyclic dinucleotides [80]. With this finding, there is now an urgent need to understand how DDX41 may sense both DNA and cyclic dinucleotides and also to dissect whether DDX41 is an essential factor in STING-dependent signaling rather than a PRR.

Other DExD/H-box helicases. In addition to DDX41, three other DExD/H-box helicases have been ascribed roles in innate DNA sensing. In a screen for cytosolic DNA-interacting proteins in pDCs, Liu and associates identified DHX9 and DHX36, which bind CpG DNA and interact with MyD88 [81]. *In vitro* infection with HSV-1 evoked MyD88 dependent TNF-α and IFN-α responses, which were partly dependent on DHX9 and DHX36, respectively. Recent studies have demonstrated that DHX9 interacts not only with DNA but also with RNA, which induces MAVS-dependent IFN and cytokine expression in myeloid DCs [73,82]. This raises important questions about the emerging roles of the DExD/H-box helicase superfamily in intracellular sensing of nucleic acids [73,81,82]. For instance, are DHX9/36 actual DNA sensors or rather downstream signaling molecules in the nucleic acid recognition pathways, as is the case for DDX3 [83]? If DHX9 is indeed a sensor for both RNA and DNA, what determines whether DHX9 acts as an RNA or DNA sensor in a given cell? Does the chemical nature of the nucleic acid agonist have an impact on whether DHX9 engages in downstream signaling through MyD88 or MAVS?

DDX60 is another DDX protein proposed to have a role in DNA sensing [84]. Like DHX9, DDX60 was found to bind both dsRNA and dsDNA, and to interact with RIG-I and MDA5. Importantly, DDX60 was essential for a full type I IFN response to RNA ligands and also to HSV-1 infection in HeLa cells [84]. Since HSV-1 infection has been reported to be sensed both by the Pol IIIRIG-I pathway and by MDA5 [61,62], and dsRNA does accumulate during productive HSV-1 infection [85] it is still not clarified whether DDX60 acts as a sensor of viral RNA or DNA during herpesvirus infection. Nevertheless, the data available on DHX9, DHX36, and DDX60 raise the question as to whether DNA sensors in certain instances act in sequential manners: Primary DNA sensors capable of initiating IFN responses (DAI, IFI16, DDX41, DNA-PK) and secondary DNA sensors working mainly as signaling amplifiers (DHX9, DHX36, DDX60). This issue needs further attention in the future. It is also possible that – analogous to what has been proposed for RNA sensing by RIG-I [86] – there are DNA binding proteins upstream of the signaling DNA sensors that play important roles in initiating and sustaining DNA-activated innate immune responses.

ROLE FOR STING IN THE RESPONSE TO INTRACELLULAR DNA SENSING

As discussed above and illustrated in Figure 8.3, STING is a central molecule for IFN induction by many DNA sensors [57,58,73,87], and STING$^{-/-}$ mice are highly susceptible to intravenous and intravaginal HSV-1 infections [87]. However, at this stage there is very limited knowledge on how DNA sensors actually 'activate' STING. STING was identified independently by several groups, and demonstrated to have functions beyond DNA-activated signaling [88–90]. Moreover, the STING–TBK1–IRF3 pathway is not activated selectively by DNA but also by cyclic-dinucleotides and virus-cell membrane fusion [91,92].

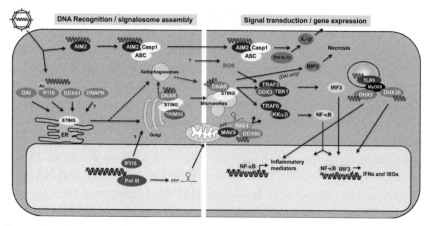

Figure 8.3 *Activation of signaling pathways by DNA.* Left: DNA recognition and signalosome assembly. Cytosolic DNA is recognized by DNA sensors and stimulates assembly of a STING signalsome and an AIM2 inflammasome. STING resides in the ER and proceeds through the Golgi to vesicular structures isolated together with mitochondria-associated ER membranes. The subcellular localization of the AIM2 inflammasome is not well described. RNA polymerase III can transcribe AT-rich dsDNA to induce immunostimulatory RNA species. Right: The STING signalsome and RIG-I-driven MAVS signaling stimulate induction of TBK1–IRF-3-dependent IFN responses. DAI specifically signals through RIP3 to induce necrotic cell death. The AIM2–ASC–caspase1 inflammasome stimulates proteolytic maturation of pro-IL-1β and pro-IL-18. Please see color plate at the back of the book.

STING is an ER-resident protein that after DNA stimulation of cells relocates via the Golgi apparatus to perinuclear mitochondria-associated ER membranes which can be isolated together with microsomes [73,87,88,93]. The nature of these vesicles remains to be characterized, but they seem to contain autophagy-associated proteins including LC3 and ATG9a [93]. In line with this, it was recently demonstrated that cytoplasmic localization of herpesvirus DNA induces autophagy [15,16], and this proceeds through a STING-dependent mechanism [16]. The interaction between STING-dependent signaling and autophagy is still not clarified. Work in fibroblasts has demonstrated that ATG9a-deficiency leads to elevated IFN-β responses [93]. In DCs, pharmacological inhibition of autophagy and deletion of ATG7 strongly inhibits HSV-1-induced IFN-β expression [16]. Thus, modulation of autophagy results in both elevated and decreased IFN responses to cytosolic DNA and herpesvirus infection. It still remains to be determined whether these apparently contradictory data are due to cell type specific roles of the autophagy machinery in

DNA responses or rather that different factors in the autophagy pathway differentially influence STING-dependent signaling from DNA sensors.

Most work on DNA-activated signaling has focused on IFN responses, and some work even suggests that cytosolic DNA exclusively stimulates the IRF-3 pathway [9]. The ability of STING to promote this pathway is dependent on ubiquitination by TRIM56, which stimulates STING-dimerization and association with TBK1 [94]. During HSV infection, p204-STING dependent induction of ISGs has been reported to rely on virus-induced production of ROS, which leads to S-glutathionylation of TRAF3 [95]. The potential role for ROS in STING dependent signaling, however, remains unclarified, since other studies have demonstrated that high concentrations of ROS lead to covalent linkage of STING dimers with impaired function [96]. With respect to DNAR-mediated activation of non-IRF3 pathways there is now evidence for NF-κB-activation by ZBP1/DAI and p204 [12,13,58]. It remains to be demonstrated whether these pathways proceed through STING. ZBP1/DAI interacts with receptor-interacting protein (RIP) 1 and 3 through RIP homotypic interaction motifs and activates NF-κB [12,13]. RNAi-mediated knockdown of RIP1 and 3 inhibits NF-κB activation. CMV activates DAI and the RIP3 pathway and stimulates not only NF-κB but also necroptosis [12,13,18]. One important question in this field is whether STING is required for all DNA-activated signaling events or whether it is only a subset of responses, such as the TBK1–IRF3 pathway, that proceed via STING [97].

Human STING has single nucleotide polymorphisms leading to a defective IFN-β response to DNA [98]. These SNPs are present in the population with a homozygosity frequency ranging from about 1.7% in Europeans to about 15% in Asians, suggesting that the haplotype is under natural selection [98]. Data on susceptibility to infectious and inflammatory diseases in these populations may reveal important information on the role of the intracellular DNA sensing machinery *in natura*.

ROLES OF KNOWN DNA SENSORS IN HERPESVIRUS INFECTION INDEPENDENT OF THEIR ROLE AS PRRs

As the list of proposed DNA sensors has expanded, we have learned that several of the sensors are proteins with key roles in other well-described cellular processes such as DNA repair and gene transcription. In addition, several of the DNARs also play positive and negative roles in the replication cycle of herpesviruses independent of their role as PRRs [64,99,100].

This illustrates the close evolutionary interaction between herpesvirus replication, basal cellular processes, and the innate DNA sensing system.

The DDR is a nuclear machinery capable of sensing and repairing ss- and dsDNA breaks [76]. The system involves sensors of DNA breaks such as Ku70/Ku80, as described above, and downstream effector signaling molecules, such as DNA-PK. Three key pathways in the DDR are the ATR (ssDNA breaks), ATM (dsDNA breaks) and DNA-PK (dsDNA breaks) pathways, which are activated by different types of DNA damage and different sensors of DNA breaks. HSV-1 and EBV infections activate the ATM DNA damage response [101,102], and for HSV-1 this is required for formation of replication centers and optimal viral replication [99]. Concerning the role of the DNA-PK DDR in herpesvirus replication, there is less evidence. It has been reported that Ku70 and Ku80 are recruited to the HSV origin of replication (oriS) but a mechanistic involvement in the replication process has not been reported [103]. Finally, the ATR pathway seems not to be activated but in fact inhibited by HSV-1 and CMV [104,105], suggesting a role for this pathway in restriction of virus replication. In addition to the above-described function of several DDR proteins in DNA sensing, there is now also evidence for involvement of the DDR in cell-intrinsic antiviral defense against incoming HSV-1 [106].

Several key herpesvirus genes are transcribed by RNA polymerase III. This includes for instance the EBV gene RNA product EBERs, which are involved in the latent cycle of EBV infection, and several of the MHV68-encoded microRNAs [64]. In one of the original papers describing Pol III as an innate DNA sensor, it was demonstrated that EBER gene transcription by Pol III generates a RIG-I agonist [60]. For other herpesvirus transcripts driven by Pol III, it remains to be demonstrated whether they actually are sensed by the RIG-I pathway. If the herpesvirus genes transcribed by Pol III lead to immune recognition, we still have to learn why the virus uses this system for a specific subset of genes instead of e.g. Pol II. In order to fully understand Pol III as an innate immune sensor, we also need to know why host RNAs transcribed by Pol III do not stimulate innate immune responses, and whether herpesviruses have adapted a strategy to evade Pol III-dependent immune activation.

The published evidence for IFI16 as a PRR in the cytoplasm and nucleus was described above. Recent data now suggest that IFI16 can also act as a restriction factor for HSV-1, HSV-2, and HCMV [100]. This occurs through a mechanism where IFI16 binds directly to the promoter of the viral DNA polymerase, displaces the essential SP-1-like transcription factors from the promoter, and thus prevents transcription. Since this occurs in the nucleus, there is now evidence for antiviral effects of IFI16 against α- and β-herpesviruses both in the cytoplasm and the nucleus.

Thus, several reported DNA sensors have roles in both support and restriction of herpesvirus replication beyond their function as DNA sensor. This illustrates the close interaction between herpesviruses and the host and also highlights potential challenges in our research efforts to specifically understand the role of the innate DNA sensing machinery when using herpesviruses as model pathogens.

VIRAL EVASION OF DNA-DRIVEN INNATE IMMUNE RESPONSES

Given the ability of herpesviruses to establish life-long infections, it is no surprise that these viruses actively counteract both the innate and adaptive immune responses [27,107]. The inhibition of innate immune responses includes both pathway-specific mechanisms as well as inhibition of components shared by many recognition pathways, e.g. IRF-3 [108–115]. The discussion here will focus on the current knowledge on specific evasion of DNA recognition and signaling by herpesviruses (Figure 8.4).

One potential way to evade the activities of the DNA sensing machinery is to impair exposure of the viral genome to DNA sensors. This strategy is adapted by HSV-1, which uses ICP34.5 to bind beclin-1 and prevent autophagy [116], which would otherwise expose the viral DNA to TLR9 in endosomes [31]. Interestingly, ICP34.5 also forms complexes with TBK1 and disrupts the interaction of TBK1 and IRF3, which prevents the induction of IFNs and ISGs [117]. There are currently no reports of herpesviruses inhibiting release of viral DNA into the cytoplasm. With the recent identification of this process proceeding via the ubiquitin–proteasome pathway, the information is now available to start looking for viruses evading this pathway [32]. Interestingly, herpesviruses contain a conserved deubiquitinase, with dual specificity (K48 and K63) in the tegument [118–120], and this enzyme constitutes an interesting candidate for a viral protein inhibiting proteasomal degradation in the cytoplasm.

An alternative way to evade DNA-driven immune responses is by rendering the viral DNA less stimulatory. A genome-wide analysis of the MHV68 genome has demonstrated that TLR9-stimulating CpG motifs,

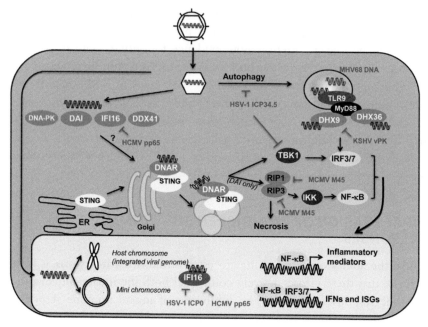

Figure 8.4 *Evasion of DNA sensing and signaling by herpesviruses.* Herpesviruses evade DNA-driven innate immune responses by inhibiting exposure of viral DNA to DNA sensors (e.g. HSV-1 ICP34.5), modulating the genome composition to become less immunostimulatory (MHV68), and by encoding proteins that specifically target DNA sensors or DNA-activated signaling molecules (e.g. MCMV, M45). The figure only includes herpesvirus evasion strategies targeting the upstream steps in DNA recognition and signaling. Viral strategies to evade innate response at the more downstream level are described elsewhere [27]. Please see color plate at the back of the book.

which are abundantly present in e.g. the MCMV genome, are under-represented in the MHV68 genome [121]. This is associated with lower stimulation of the TLR9 pathway by MHV68 as compared to MCMV, suggesting that this γ-herpesvirus specifically seeks to evade recognition by TLR9.

Pp65 represents the major tegument protein of HCMV, and it has long been known to have immune evasion activities, and was described to inhibit HCMV-induced activation of IRF-3 by HCMV infection [122]. It has now been described that pp65 can physically interact with IFI16 and this leads to activation of the viral immediate early gene promoter [123]. The work by Cristea et al. was performed in fibroblasts, where both IFI16 and pp65 are localized nearly exclusively to the nucleus, hence suggesting the interaction to occur in the nucleus. It remains to be formally demonstrated that interaction between pp65 and IFI16 inhibits DNA signaling.

M45 is a tegument protein of MCMV [124]. It contains 2 RIP homotypic interaction motifs, and disrupts the recruitment of RIP1 and 3 to DAI, hence impairing signaling to NF-κB [12]. Furthermore, using M45-deficient virus, it was reported that this protein specifically inhibits RIP3-dependent necrosis [18]. Importantly, it has recently been reported that this necrosis pathway proceeds through DAI [19], and hence adds necrosis to the list of biological processes stimulated by DNA and places M45 as an even more important inhibitor of DNA-activated signaling.

KSHV encodes a kinase, vPK, from ORF36, which has been proposed to be packed into the virus particle as a tegument protein [125]. One identified target of vPK is DHX9, and this is associated with regulation of transcriptional activity [126]. However, we still have very limited evidence for vPK as a viral protein mediating evasion of DNA recognition. For instance, vPK is described mainly as a nuclear protein, so the current knowledge on vPK does not formally demonstrate this kinase as an inhibitor of DHX9 as a cytosolic DNA sensor.

It is interesting to note that all proposed inhibitors of DNA recognition and signaling can be found in the viral tegument, and hence released into the cell upon entry. Since DNA recognition is an event occurring early after viral entry often prior to viral gene expression [32], it makes sense that efficient inhibition of this process is dependent on delivery of the inhibitory protein to the target cell with the virion. With this in mind, the prediction would be that other viral inhibitors of DNA recognition and signaling should be found among virion-associated proteins.

If the infecting herpesvirus makes it to the nucleus, it is protected from many of the known sensing systems, but we are now starting to learn that DNA sensing also takes place in the nucleus [127], suggesting a need for viral evasion strategies. One such strategy includes viral means to make the genome look as similar to the host DNA as possible. Herpesvirus genomes become circularized in the nucleus, and in addition to this, the genomes of some herpesviruses, including HHV6 and Marek's disease virus, become integrated into the host genome [128,129]. Finally, the viral genomes associate with histones to form mini-chromosomes, the transcription from which is regulated by histone acetylation [130]. These measures render the viral DNA difficult to distinguish from host DNA, and hence may protect it from degradation and recognition by PRRs in the nucleus. A second viral strategy includes active inhibition of the nuclear

DNA sensing machinery. It has recently been reported that HSV-1 ICP0 localizes to the nucleus with IFI16 in human foreskin fibroblasts and promotes IFI16 relocalization and degradation, hence inhibiting IFN gene expression [24]. In the same cell type, infection with HCMV leads to specific exclusion of Ku70 and DNA-PKsc from the areas of replication in the nucleus [105].

In addition to the above-described specific evasion and inhibition of the DNA recognition machinery, there are numerous examples of herpesvirus inhibition of signaling at a downstream level shared between DNA sensors and other classes of PRRs. This has been reviewed elsewhere [27]. As an example, all subfamilies of herpesviruses target IRF-3 to inhibit IFN induction, e.g. through proteasomal degradation [108–115]. Thus, there is ample evidence for herpesviruses evading DNA-driven innate immune responses. However, it is most likely that we still have a lot to learn on how herpesviruses evade activation of immune responses through DNA. For instance, it would be interesting to learn whether DNA viruses exploit DNA sensing to promote their own replication. In addition, given the important role for STING in signaling by DNA sensors, it seems likely that herpesviruses have evolved strategies to inhibit STING function. To fully understand the impact of the viral evasion strategies uncovered, viral mutants should be generated when technically possible and tested in relevant mouse models.

DNA SENSING BEYOND HERPESVIRUSES

In addition to herpesvirus infections, DNA recognition is likely to be important for the host response to many other microbial infections including viruses such as adenoviruses, poxviruses, HIV, and papillomaviruses [33,58,65,131]; obligate and facultative intracellular bacteria such as *Listeria*, *Francisella*, and *Chlamydia* species [9,69,132]; intracellular parasites such as *Plasmodium falciparum* [34]. In addition, many extracellular bacteria are recognized by TLR9 [133,134], and may even be sensed by intracellular DNA sensors despite their extracellular life cycle [135]. Finally, although not discussed in this chapter it has also been reported that the DNA sensing machinery potently stimulates adaptive immune responses, which may be relevant for the efficacy of DNA vaccines [10,11].

Among viruses, HIV is unique with a replication cycle that involves RNA, ssDNA, RNA:DNA hybrids, and dsDNA, and there is thus a potential role for several classes of nucleic acid sensors of RNA and DNA in HIV recognition [33,136–138]. The Warner Greene laboratory reported that incomplete reverse transcripts (ssDNA) accumulate in abortively infected CD4$^+$ T cells leading to a coordinated proapoptotic and pro-inflammatory response involving caspase-1 and caspase-3 [138]. The authors argued that this mechanism could be involved in the pronounced bystander cell death in HIV infection. A separate study focusing on IFN responses found that elimination of expression of the cytoplasmic DNase Trex1 augmented HIV-induced type I IFN responses [33]. Trex1 is a 5'-3' exonuclease and the IFN response to ssDNA was more sensitive to the presence of Trex1 than the IFN response induced by dsDNA. This suggests that the HIV-induced IFN response involves a PRR sensing ssDNA. The sensor driving this response has not been identified, but was found to signal through the STING–TBK1–IRF3 pathway [33]. Thus, in cells productively infected with HIV, excess viral DNA does not seem to accumulate, hence preventing innate DNA recognition and activation of antiviral, inflammatory and apoptotic responses. This occurs if Trex1 is defective or the virus cannot complete its replication cycle. Key questions facing the investigators in this field include: what is/are the sensor(s) recognizing HIV DNA, and which DNA forms are recognized? Does HIV actively counteract DNA sensing/signaling and what are the mechanisms? Given the predominant role for CD4$^+$ T cells as a source of HIV, it will be interesting to learn which DNA sensors are expressed in T cells and whether they contribute to defense against HIV infection.

Early after identification of AIM2, it was reported that *L. monocytogenes* and *Francisella tularensis* induce IL-1β via the AIM2 inflammasome [69,139,140]. The importance of AIM2-driven innate immune responses is underscored by the active antagonism of AIM2 by *F. tularensis* through the mviN protein [141]. The IFN response to some bacteria can also be driven by intracellular DNA [9], and DDX41, LRRFIP1 and p204 have been reported to be essential for optimal IFN-β responses to *L. monocytogenes* (DDX41, and LRRFIP1) and *Mycobacterium tuberculosis* (p204) in different cells [35,73,142]. However, cyclic-dinucleotides, which are second messenger signaling molecules produced by many bacteria, are also potent inducers of type I IFN responses. It has recently been proposed by Vance and associates that cyclic-dinucleotides are sensed directly by STING and induce IFN-β [91], and the relative importance of cyclic-dinucleotides versus DNA in triggering IFN responses to bacterial infections in different cell types is a key issue that needs to be addressed.

We are now starting to learn how bacterial DNA is released into the cytoplasm for host recognition. It was reported that *M. tuberculosis* DNA is delivered into the cytosol through ESX-1 secretion, although the bacterium remains in the phagosome [142]. Whether this mechanism of DNA release is specific for *M. tuberculosis* remains to be determined; we also have incomplete knowledge on whether there is interaction between the induction of IFN by cyclic-dinucleotides and cytoplasmic DNA. Finally, the work on *M. tuberculosis* revealed that this bacterium actually exploits DNA-activated IRF3-driven responses to establish long-term infection, hence providing yet another example of microbial exploitation of innate recognition [142].

Some parasites pass through an intracellular stage during their life cycle, and hence may be recognized by the innate immune system via DNA sensors. TLR9 has been convincingly demonstrated to be able to detect *Toxoplasma gondii*, *Plasmodium falciparum*, and *Trypanosoma cruzi* [143–145]. *In vivo*, TLR9-deficiency decreased the susceptibility to LPS-induced lethality in mice infected with a rodent malaria parasite [146]. In addition, an association between a TLR9 polymorphism and malaria during pregnancy has been reported [147]. Patients with malaria exhibit elevated type I IFN response in the blood and murine studies indicate a role for the IFNs in the pathogenesis of malaria [34,144,146]. Recent studies have demonstrated that the *P. falciparum* genome, which has an A/T content of about 80%, potently stimulates STING-dependent type I IFN responses through an unidentified DNA sensor [34]. Interestingly, the immunostimulatory DNA was not classical B-form dsDNA but rather hairpin loop DNA. Although it remains to be formally demonstrated that it is the stem–loop DNA that stimulate the IFN response during a *P. falciparum* infection, these findings suggest that innate sensing of DNA is not limited to dsDNA and also that the structure of the microbial DNA being recognized by the immune system may be different from the structure of the functional genome.

In addition to microbial DNA, self-DNA also has the potential to trigger innate immune responses. Soon after the discovery of TLR9 as a sensor of DNA, data were published supporting a role for TLR9 in the pathogenesis of SLE [20], which is a systemic autoimmune disease characterized by production of autoantibodies and nucleic-acid-associated proteins. Another key feature of SLE is a signature of ISG expression in the peripheral blood [148,149]. Studies in a murine model for SLE have demonstrated a key role for the type I IFN system in the pathogenesis of SLE-like diseases [150], and there are data indicating a role for TLR9 in induction of the IFN response, stimulation of autoantibody production, and development of disease [151]. However, the role of TLR9-dependent DNA sensing in SLE pathogenesis is still controversial [152,153].

More recently, data have emerged on roles for immune-stimulatory DNA in the pathogenesis of inflammatory diseases other than SLE. Lack of Trex1 leads to cytosolic accumulation of DNA originating from e.g. endogenous retroelements [21]. This leads to development of IFN-dependent autoimmune diseases [21]. Interestingly, in humans, Trex1 mutants are associated with Aicardi–Goutières syndrome, which is an immune-mediated neurodevelopmental disorder where IFNs play an essential role in the pathogenesis. The DNA-driven and Trex1-restricted autoinflammation is dependent on STING [154]. It will be interesting to learn which DNA sensors are involved in detection of DNA in Trex1-insufficient individuals, and also to get a full understanding of how cells normally keep the cytoplasm clear of DNA.

Psoriasis is an inflammatory skin disease of unknown origin. Recently, abundant cytosolic DNA and increased AIM2 expression in keratinocytes in psoriatic lesions was reported, and cytoplasmic DNA was found to be a potent activator of inflammasomes in keratinocytes [155]. In addition, there is evidence for an IFN gene expression profile in early psoriasis plaques [156], and given the presence of intracellular DNA in psoriaris plaques, there may be a role for intracellular DNA sensors in driving this response and also promoting infiltration of pathogenic Th17 cells.

All together, the above discussion illustrates that although much of our basic knowledge on DNA sensing has first been generated in herpesvirus systems, this information has been used extensively to generate new insight into other systems, most notably non-viral intracellular infections and anti-inflammatory diseases.

CONCLUDING REMARKS

DNA is an important molecule capable of carrying genetic information. The presence of foreign DNA inside a cell can therefore potentially result in altered composition of the host genome if integrated; hence foreign intracellular DNA represents a threat to the integrity of our genome. Maybe for that reason, it is particularly important for cells to detect foreign DNA and mount appropriate responses. The field of innate DNA recognition has developed rapidly in the past few years with identification of more than 10 sensors, and there is now a need to describe their roles in defense and disease, their mechanisms of action, and interplay with other

PRRs. Two purposes of this chapter have been to critically assess the current literature on innate DNA recognition and to identify some of the key questions that now need to be addressed in this field.

Innate DNA recognition is at the heart of innate immunology and captures maybe the most important question in immunology, namely how to distinguish self from non-self? DNA is actively utilized and also protected by microbes and cells, and we now know that innate DNA recognition is closely connected to the process of microbial replication – and may in some situations even be exploited by the microbe to promote infection. It is therefore not surprising that innate DNA sensing is interacting with basal cellular processes such as DDR and polymerase III-driven transcription. It is likely that future discoveries on DNA-driven innate and adaptive immune responses may reveal key roles of this part of the immune system in infections and inflammatory diseases.

ACKNOWLEDGMENTS

The work in the SRP laboratory is supported by grants from The Danish Medical Research Council (09-072636, 12-124330), The Lundbeck Foundation (R83-A7598), The Novo Nordisk Foundation, The Velux Foundation, Aase og Ejnar Danielsens Fond, and Aarhus University Research Foundation. AGB is funded by Science Foundation Ireland and the NIH (AI093752).

REFERENCES

[1] Medzhitov R. Origin and physiological roles of inflammation. Nature 2008;454:428–35.
[2] Lemaitre B, Nicolas E, Michaut L, Reichhart JM, Hoffmann JA. The dorsoventral regulatory gene cassette Spatzle/Toll/cactus controls the potent antifungal response in *Drosophila* adults. Cell 1996;86:973–83.
[3] Medzhitov R, Preston-Hutlburt P, Janeway Jr CA. A human homologue of the *Drosophila* Toll protein signals activation of adaptive immunity. Nature 1997;388:394–7.
[4] Poltorak A, He X, Smirnova I, Liu M, van Huffel C, Du X, et al. Defective LPS signaling in C3H/HeJ and C57BL/10ScCr mice: Mutations in Tlr4 gene. Science 1999;282:2085–8.
[5] Franchi L, Wamer N, Viani K, Nunez G. Function of Nod-like receptors in microbial recognition and host defense. Immunol Rev 2009;227:106–28.
[6] Kato H, Takahasi K, Fujita T. RIG-I-like receptors: cytoplasmic sensors for non-self RNA. Immunol Rev 2011;243:91–8.
[7] Keating SE, Baran M, Bowie AG. Cytosolic DNA sensors regulating type I interferon induction. Trends Immunol 2011;32:574–81.
[8] Ishii KJ, Coban C, Kato H, Takahashi K, Torii Y, Takeshita F, et al. A Toll-like receptor-independent antiviral response induced by double-stranded B-form DNA. Nat Immunol 2006;7:40–8.

[9] Stetson DB, Medzhitov R. Recognition of cytosolic DNA activates an IRF3-dependent innate immune response. Immunity 2006;24:93–103.

[10] Ishii KJ, Kawagoe T, Koyama S, Matsui K, Kumar H, Kawai T, et al. TANK-binding kinase-1 delineates innate and adaptive immune responses to DNA vaccines. Nature 2008;451:725–9.

[11] Kis-Toth K, Szanto A, Thai TH, Tsokos GC. Cytosolic DNA-activated human dendritic cells are potent activators of the adaptive immune response. J Immunol 2011;187:1222–34.

[12] Rebsamen M, Heinz LX, Meylan E, Michallet MC, Schroder K, Hofmann K, et al. DAI/ZBP1 recruits RIP1 and RIP3 through RIP homotypic interaction motifs to activate NF-kappaB. EMBO Rep 2009;10:916–22.

[13] Kaiser WJ, Upton JW, Mocarski ES. Receptor-interacting protein homotypic interaction motif-dependent control of NF-kappa B activation via the DNA-dependent activator of IFN regulatory factors. J Immunol 2008;181:6427–34.

[14] Muruve DA, Petrilli V, Zaiss AK, White LR, Clark SA, Ross PJ, et al. The inflammasome recognizes cytosolic microbial and host DNA and triggers an innate immune response. Nature 2008;452:103–7.

[15] McFarlane S, Aitken J, Sutherland JS, Nicholl MJ, Preston VG, Preston CM. Early induction of autophagy in human fibroblasts after infection with human cytomegalovirus or herpes simplex virus 1. J Virol 2011;85:4212–21.

[16] Rasmussen SB, Horan KA, Holm CK, Stranks AJ, Mettenleiter TC, Simon AK, et al. Activation of autophagy by alpha-herpesviruses in myeloid cells is mediated by cytoplasmic viral DNA through a mechanism dependent on stimulator of IFN genes. J Immunol 2011;187:5268–76.

[17] Wenzel M, Wunderlich M, Besch R, Poeck H, Willms S, Schwantes A, et al. Cytosolic DNA triggers mitochondrial apoptosis via DNA damage signaling proteins independently of AIM2 and RNA polymerase III. J Immunol 2012;188:394–403.

[18] Upton JW, Kaiser WJ, Mocarski ES. Virus inhibition of RIP3-dependent necrosis. Cell Host Microbe 2010;7:302–13.

[19] Upton JW, Kaiser WJ, Mocarski ES. DAI/ZBP1/DLM-1 Complexes with RIP3 to mediate virus-induced programmed necrosis that is targeted by murine cytomegalovirus vIRA. Cell Host & Microbe 2012;11:290–7.

[20] Leadbetter EA, Rifkin IR, Hohlbaum AM, Beaudette BC, Shlomchik MJ, Marshak-Rothstein A. Chromatin-IgG complexes activate B cells by dual engagement of IgM and Toll-like receptors. Nature 2002;416:603–7.

[21] Stetson DB, Ko JS, Heidmann T, Medzhitov R. Trex1 prevents cell-intrinsic initiation of autoimmunity. Cell 2008;134:587–98.

[22] Kerur N, Veettil MV, Sharma-Walia N, Bottero V, Sadagopan S, Otageri P, et al. IFI16 acts as a nuclear pathogen sensor to induce the inflammasome in response to Kaposi sarcoma-associated herpesvirus infection. Cell Host Microbe 2011;9:363–75.

[23] Li T, Diner BA, Chen J, Cristea IM. Acetylation modulates cellular distribution and DNA sensing ability of interferon-inducible protein IFI16. PNAS 2012;109:10558–63.

[24] Orzalli MH, DeLuca NA, Knipe DM. HSV-1 ICP0 redistributes the nuclear IFI16 pathogen sensor and promotes its degradation. PNAS 2012;109:3008–17.

[25] Hemmi H, Takeuchi O, Kawai T, Kaisho T, Sato S, Sanjo H, et al. A Toll-like receptor recognizes bacterial DNA. Nature 2000;408:740–5.

[26] Takaoka A, Wang Z, Choi MK, Yanai H, Negishi H, Ban T, et al. DAI (DLM-1/ZBP1) is a cytosolic DNA sensor and an activator of innate immune response. Nature 2007;448:501–5.

[27] Paludan SR, Bowie AG, Horan KA, Fitzgerald KA. Recognition of herpesviruses by the innate immune system. Nat Rev Immunol 2011;11:143–54.

[28] Pellett PE, Roizman B. The family Herpesviridae: A brief introduction. In: Knipe DM, Howley PM, editors. Fields Virology. Philadelphia: Lippincott, Williams, Wilkins; 2007. p. 2579-00.

[29] Lund J, Sato A, Akira S, Medzhitov R, Iwasaki A. Toll-like receptor 9-mediated recognition of herpes simplex virus-2 by plasmacytoid dendritic cells. J Exp Med 2003;198:513–20.

[30] Barton GM, Kagan JC, Medzhitov R. Intracellular localization of Toll-like receptor 9 prevents recognition of self DNA but facilitates access to viral DNA. Nat Immunol 2006;7:49–56.

[31] Lee HK, Lund JM, Ramanathan B, Mizushima N, Iwasaki A. Autophagy-dependent viral recognition by plasmacytoid dendritic cells. Science 2007;315:1398–401.

[32] Horan KA, Hansen K, Jakobsen MR, Holm CK, Waggoner L, West JA, et al. Proteasomal degradation of herpes simplex virus capsids in macrophage releases DNA to the cytosol for recognition by DNA sensors. J Immunol 2012;190:2311–9.

[33] Yan N, Regalado-Magdos AD, Stiggelbout B, Lee-Kirsch MA, Lieberman J. The cytosolic exonuclease TREX1 inhibits the innate immune response to human immunodeficiency virus type 1. Nat Immunol 2010;11:1005–13.

[34] Sharma S, DeOliveira RB, Kalantari P, Parroche P, Goutagny N, Jiang ZZ, et al. Innate immune recognition of an AT-rich stem-loop DNA motif in the *Plasmodium falciparum* genome. Immunity 2011;35:194–207.

[35] Yang P, An H, Liu X, Wen M, Zheng Y, Rui Y, et al. The cytosolic nucleic acid sensor LRRFIP1 mediates the production of type I interferon via a beta-catenin-dependent pathway. Nat Immunol 2010;11:487–94.

[36] Kadowaki N, Ho S, Antonenko S, Malefyt RW, Kastelein RA, Bazan F, et al. Subsets of human dendritic cell precursors express different toll-like receptors and respond to different microbial antigens. J Exp Med 2001;194:863–9.

[37] Ahmad-Nejad P, Hacker H, Rutz M, Bauer S, Vabulas RM, Wagner H. Bacterial CpG-DNA and lipopolysaccharides activate Toll-like receptors at distinct cellular compartments. Eur J Immunol 2002;32:1958–68.

[38] Ewald SE, Lee BL, Lau L, Wickliffe KE, Shi GP, Chapman HA, et al. The ectodomain of Toll-like receptor 9 is cleaved to generate a functional receptor. Nature 2008;456 658–88.

[39] Yasuda K, Richez C, Uccellini MB, Richards RJ, Bonegio RG, Akira S, et al. Requirement for DNA CpG content in TLR9-dependent dendritic cell activation induced by DNA-containing immune complexes. J Immunol 2009;183:3109–17.

[40] Latz E, Verma A, Visintin A, Gong M, Sirois CM, Klein DC, et al. Ligand-induced conformational changes allosterically activate Toll-like receptor 9. Nat Immunol 2007;8:772–9.

[41] Latz E, Schoenemeyer A, Visintin A, Fitzgerald KA, Monks BG, Knetter CF, et al. TLR9 signals after translocating from the ER to CpG DNA in the lysosome. Nat Immunol 2004;5:190–8.

[42] Krug A, Luker GD, Barchet W, Leib DA, Akira S, Colonna M. Herpes simplex virus type 1 activates murine natural interferon-producing cells through toll-like receptor 9. Blood 2004;103:1433–7.

[43] Varani S, Cederarv M, Feld S, Tammik C, Frascaroli G, Landini MP, et al. Human cytomegalovirus differentially controls B cell and T cell responses through effects on plasmacytoid dendritic cells. J Immunol 2007;179:7767–76.

[44] Lim WH, Kireta S, Russ GR, Coates PT. Human plasmacytoid dendritic cells regulate immune responses to Epstein-Barr virus (EBV) infection and delay EBV-related mortality in humanized NOD-SCID mice. Blood 2007;109:1043–50.

[45] Fiola S, Gosselin D, Takada K, Gosselin J. TLR9 contributes to the recognition of EBV by primary monocytes and plasmacytoid dendritic cells. J Immunol 2010;185:3620–31.

[46] Tabeta K, Georgel P, Janssen E, Du X, Hoebe K, Crozat K, et al. Toll-like receptors 9 and 3 as essential components of innate immune defense against mouse cytomegalovirus infection. Proc Natl Acad Sci USA 2004;101:3516–21.

[47] Krug A, French AR, Barchet W, Fischer JA, Dzionek A, Pingel JT, et al. TLR9-dependent recognition of MCMV by IPC and DC generates coordinated cytokine responses that activate antiviral NK cell function. Immunity 2004;21:107–19.

[48] Guggemoos S, Hangel D, Hamm S, Heit A, Bauer S, Adler H. TLR9 contributes to antiviral immunity during gammaherpesvirus infection. J Immunol 2008;180:438–43.

[49] Wuest T, Austin BA, Uematsu S, Thapa M, Akira S, Carr DJ. Intact TRL 9 and type I interferon signaling pathways are required to augment HSV-1 induced corneal CXCL9 and CXCL10. J Neuroimmunol 2006;179:46–52.

[50] Sarangi PP, Kim B, Kurt-Jones E, Rouse BT. Innate recognition network driving herpes simplex virus-induced corneal immunopathology: role of the Toll pathway in early inflammatory events in stromal keratitis. J Virol 2007;81:11128–38.

[51] Sorensen LN, Reinert LS, Malmgaard L, Bartholdy C, Thomsen AR, Paludan SR. TLR2 and TLR9 synergistically control herpes simplex virus infection in the brain. J Immunol 2008;181:8604–12.

[52] Smiley JR. Herpes simplex virus virion host shutoff protein: immune evasion mediated by a viral RNase? J Virol 2004;78:1063–8.

[53] Delale T, Paquin A, Asselin-Paturel C, Dalod M, Brizard G, Bates EE, et al. MyD88-dependent and -independent murine cytomegalovirus sensing for IFN-alpha release and initiation of immune responses in vivo. J Immunol 2005;175:6723–32.

[54] Netea MG, Sutmuller R, Hermann C, Van der Graaf CA, Van der Meer JW, van Krieken JH, et al. Toll-like receptor 2 suppresses immunity against Candida albicans through induction of IL-10 and regulatory T cells. J Immunol 2004;172:3712–8.

[55] Zucchini N, Bessou G, Traub S, Robbins SH, Uematsu S, Akira S, et al. Cutting edge: Overlapping functions of TLR7 and TLR9 for innate defense against a herpesvirus infection. J Immunol 2008;180:5799–803.

[56] Yang K, Puel A, Zhang S, Eidenschenk C, Ku CL, Casrouge A, et al. Human TLR-7-, -8-, and -9-mediated induction of IFN-alpha/beta and -lambda is IRAK-4 dependent and redundant for protective immunity to viruses. Immunity 2005;23:465–78.

[57] Defilippis VR, Alvarado D, Sali T, Rothenburg S, Fruh K. Human cytomegalovirus induces the interferon response via the DNA sensor ZBP1. J Virol 2010;84:585–98.

[58] Unterholzner L, Keating SE, Baran M, Horan KA, Jensen SB, Sharma S, et al. IFI16 is an innate immune sensor for intracellular DNA. Nat Immunol 2010;11:997–1004.

[59] Furr SR, Chauhan VS, Moerdyk-Schauwecker MJ, Marriott I. A role for DNA-dependent activator of interferon regulatory factor in the recognition of herpes simplex virus type 1 by glial cells. J Neuroinflamm 2011;8:99.

[60] Ablasser A, Bauernfeind F, Hartmann G, Latz E, Fitzgerald KA, Hornung V. RIG-I-dependent sensing of poly(dA:dT) through the induction of an RNA polymerase III-transcribed RNA intermediate. Nat Immunol 2009;10:1065–72.

[61] Chiu YH, Macmillan JB, Chen ZJ. RNA polymerase III detects cytosolic DNA and induces type I interferons through the RIG-I pathway. Cell 2009;138:576–91.

[62] Melchjorsen J, Rintahaka J, Søby S, Horan KA, Ostergaard L, Paludan SR, et al. Innate recognition of HSV in human primary macrophages is mediated via the MDA5/MAVS pathway and MDA5/MAVS/Pol III independent pathways. J Virol 2010;84:11350–8.

[63] Howe JG, Shu MD. Epstein–Barr virus small RNA (Eber) genes – unique transcription units that combine RNA polymerase-II and polymerase-III promoter elements. Cell 1989;57:825–34.

[64] Pfeffer S, Sewer A, Lagos-Quintana M, Sheridan R, Sander C, Grasser FA, et al. Identification of microRNAs of the herpesvirus family. Nature Methods 2005;2:269–76.

[65] Hornung V, Ablasser A, Charrel-Dennis M, Bauernfeind F, Horvath G, Caffrey DR, et al. AIM2 recognizes cytosolic dsDNA and forms a caspase-1-activating inflammasome with ASC. Nature 2009;458:514–8.

[66] Burckstummer T, Baumann C, Bluml S, Dixit E, Durnberger G, Jahn H, et al. An orthogonal proteomic-genomic screen identifies AIM2 as a cytoplasmic DNA sensor for the inflammasome. Nat Immunol 2009;10:266–72.

[67] Fernandes-Alnemri T, Yu JW, Datta P, Wu J, Alnemri ES. AIM2 activates the inflammasome and cell death in response to cytoplasmic DNA. Nature 2009;458:509–13.

[68] Roberts TL, Idris A, Dunn JA, Kelly GM, Burnton CM, Hodgson S, et al. HIN-200 proteins regulate caspase activation in response to foreign cytoplasmic DNA. Science 2009;323:1057–60.

[69] Rathinam VA, Jiang Z, Waggoner SN, Sharma S, Cole LE, Waggoner L, et al. The AIM2 inflammasome is essential for host defense against cytosolic bacteria and DNA viruses. Nat Immunol 2010;11:395–402.

[70] Nour AM, Reichelt M, Ku CC, Ho MY, Heineman TC, Arvin AM. Varicella-zoster virus infection triggers formation of an interleukin-1 beta (IL-1 beta)-processing inflammasome complex. J Biol Chem 2011;286:17921–33.

[71] Jin TC, Perry A, Jiang JS, Smith P, Curry JA, Unterholzner L, et al. Structures of the HIN domain: DNA complexes reveal ligand binding and activation mechanisms of the AIM2 inflammasome and IFI16 receptor. Immunity 2012;36:561–71.

[72] Goubau D, Rehwinkel J. Sousa CRE. PYHIN proteins: center stage in DNA sensing. Nat Immunol 2010;11:984–6.

[73] Zhang ZQ, Yuan B, Bao MS, Lu N, Kim T, Liu YJ. The helicase DDX41 senses intracellular DNA mediated by the adaptor STING in dendritic cells. Nat Immunol 2011;12:959–62.

[74] Conrady CD, Zheng M, Fitzgerald KA, Lui C, Carr DJ. Resistance to HSV-1 infection in the epithelium resides with the novel innate sensor, IFI-16. Mucosal Immunol 2012;5:173–83.

[75] Brunette RL, Young JM, Whitley DG, Brodsky IE, Malik HS, Stetson DB. Extensive evolutionary and functional diversity among mammalian AIM2-like receptors. J Exp Med 2012;209(11):1969-83.

[76] Ciccia A, Elledge SJ. The DNA damage response: making it safe to play with knives. Mol Cell 2010;40:179–204.

[77] Zhang X, Brann TW, Zhou M, Yang J, Oguariri RM, Lidie KB, et al. Cutting edge: Ku70 is a novel cytosolic DNA sensor that induces type III rather than type I IFN. J Immunol 2011;186:4541–5.

[78] Ferguson B, Mansur D, Peters N, Ren H, Smith GL. DNA-PK is a DNA sensor for IRF-3-dependent innate immunity. eLIFE 2012;1:e00047.

[79] Taylor TJ, Knipe DM. Proteomics of herpes simplex virus replication compartments: Association of cellular DNA replication, repair, recombination, and chromatin remodeling proteins with ICP8. J Virol 2004;78:5856–66.

[80] Parvatiyar K, Zhang Z, Teles RM, Ouyang S, Jiang Y, Iyer SS, et al. The helicase DDX41 recognizes the bacterial secondary messenger cyclic di-GMP and cyclic di-AMP to activate a type I interferon immune response. Nat Immunol 2012;13:1155–61.

[81] Kim T, Pazhoor S, Bao M, Zhang Z, Hanabuchi S, Facchinetti V, et al. Aspartate-glutamate-alanine-histidine box motif (DEAH)/RNA helicase A helicases sense microbial DNA in human plasmacytoid dendritic cells. Proc Natl Acad Sci USA 2010;107:15181–6.

[82] Zhang ZQ, Yuan B, Lu N, Facchinetti V, Liu YJ. DHX9 pairs with IPS-1 to sense double-stranded RNA in myeloid dendritic cells. J Immunol 2011;187:4501–8.

[83] Schroder M, Baran M, Bowie AG. Viral targeting of DEAD box protein 3 reveals its role in TBK1/IKKepsilon-mediated IRF activation. EMBO J 2008;27:2147–57.

[84] Miyashita M, Oshiumi H, Matsumoto M, Seya T. DDX60, a DEXD/H box helicase, is a novel antiviral factor promoting RIG-I-like receptor-mediated signaling. Mol Cell Biol 2011;31:3802–19.

[85] Weber F, Wagner V, Rasmussen SB, Hartmann R, Paludan SR. Double-stranded RNA is produced by positive-strand RNA viruses and DNA viruses but not in detectable amounts by negative-strand RNA viruses. J Virol 2006;80:5059–64.

[86] Kok KH, Lui PY, Ng MHJ, Siu KL, Au SWN, Jin DY. The double-stranded RNA-binding protein PACT functions as a cellular activator of RIG-I to facilitate innate antiviral response. Cell Host & Microbe 2011;9:299–309.

[87] Ishikawa H, Ma Z, Barber GN. STING regulates intracellular DNA-mediated, type I interferon-dependent innate immunity. Nature 2009;461:788–92.

[88] Ishikawa H, Barber GN. STING is an endoplasmic reticulum adaptor that facilitates innate immune signalling. Nature 2008;455:674–8.

[89] Jin L, Waterman PM, Jonscher KR, Short CM, Reisdorph NA, Cambier JC. MPYS a novel membrane tetraspanner, is associated with major histocompatibility complex class II and mediates transduction of apoptotic signals. Mol Cell Biol 2008;28:5014–26.

[90] Zhong B, Yang Y, Li S, Wang YY, Li Y, Diao F, et al. The adaptor protein MITA links virus-sensing receptors to IRF3 transcription factor activation. Immunity 2008;29:538–50.

[91] Burdette DL, Monroe KM, Sotelo-Troha K, Iwig JS, Eckert B, Hyodo M, et al. STING is a direct innate immune sensor of cyclic di-GMP. Nature 2011;478:515–8.

[92] Holm CK, Jensen SB, Jakobsen MR, Cheshenko N, Horan KA, Moller HB, et al. Virus-cell fusion as a trigger of innate immunity dependent on the adaptor STING. Nat Immunol 2012;13:737–43.

[93] Saitoh T, Fujita N, Yoshimori T, Akira S. Regulation of dsDNA-induced innate immune responses by membrane trafficking. Autophagy 2010;6:430–2.

[94] Tsuchida T, Zou JA, Saitoh T, Kumar H, Abe T, Matsuura Y, et al. The ubiquitin ligase TRIM56 regulates innate immune responses to intracellular double-stranded DNA. Immunity 2010;33:765–76.

[95] Gonzalez Dosal R, Horan KA, Rahbek SH, Ichijo H, Chen ZJ, Mieyal JJ, et al. HSV infection induces production of ROS, which potentiate signaling from pattern recognition receptors: role for S-glutathionylation of TRAF3 and 6. PLoS Pathog 2011;7:e1002250.

[96] Jin L, Lenz LL, Cambier JC. Cellular reactive oxygen species inhibit MPYS induction of IFN beta. PLoS One 2010:5.

[97] Tanaka Y, Chen ZJ. STING specifies IRF3 phosphorylation by TBK1 in the cytosolic DNA signaling pathway. Sci Signal 2012;5:ra20.

[98] Jin L, Xu LG, Yang IV, Davidson EJ, Schwartz DA, Wurfel MM, et al. Identification and characterization of a loss-of-function human MPYS variant. Genes Immun 2011;12:263–9.

[99] Lilley CE, Carson CT, Muotri AR, Gage FH, Weitzman MD. DNA repair proteins affect the lifecycle of herpes simplex virus 1. Proc Natl Acad Sci USA 2005;102:5844–9.

[100] Gariano GR, Dell'Oste V, Bronzini M, Gatti D, Luganini A, De Andrea M, et al. The intracellular DNA sensor IFI16 gene acts as restriction factor for human cytomegalovirus replication. Plos Pathogens 2012:8.

[101] Shirata N, Kudoh A, Daikoku T, Tatsumi Y, Fujita M, Kiyono T, et al. Activation of ataxia telangiectasia-mutated DNA damage checkpoint signal transduction elicited by herpes simplex virus infection. J Biol Chem 2005;280:30336–41.

[102] Kudoh A, Fujita M, Zhang LM, Shirata N, Daikoku T, Sugaya Y, et al. Epstein–Barr virus lytic replication elicits ATM checkpoint signal transduction while providing an S-phase-like cellular environment. J Biol Chem 2005;280:8156–63.

[103] Han KJ, Yang Y, Xu LG, Shu HB. Analysis of a TIR-less splice variant of TRIF reveals an unexpected mechanism of TLR3-mediated signaling. J Biol Chem 2010;285:12543–50.

[104] Wilkinson DE, Weller SK. Herpes simplex virus type I disrupts the ATR-dependent DNA-damage response during lytic infection. J Cell Sci 2006;119:2695–703.

[105] Luo MH, Rosenke K, Czornak K, Fortunato EA. Human cytomegalovirus disrupts both ataxia telangiectasia mutated protein (ATM)- and ATM-Rad3-related kinase-mediated DNA damage responses during lytic infection. J Virol 2007;81:1934–50.

[106] Lilley CE, Chaurushiya MS, Boutell C, Everett RD, Weitzman MD. The intrinsic antiviral defense to incoming HSV-1 genomes includes specific DNA repair proteins and is counteracted by the viral protein ICP0. Plos Pathogens 2011:7.

[107] Hansen TH, Bouvier M. MHC class I antigen presentation: learning from viral evasion strategies. Nat Rev Immunol 2009;9:503–13.

[108] Melchjorsen J, Siren J, Julkunen I, Paludan SR, Matikainen S. Induction of cytokine expression by herpes simplex virus in human monocyte-derived macrophages and dendritic cells is dependent on virus replication and is counteracted by ICP27 targeting NF-kappaB and IRF-3. J Gen Virol 2006;87:1099–108.

[109] Zhu FX, King SM, Smith EJ, Levy DE, Yuan YA. Kaposi's sarcoma-associated herpesviral protein inhibits virus-mediated induction of type I interferon by blocking IRF-7 phosphorylation and nuclear accumulation. Proc Natl Acad Sci USA 2002;99:5573–8.

[110] Lubyova B, Pitha PM. Characterization of a novel human herpesvirus 8-encoded protein, vIRF-3, that shows homology to viral and cellular interferon regulatory factors. J Virol 2000;74:8194–201.

[111] Cloutier N, Flamand L. Kaposi sarcoma-associated herpesvirus latency-associated nuclear antigen inhibits interferon (IFN) beta expression by competing with IFN regulatory factor-3 for binding to IFNB promoter. J Biol Chem 2010;285:7208–21.

[112] Sen N, Sommer M, Che X, White K, Ruyechan WT, Arvin AM. Varicella zoster virus immediate early protein 62 blocks IRF3 phosphorylation at key serine residues: a novel mechanism of IRF3 inhibition among herpesviruses. J Virol 2010;84:9240–53.

[113] Melroe GT, DeLuca NA, Knipe DM. Herpes simplex virus 1 has multiple mechanisms for blocking virus-induced interferon production. J Virol 2004;78:8411–20.

[114] Paladino P, Collins SE, Mossman KL. Cellular localization of the herpes simplex virus ICP0 protein dictates its ability to block IRF3-mediated innate immune responses. PLoS One 2010;5:e10428.

[115] Melroe GT, Silva L, Schaffer PA, Knipe DM. Recruitment of activated IRF-3 and CBP/p300 to herpes simplex virus ICP0 nuclear foci: Potential role in blocking IFN-beta induction. Virology 2007;360:305–21.

[116] Orvedahl A, Alexander D, Talloczy Z, Sun Q, Wei Y, Zhang W, et al. HSV-1 ICP34.5 confers neurovirulence by targeting the Beclin 1 autophagy protein. Cell Host Microbe 2007;1:23–35.

[117] Verpooten D, Ma Y, Hou S, Yan Z, He B. Control of TANK-binding kinase 1-mediated signaling by the gamma(1)34.5 protein of herpes simplex virus 1. J Biol Chem 2009;284:1097–105.

[118] Gonzalez CM, Wang L, Damania B. Kaposi's sarcoma-associated herpesvirus encodes a viral deubiquitinase. J Virol 2009;83:10224–33.

[119] Kim ET, Oh SE, Lee YO, Gibson W, Ahn JH. Cleavage specificity of the UL48 deubiquitinating protease activity of human cytomegalovirus and the growth of an active-site mutant virus in cultured cells. J Virol 2009;83:12046–56.

[120] Schlieker C, Korbel GA, Kattenhorn LM, Ploegh HL. A deubiquitinating activity is conserved in the large tegument protein of the Herpesviridae. J Virol 2005;79:15582–5.

[121] Pezda AC, Penn A, Barton GM, Coscoy L. Suppression of TLR9 immunostimulatory motifs in the genome of a gammaherpesvirus. J Immunol 2011;187:887–96.

[122] Abate DA, Watanabe S, Mocarski ES. Major human cytomegalovirus structural protein pp65 (ppUL83) prevents interferon response factor 3 activation in the interferon response. J Virol 2004;78:10995–1006.

[123] Cristea IM, Moorman NJ, Terhune SS, Cuevas CD, O'Keefe ES, Rout MP, et al. Human cytomegalovirus pUL83 stimulates activity of the viral immediate-early promoter through its interaction with the cellular IFI16 protein. J Virol 2010;84:7803–14.

[124] Kattenhorn LM, Mills R, Wagner M, Lomsadze A, Makeev V, Borodovsky M, et al. Identification of proteins associated with murine cytomegalovirus virions. J Virol 2004;78:11187–97.

[125] Gershburg E, Pagano JS. Conserved herpesvirus protein kinases. Biochim Biophys Acta-Proteins Proteom 2008;1784:203–12.

[126] Jong JE, Park J, Kim S, Seo T. Kaposi's sarcoma-associated herpesvirus viral protein kinase interacts with RNA helicase a and regulates host gene expression. J Microbiol 2010;48:206–12.

[127] Li J, Hu S, Zhou L, Ye L, Wang X, Ho J, et al. Interferon lambda inhibits herpes simplex virus type I infection of human astrocytes and neurons. Glia 2011;59:58–67.

[128] Arbuckle JH, Medveczky MM, Luka J, Hadley SH, Luegmayr A, Ablashi D, et al. The latent human herpesvirus-6A genome specifically integrates in telomeres of human chromosomes in vivo and in vitro. Proc Natl Acad Sci USA 2010;107:5563–8.

[129] Delecluse HJ, Hammerschmidt W. Status of Mareks-disease virus in established lymphoma cell-lines – herpesvirus integration is common. J Virol 1993;67:82–92.

[130] Knipe DM, Cliffe A. Chromatin control of herpes simplex virus lytic and latent infection. Nat Rev Microbiol 2008;6:211–21.

[131] Nociari M, Ocheretina O, Schoggins JW, Falck-Pedersen E. Sensing infection by adenovirus: toll-like receptor-independent viral DNA recognition signals activation of the interferon regulatory factor 3 master regulator. J Virol 2007;81:4145–57.

[132] Prantner D, Darville T, Nagarajan UM. Stimulator of IFN gene is critical for induction of IFN-beta during *Chlamydia muridarum* infection. J Immunol 2010;184:2551–60.

[133] Mogensen TH, Paludan SR, Kilian M, Ostergaard L. Live *Streptococcus pneumoniae, Haemophilus influenzae,* and *Neisseria meningitidis* activate the inflammatory response through Toll-like receptors 2, 4, and 9 in species-specific patterns. J Leukoc Biol 2006;80:267–77.

[134] Bafica A, Scanga CA, Feng CG, Leifer C, Cheever A, Sher A. TLR9 regulates Th1 responses and cooperates with TLR2 in mediating optimal resistance to *Mycobacterium tuberculosis.* J Exp Med 2005;202:1715–24.

[135] Charrel-Dennis M, Latz E, Halmen KA, Trieu-Cuot P, Fitzgerald KA, Kasper DL, et al. TLR-independent type I interferon induction in response to an extracellular bacterial pathogen via intracellular recognition of its DNA. Cell Host Microbe 2008;4:543–54.

[136] Solis M, Nakhaei P, Jalalirad M, Lacoste J, Douville R, Arguello M, et al. RIG-I-mediated antiviral signaling is inhibited in HIV-1 infection by a protease-mediated sequestration of RIG-I. J Virol 2011;85:1224–36.

[137] Berg RK, Melchjorsen J, Rintahaka J, Diget E, Soby E, Horan KA, et al. Genomic HIV RNA induces innate immune responses through RIG-I-dependent sensing of secondary-structured RNA. PLoS One 2012;7:e29291.

[138] Doitsh G, Cavrois M, Lassen KG, Zepeda O, Yang ZY, Santiago ML, et al. Abortive HIV infection mediates CD4 T cell depletion and inflammation in human lymphoid tissue. Cell 2010;143:789–801.

[139] Warren SE, Armstrong A, Hamilton MK, Mao DP, Leaf IA, Miao EA, et al. Cutting edge: cytosolic bacterial DNA activates the inflammasome via AIM2. J Immunol 2010;185:818–21.

[140] Fernandes-Alnemri T, Yu JW, Juliana C, Solorzano L, Kang S, Wu J, et al. The AIM2 inflammasome is critical for innate immunity to *Francisella tularensis*. Nat Immunol 2010;11:385–93.

[141] Ulland TK, Buchan BW, Ketterer MR, Fernandes-Alnemri T, Meyerholz DK, Apicella MA, et al. Cutting edge: mutation of *Francisella tularensis* mviN leads to increased macrophage absent in melanoma 2 inflammasome activation and a loss of virulence. J Immunol 2010;185:2670–4.

[142] Manzanillo PS, Shiloh MU, Portnoy DA, Cox JS. *Mycobacterium tuberculosis* activates the DNA-dependent cytosolic surveillance pathway within macrophages. Cell Host Microbe 2012;11:469–80.

[143] Minns LA, Menard LC, Foureau DM, Darche S, Ronet C, Mielcarz DW, et al. TLR9 is required for the gut-associated lymphoid tissue response following oral infection of *Toxoplasma gondii*. J Immunol 2006;176:7589–97.

[144] Pichyangkul S, Yongvanitchit K, Kum-arb U, Hemmi H, Akira S, Krieg AM, et al. Malaria blood stage parasites activate human plasmacytoid dendritic cells and murine dendritic cells through a Toll-like receptor 9-dependent pathway. J Immunol 2004;172:4926–33.

[145] Bafica A, Santiago HC, Goldszmid R, Ropert C, Gazzinelli RT, Sher A. Cutting edge: TLR9 and TLR2 signaling together account for MyD88-dependent control of parasitemia in *Trypanosoma cruzi* infection. J Immunol 2006;177:3515–9.

[146] Franklin BS, Parroche P, Ataidea MA, Lauw F, Ropert C, de Oliveira RB, et al. Malaria primes the innate immune response due to interferon-gamma induced enhancement of toll-like receptor expression and function. Proc Natl Acad Sci USA 2009;106:5789–94.

[147] Mockenhaupt FP, Hamann L, von Gaertner C, Bedu-Addo G, von Kleinsorgen C, Schumann RR, et al. Common polymorphisms of Toll-like receptors 4 and 9 are associated with the clinical manifestation of malaria during pregnancy. J Infect Dis 2006;194:184–8.

[148] Bennett L, Palucka AK, Arce E, Cantrell V, Borvak J, Banchereau J, et al. Interferon and granulopoiesis signatures in systemic lupus erythematosus blood. J Exp Med 2003;197:711–23.

[149] Baechler EC, Batliwalla FM, Karypis G, Gaffney PM, Ortmann WA, Espe KJ, et al. Interferon-inducible gene expression signature in peripheral blood cells of patients with severe lupus. Proc Natl Acad Sci USA 2003;100:2610–5.

[150] Agrawal H, Jacob N, Carreras E, Bajana S, Putterman C, Turner S, et al. Deficiency of type I IFN receptor in lupus-prone New Zealand mixed 2328 mice decreases dendritic cell numbers and activation and protects from disease. J Immunol 2009;183:6021–9.

[151] Christensen SR, Shupe J, Nickerson K, Kashgarian M, Flavell RA, Shlomchik MJ. Toll-like receptor 7 and TLR9 dictate autoantibody specificity and have opposing inflammatory and regulatory roles in a murine model of lupus. Immunity 2006;25:417–28.

[152] Yu P, Wellmann U, Kunder S, Quintanilia-Martinez L, Jennen L, Dear N, et al. Toll-like receptor 9-independent aggravation of glomerulonephritis in a novel model of SLE. Internat Immunol 2006;18:1211–9.

[153] Santiago-Raber ML, Dunand-Sauthier I, Wu TF, Li QZ, Uematsu S, Akira S, et al. Critical role of TLR7 in the acceleration of systemic lupus erythematosus in TLR9-deficient mice. J Autoimmun 2010;34:339–48.

[154] Gall A, Treuting P, Elkon KB, Loo YM, Gale M, Barber GN, et al. Autoimmunity initiates in nonhematopoietic cells and progresses via lymphocytes in an interferon-dependent autoimmune disease. Immunity 2012;36:120–31.

[155] Dombrowski Y, Peric M, Koglin S, Kammerbauer C, Goss C, Anz D, et al. Cytosolic DNA triggers inflammasome activation in keratinocytes in psoriatic lesions. Sci Translat Med 2011;3:168.

[156] Albanesi C, Scarponi C, Bosisio D, Sozzani S, Girolomoni G. Immune functions and recruitment of plasmacytoid dendritic cells in psoriasis. Autoimmunity 2010;43:215–9.

CHAPTER 9

Cancer Pathogenesis and DNA Sensing[*]

Y.J. Shen[1,2], A.R. Lam[1,2], S.W.S. Ho[1], C.X. Koo[1,2,3],
N. Le Bert[1] and S. Gasser[1,2]

[1]Immunology Programme and Department of Microbiology, Centre for Life Sciences, National University of Singapore, Singapore
[2]NUS Graduate School for Integrative Sciences and Engineering, National University of Singapore, Singapore
[3]Laboratory of Adjuvant Innovation, National Institute of Biomedical Innovation (NIBIO), Ibaraki, Osaka, Japan

The immune system was suggested to represent a barrier that prevents tumor formation and to act in addition to cell intrinsic mechanisms such as the DNA damage response [1,2]. The mechanisms that allow the immune system to recognize tumor cells are a topic of intense research interest. Optimal immune responses to self-cells generally only occur when the cells are first exposed to pathogen-associated molecular patterns, or when the cells are subjected to disease-related stress. During infections, pattern recognition receptors (PRR) such as Toll-like receptors (TLRs) and numerous intracellular sensors that detect nucleotides can initiate inflammatory signals [3,4]. Recognition of cancer cells, on the other hand, is thought to depend on the activation of disease-associated stress [5–8]. Here we review cytosolic DNA sensors and discuss their role in pathways that are associated with tumorigenesis.

IFN-INDUCING CYTOSOLIC DNA SENSORS

ZBP1 (DAI/DLM-1)

ZBP1 (Z-DNA binding protein 1, also named DAI/DLM-1) was the first reported cytosolic DNA sensor that functions independently of TLRs [9]. Upon transfection of cells with double-stranded (ds) DNA, ZBP1 mediates a type I interferon (IFN) response by forming a complex with TANK binding kinase 1 (TBK1) and IFN regulatory factor 3 (IRF3) [9]. In addition, ZBP1 can interact with receptor interacting protein kinase 1

[*]Authors contributed equally.

Biological DNA Sensor.
DOI: http://dx.doi.org/10.1016/B978-0-12-404732-7.00009-5

(RIP1) and RIP3 leading to nuclear factor kappa-light-chain-enhancer of activated B cells (NF-κB) activation [10]. ZBP1 senses dsDNA irrespective of the sequence. However, DNA fragments shorter than 100 bp induce little type I IFN suggesting a minimum length requirement for ZBP1 binding to DNA. ZBP1 contains two N-terminal Z–DNA binding domains Zα and Zβ, which are known to induce and stabilize uncommon left-handed Z-form DNA, and a putative C-terminal DNA binding domain [9,11]. The Zα domain was shown to regulate the cytosolic distribution of ZBP1, since ZBP1 lacking the Zα domain accumulated in large cytoplasmic granules, which interact with stress granules (SGs) and processing bodies (PDs) [12]. SGs and PDs regulate RNA metabolism and usually occur in response to stress conditions like heat shock, oxidative stress, hyperosmolarity, viral infection and UV irradiation, indicating that ZBP1 may be involved in mRNA sorting and metabolism [12–14]. A link of ZBP1 to cancer was suggested by the finding that ZBP1 is highly upregulated in the peritoneal lining tissue of mice bearing ascitic tumors [15]. Z-DNA-forming sequences in the genome can generate large-scale deletions possibly due to DNA repair processes and therefore can pose a danger to genome integrity in proliferating cells such as cancer cells [16].

PYHIN Family/p200 Protein Family

The PYHIN (pyrin and HIN200 domain-containing proteins) family consists of the mouse IFN-inducible genes *Ifi200, p202a, p202b, p203, p204, myeloid cell nuclear differentiation antigen 1 (Mnda1)*, and *absent in melanoma 2 (Aim2*; see below), and the human HIN-200 genes *IFI16, MNDA, AIM2* and *PYHIN1* [16]. PYHIN family proteins are characterized by one N-terminal pyrin domain, and at least one C-terminal HIN domain [18–21]. AIM2 binds to DNA via its HIN200 domain and oligomerizes with adaptor protein apoptosis-associated speck-like protein containing a CARD (ASC) to form an AIM2-inflammasome promoting maturation of the pro-inflammatory cytokines IL-1β and IL-18 (see below) [19,20]. In contrast to other members of the PYHIN family, IFI16 and the mouse homolog p204 contain two HIN domains and induce IFN-β production upon DNA transfection [21,22]. Many PYHIN family members were also shown to modulate proliferation, differentiation, and transcriptional regulation [23–25].

IFI16 and IFI204

IFI16 was identified as an IFN-γ-inducible nuclear gene in human myeloid cell lines [26]. Human IFI16 is predominantly expressed in lymphocytes, monocytes, and epithelial cells [26,27]. In contrast, the mouse homolog IFI204 is expressed in many tissues. The expression of IFI16 and IFI204 is upregulated by type I and II IFNs [28]. IFI16 and IFI204 activate type I IFN through stimulator of IFN genes (STING) signaling in response to DNA virus infection or transfection of cells with dsDNA [21]. Upon infection of HMVEC-d endothelial cells with Kapososi's sarcoma-associated herpes virus (KSHV), IFI16 interacts with the adaptor molecule ASC and pro-caspase-1 to form a functional inflammasome [22].

Increased expression of many p200 family proteins, including IFI16 and IFI204, inhibits cell cycle progression [28–33]. IFI204 interacts with both wild type and oncogenic RAS-GTP and prevents binding of the RAS effectors RAF-1, PI3K, and RAL [30]. Oncogenic K-RAS induces the expression and translocation of IFI204 from the nucleus to the cytoplasm in murine fibroblasts [30]. p53, which is activated by oncogenic RAS [34], was found to directly regulate the expression of IFI16 [32] and the activity of p53 is further stimulated by IFI16 and IFI204 [35–38]. Consistent with the idea that IFI204 functions as part of a negative feedback mechanism for RAS-mediated proliferation signals, IFI16 expression is often downregulated in breast cancer tissue [36] and inversely correlated with proliferation activity and grade in head and neck squamous cell carcinomas [39,40]. Furthermore, transcriptional silencing of IFI16 via histone deacetylase (HDAC) was also found in human prostate cancer cell lines [41]. However, IFI16 is highly expressed in the basal epithelia that contains cells with high proliferative potential suggesting the existence of other modes of regulation [33]. Furthermore, IFI16 protein expression is regulated by post-translational mechanisms that are independent of p53 [37] and IFI204 expression was associated with myoblast, macrophage, myeloid and lymphocytic differentiation [28, 42–44].

Mouse p202, another HIN-200 pyrin family protein, binds to DNA and is thought to negatively regulate innate immune responses [45,46]. Similar to IFI204, p202 impairs proliferation of cells and inhibits the transcriptional activity of genes involved in cell cycle progression including *E2f1*, *Ap-1*, *C-Fos* and *C-Jun*. In contrast to IFI204, p202 inhibits the transcriptional activity of p53 [47].

DEXD/H-Box Helicase Superfamily DNA Sensors

Human RNA helicases represent a large family of proteins that play important roles in RNA processing and DNA damage repair [48,49]. Many DEAD-box (DDX) or DEAH-box (DHX) containing members of the RNA helicases family have been suggested to act as cytosolic DNA and RNA sensors [50].

DHX9 and DHX36

DHX9 (RNA helicase A) and DHX36 (also referred to as G4 resolvase (G4R1) or RNA helicase associated with AU-rich elements (RHAU)) recognize cytosolic DNA containing CpG motifs and trigger IFN-β production via MyD88 in plasmacytoid dendritic cells (pDCs) [51,52]. DHX9 and DHX36 play important roles in RNA metabolism, DNA repair and genome stability [48,53–57].

DHX9 is involved in the nuclear export of retroviruses including human immunodeficiency virus (HIV) and type D retroviruses [54,57–63]. Moreover, DHX9 unwinds mutagenic structures like triplex DNA by displacing the third strand of DNA triplex complexes [64]. It cooperates with another DHX family member, the Werner gene (*Wrn*), to resolve Holiday junction-like intermediates at stalled replication forks and to stimulate the activity of topoisomerase II alpha, an enzyme important for DNA repair [65]. DHX9 is overexpressed in small cell lung cancers and non-small cell carcinomas suggesting a role of DHX9 in tumorigenesis [66]. The role of DHX9 in tumorigenesis is probably complex as DHX9 was found to upregulate the expression of the tumor suppressor gene *p16INK4a* [67] and to impair the activity of the breast cancer type 1 susceptibility protein BRCA1, another tumor suppressor gene involved in DNA repair [68,69].

DHX36 binds to messenger RNAs and overexpression of DHX36 accelerates degradation of certain AU-containing mRNAs [70–72]. DHX36 is mostly present in the nucleus, but re-localizes to stress granules upon inhibition of translation [70,72]. DHX36 exhibits an ATP-dependent guanine-quadruplex (G4) resolvase activity and specificity *in vitro* [73,74]. G4-DNA is a highly stable alternative DNA structure that can form spontaneously in guanine-rich regions of single-stranded DNA under physiological condition [73,75]. Telomeres and genomic regions that encode a number of oncogenes, ribosomal DNA (rDNA), repetitive G-rich microsatellites and the immunoglobulin heavy-chain switch regions have a high potential to form G4-DNA structures. In contrast, genomic sequences around tumor-suppressor genes, which maintain genomic stability and are commonly haploinsufficient, often have a low potential to form G4-DNA structures. G4-DNA structures are highly stable and can promote arrest of replication forks, recombination and genomic instability [75]. It is therefore possible that DHX36 plays a role in genome stability, although no report has shown a direct role for DHX36 in tumorigenesis.

DDX1 and DDX21

DHX36 forms a complex with the DEAD family helicases DDX1 and DDX21 to sense dsRNA in myeloid dendritic cells (mDCs) [51]. DDX1 is a nuclear protein that is widely expressed in the early stages of development, but displays a restricted expression pattern in later stages [76]. DDX1 interacts with factors involved in RNA metabolism [77,78]. The *DDX1* gene is located in chromosomal region 2p24 that is often overexpressed in retinoblastomas and *DDX1* was found to be co-amplified with *MYCN* in neuroblastoma cell lines and some primary tumors [48,79–81]. Conflicting results have been reported regarding the prognostic significance of DDX1 and it is unclear whether and how DDX1 contributes to tumorigenesis of neuroblastomas [48].

DDX21, also called RHII/GUα, is important for *C-JUN* transcription activation and processing of 20S to 18S rRNA [82–85]. No evidence for a role of DDX21 in cancer has been reported.

DDX41

Recently, DDX41 was identified as a new DNA sensor [86]. DDX41 directly associates with the membrane bound adaptor STING and activates TBK1 and NF-κB. A function of DDX41 in tumorigenesis has not been described, but a drosophila homolog of DDX41, ABSTRACT, plays a role in the development of drosophila [87–89].

LRRFIP1

Leucine-rich repeat containing protein (LRRFIP1) was initially cloned as a GC-binding factor called GCF2 [90]. LRRFIP1 was later shown to recognize both cytosolic dsDNA and dsRNA in macrophages infected by *Listeria monocytogenes* and vesicular stomatitis virus [91]. LRRFIP1 interacts with β-catenin and promotes β-catenin activation. Activated nuclear β-catenin enhances recruitment of the p300-histone acteyltransferase complex to the *Ifnb1* promoter through IRF3 leading to the transcription of IFN-β. β-catenin is an important regulator of genes implicated in cell proliferation, inhibition of apoptosis and tumor progression [92,93]. Mutations that constitutively stabilize β-catenin can cause colorectal carcinomas and other forms of cancer [94,95]. LRRFIP1 also acts as a transcriptional repressor and was shown to directly bind to the GC-rich sequences of the epidermal growth factor receptor (EGFR) and to the tumor necrosis factor alpha (TNF-α) promoter [96–98]. LRRFIP1 is expressed in many tissues and is upregulated in breast cancers and Burkitt lymphomas [99–101]. It promotes colorectal cancer metastasis and liver invasion by regulating RhoA-induced cell adhesion, migration, and invasion of cancer cells [100,101]. LRRFIP1 also contributes to the resistance of cells to the topoisomerase II inhibitor VM-26 and the two genotoxic drugs cisplatin and doxorubicin [102,103]. However, the microRNA-21 (miR-21), which is overexpressed in many cancers, targets LRRFIP1 suggesting a complex regulation of LRRFIP1 in tumor cells [102].

RIG-I

Retinoic acid-induced gene I (RIG-I), also called DDX58, is part of the RIG-I-like receptor (RLR) family and contains a RNA helicase-DEAD-box motif and a caspase recruitment domain (CARD) [104]. RIG-I recognizes viral dsRNA that carries a 5′-triphosphate moiety (5′-ppp) [105,106]. Recently it was suggested that RIG-I also indirectly recognizes AT-rich dsDNA, which is transcribed by RNA polymerase III into 5′-triphosphate RNA [107].

RIG-I is highly expressed in differentiated skin and colon mucosal tissues, while its expression is decreased in many cancer tissues including skin and colorectal cancer [104,108]. A number of studies suggest that RIG-I may act as a tumor suppressor gene in some cancers [109–111]. Overexpression of RIG-I in melanoma cells induced apoptosis in a NOXA, CASPASE-9 and APAF-1, but not p53-dependent manner [109]. In leukemia cells, induction of RIG-I contributes to IFN-α and retinoic acid–mediated inhibition of proliferation of acute myelogenous leukemia (AML) cells [110]. RIG-I also impairs the activation of ERK, JNK, and the p38 MAPK by RAS in cervical and gastric cancer cells [111].

HMBG1

The high mobility group protein B1 (HMGB1) is an abundant protein that binds to distorted and damaged DNA [112,113]. It is involved in diverse functions, including recognition of cytosolic DNA, regulation of chromatin structure, transcription and DNA repair.

HMGB1 expression is upregulated in several tumors and tumor endothelium [114]. The metastatic potential of cancer cells was found to correlate with HMGB1 expression, possibly due to its ability to modulate the adhesion of cells, the extracellular matrix and angiogenesis [115–118]. HMGB1 can also stimulate endothelial cell proliferation *in vitro* and neovascularization *in vivo*[119]. In agreement with a role for HMGB1 in cancer, blocking of HMGB1 impairs proliferation of tumors and the ability to form metastasis in a mouse model for lung cancer [120]. HMGB1 is also released from tumor cells treated with different genotoxic agents and soluble HMGB1 was found to enhance processing and cross-presentation of tumor antigens by DCs [121].

KU70

The KU70 forms a heterodimer with KU80 that interacts with the helicase WRN and binds to dsDNA breaks [122,123]. KU70 plays an important role in the non-homologous end joining (NHEJ) DNA repair pathway, V(D)J recombination, and telomere maintenance. A recent report suggests that KU70 also functions as a DNA sensor that induces the production of type III IFN in an IRF1- and IRF7-dependent manner [124].

MEDIATORS OF IFN-INDUCING CYTOSOLIC DNA SENSORS

ZBP1, IFI16, DDX41 and RIG-I interact with stimulator of IFN genes (STING) to activate TBK1, I-kappaB kinase epsilon (IKKε), IRF3 and the IKK complex [50,125]. In contrast to the other sensors, RIG-I relays signals to members of the interferon regulatory factor (IRF) and NF-κB family through the IFNB-promoter stimulator 1 (IPS1).

DDX1, DDX21 and DHX36 were suggested to form a TRIF-interacting complex, which induces the IRF3-dependent production of type I IFNs and the activation of the IKK complex.

STING AND TRIF PATHWAYS
STING

STING (also known as MITA, MPYS and ERIS) is encoded by *Tmem173* and comprises putative transmembrane regions. STING is mainly expressed in the endoplasmic reticulum (ER) and interacts with SSR2 signal sequence receptor 2 (SSR2), a member of the translocon-associated protein (TRAP) complex [126,127]. TRAP is associated with the exocyst complex and both complexes facilitate protein synthesis, folding and secretion [128]. Consistent with a role of the exocyst complex in STING function, Ishikawa et al. reported that, in the presence of cytosolic DNA, STING relocalizes with TBK1 from the ER to perinuclear vesicles that contain the exocyst component Sec5 A [129]. The exocyst is an effector complex for activated v-ral simian leukemia viral oncogene homolog (RAL) proteins, which can support tumorigenic progression and invasion [130–132]. However, the role of STING in tumorigenesis has not been elucidated, but STING was shown to induce apoptosis and possibly inhibit proliferation of B cell lymphoma cells [130–133]. Interestingly, oxidative stress, which is present in many types of cancer cells, inhibits STING [134]. Given the importance of STING in the activation of type I IFN expression in response to a variety of DNA virus infections, it is possible that STING also acts as a tumor suppressor by preventing infection of cells by oncoviruses [135].

TRIF

TIR-domain-containing adapter-inducing interferon-β (TRIF) recruits the canonical IKK complex, which comprises IKKα, IKKβ and IKKγ (also known as NEMO) and the two non-canonical IκB kinase homologs TBK1 and IKKε [136]. Similarly to STING, the role of TRIF in cancer has not been studied in detail, but TRIF was shown to induce RIP-1 and CASPASE-8-dependent apoptosis in various human cancer cells [137–140].

TBK1 and IKKε

The activation of TBK1 and IKKε kinases triggers a cascade of signals through phosphorylation of transcription factors including members of the IRF and the canonical NF-κB families [141,142]. Recent findings suggest a role for TBK1 and IKKε in RAS-induced transformation of cells [143–145]. TBK1 is recruited and activated by a RalB-Sec5 effector complex that is required for RAS-mediated transformation. TBK1 was necessary to inhibit apoptosis in response to RAS activation possibly by engaging the v-akt murine thymoma viral oncogene homolog (AKT) survival pathway [143,144,146,147]. Similarly, IKKε was identified as an effector of the phosphatidylinositol 3-kinase (PI3K)-AKT pathway, which cooperates with MAP kinase-ERK kinase (MEK) to promote transformation. Impairing IKKε expression by siRNA inhibited proliferation, adhesion, and invasiveness of several tumor cell lines [148,149]. In prostate cancer patients, a correlation between IKKε expression levels and malignancy was observed [150].

Interferon Regulator Factor 3 and 7

IRF3 and IRF7 are the main regulators of type I IFN expression [151]. IRF3 is localized in the cytoplasm in a latent form. Upon phosphorylation by TBK1 or IKKε, IRF3 forms a homodimer, followed by translocation into the nucleus. An IFN enhanceosome is then assembled consisting of IRF3, NF-κB and activating transcription factor 2 (ATF2)–C-JUN heterodimers that recruit the histone acetyltransferase CREB-binding protein (CBP) to the *IFNB* promoter [152–154]. IRF3 also binds to the *IFNA* promoter in humans and *Ifna4* promoter in mice [155]. IFN-β expression was abolished in *Irf3*$^{-/-}$ DCs in response to LPS and was significantly impaired upon Poly (I:C) treatment [156] or virus infection of *Irf3*$^{-/-}$ mouse embryonic fibroblasts (MEFs) [157]. Type I IFN can also be induced by IRF7, which can form dimers with IRF3, and type I IFN expression was impaired in *Irf7*$^{-/-}$ pDCs or MEFs upon virus infection [158,159].

IRFs and IFNs have been implicated in eliciting anti-tumor effects [151]. Expression of a dominant-negative mutant form of IRF3 in tumor cells increased their ability to form tumors in nude mice, while ectopic expression of IRF3 inhibited cell proliferation and increased apoptosis due to activation of p53 [160] and possibly secretion of type I IFNs [161]. 5,6-dimethylxanthenone-4-acetic acid (DMXAA), which is currently in clinical trials for use as a chemotherapeutic drug against lung, ovarian and prostate cancer, was found to activate the TBK1–IRF3 pathway [162]. However, some human lung tumors express constitutively activated IRF3 [163]. The TBK1–IRF3 axis was also implicated in the production of factors such as chemokine (C-C motif) ligand 5 (CCL5) and IL-8, which stimulate angiogenesis and tumor growth [164]. Restoration of IRF7 in tumor cells isolated from metastatic mouse primary mammary tumors restored the ability to secrete IFN and reduced the metastatic burden in a breast cancer tumor mouse model [165]. Decreased IRF7 expression levels were found to correlate with increased metastatic events in human breast cancer, supporting the notion that IRF7 plays an important role in suppressing metastasis of cancer cells. IRF7 may also act indirectly on tumor cells as macrophages transduced with a constitutively active mutant form of IRF7 blocked proliferation of human tumor cell lines via IFN in a co-culture assay [166].

NF-κB

Another major transcription factor responsible for inflammation is NF-κB, which is important for driving the expression of various cytokines, chemokines and receptors needed for antigen presentation or immune recognition [167]. As a central mediator of inflammation, NF-κB plays an important role in supporting pro-survival properties of tumor cells. In the classical NF-κB pathway, microbial infections and proinflammatory cytokines activate the IKK complex and target the NF-κB inhibitor IκBα for proteasomal degradation [168]. The released NF-κB dimers, which consist of REL-A (p65) and p50, translocate to the nucleus and mediate transcription of target genes [168]. NF-κB is constitutively activated in various tumor cells and inhibition of NF-κB can induce apoptosis of tumor cells [169,170]. The role of NF-κB in tumorigenesis has been covered by a recent review to which we would like to refer the reader for a more in depth coverage of the field [168].

PROINFLAMMATORY CYTOKINES

Several NF-κB target genes, including the proinflammatory cytokines TNF-α and IL-6, are associated with tumor development and progression in humans and mice [168]. These cytokines can promote proliferation, survival or apoptosis of tumor cells in an auto- and paracrine fashion [171]. Low concentrations of TNF-α released from tumor-associated inflammatory cells augment tumor proliferation and progression, while high concentrations of TNF-α can induce necrosis of tumor cells [168,172]. The former function of TNF-α may be more relevant *in vivo* as *Tnfa*-deficient mice and *Tnfr1*-deficient mice are resistant to carcinogen-induced tumors [173,174].

IL-6 was found to promote or inhibit the progression of lymphomas depending on their developmental stage [175]. In the bone marrow, IL-6 signaling suppresses the development of B cell lymphomas by promoting differentiation of B cells, while IL-6 stimulates survival of transformed B cells. IL-6 also supports the survival of colitis-associated cancers in mice and promotes the growth of hepatocellular carcinomas (HCCs) [176–178]. In humans, serum levels of IL-6 predict HCC development in patients with chronic hepatitis B [179].

Type 1 Interferons

Type 1 interferons (IFNs) have been linked to suppression of tumorigenesis in both hematological and solid tumors [180]. *Ifnar1*-deficient mice are more susceptible to carcinogen-induced tumors and transplanted syngeneic tumors [6]. The inactivation of IFN-consensus-sequence binding protein (ICSBP), a transcription factor involved in the regulation of IFN-stimulated genes, results in a chronic myeloid leukemia (CML)-like disease and many patients with untreated CML have low ICSBP levels [181]. In addition, type I IFNs can enhance the susceptibility of cells to p53-mediated apoptosis [182].

INFLAMMASOME-INDUCING CYTOSOLIC DNA SENSORS

The cytosolic DNA sensors AIM2 and NOD-like receptor family pyrin domain containing 3 (NLRP3) activate the inflammasome, but not the pathways leading to type I IFN secretion [18–20,46,183–187]. The inflammasome is a large multi-protein complex that assembles in response to DNA, monosodium urate (MSU) crystals, alum, and other triggers. The inflammasome controls the activity of CASPASE-1, which processes the proinflammatory cytokines interleukin-1β (IL-1β) and IL-18 to their mature forms [188].

AIM2

AIM2 is a member of the hematopoietic nuclear PYHIN protein family [23]. *AIM2* transcripts are expressed in non-lymphoid and lymphoid tissues [189]. It is the only human PYHIN protein that is predominantly expressed in the cytoplasm [186]. AIM2 binds to cytosolic DNA and forms a CASPASE-1 activating inflammasome with the adaptor protein apoptosis-associated speck-like protein containing a caspase recruitment domain (ASC) [18,19,46,190]. AIM2 was originally cloned in a functional screen to identify putative tumor suppressor genes present on chromosome 6, which is often mutated in human melanomas, colorectal carcinomas, gastric and endometrial cancers [184,191–195]. Overexpression of AIM2 suppresses proliferation and increases the rate of apoptosis of several cells, similar to the effects of other members of the HIN-200 family of proteins [23,24,26,45,189,196–199]. AIM2 was also shown to promote invasiveness of colorectal cancer [196,199]. Interestingly, inhibition of AIM2 expression results in the activation of the DNA damage response suggesting a role of AIM2 in the maintenance of genomic stability in cells [31].

NLRP3

NLRP3 forms cytosolic oligomers with ASC in DCs and macrophages in response to oxidized DNA [200–202]. The function of NLPR3 in cancer is not well studied, but NLRP3 was shown to confer protection against colorectal tumor formation [203]. Another study found that the priming of IFN-γ-producing CD8+ T cells by dying tumor cells depends on

NLRP3, suggesting that NLRP3 is essential for initiating an anti-tumor immune response [200].

ASC/PYCARD/TMS1

ASC, also called PYCARD/TMS1, is a bipartite protein that consists of a pyrin domain and a caspase recruitment domain (CARD) domain motif [188]. ASC is expressed in epithelial tissues and immune cells such as neutrophils and macrophages [204]. In the human promyelocytic leukemia cell line HL-60, ASC aggregates during apoptosis and promotes apoptosis in response to genotoxic drugs [205]. Many genotoxic agents activate p53, which was shown to induce ASC expression [206]. Consistent with a role of ASC in the p53-dependent apoptosis pathway, *ASC* is methylated and silenced in 11% of gastric carcinomas, 40–41% of small cell and non-small cell lung carcinomas, 50% of malignant melanomas and 44% of primary glioblastomas [204,207–215]. Importantly, methylation of *ASC* was specific for cancer cells as *ASC* was not methylated in the corresponding normal tissue. In support of the idea that ASC acts as a tumor suppressor, ectopic overexpression of ASC in breast cancer cells was shown to inhibit their ability to proliferate and form colonies [210].

CASPASE-1

CASPASE-1 (ICE; IL-1 concerting enzyme) is a component in all inflammasome types [216,217]. Initially synthesized as a zymogen in cells, it is autocatalyzed to its mature form upon inflammasome activation. It acts as a specific cysteine protease, which targets cleavage sites after aspartic acid residues in proteins [202].

Similar to ASC, CASPASE-1 expression is lost or reduced in many primary prostate cancers and a subset of human gastric carcinomas [218,219]. No difference at the *CASPASE-1* transcript level was found, suggesting a posttranscriptional regulation of CASPASE-1 expression in tumor cells.

IL-1β and IL-18

IL-1β and IL-18 are synthesized as inactive pro-forms, which must be cleaved by CASPASE-1, CASPASE-8 or CASPASE-11 to yield the functional product [202,220]. In response to inflammasome activation, the secretion of proinflammatory cytokines IL-1β and IL-18 results in widespread immune responses.

Chronic inflammation leading to constitutive IL-1-mediated signals was shown to promote tumorigenesis [221,222]. Expression of human

IL-1β in the stomachs of mice induces gastric inflammation and tumorigenesis [223]. The levels of IL-1β and IL-18 in the serum of carcinoma patients correlate with prognosis and can predict the therapeutic response [224]. Late-stage melanoma secrete IL-1β, which supports their proliferation [225,226]. IL-1β may also play an important role in angiogenesis and invasiveness of tumor cells [227–230].

Although IL-1β appears to promote tumorigenesis, it can also contribute to therapeutically relevant anticancer immune responses [200]. $Caspase1^{-/-}$ and $Il18^{-/-}$ mice are more susceptible to azoxymethane (AOM)-DSS-induced tumorigenesis [231]. Administration of IL-18 limits AOM-DSS-induced colorectal carcinogenesis in $caspase1^{-/-}$ mice by activating the tumor suppressor STAT1 and IFN-γ [203,232]. IL-18 also mediates anticancer effects in other tumor types, including sarcoma and melanoma. Administration of IL-18 to mice suppressed the proliferation of murine T241 fibrosarcoma and mammary carcinomas, possibly by regulating angiogenesis [233,234].

CONCLUSION

Cytosolic RNA and DNA sensors are critical for efficient innate and adaptive immunity to viral and bacterial infections. Many sensors, in particular members of the RNA helicase superfamily, also play an important role in RNA metabolism, DNA repair and possibly cancer. Some sensors appear to function as tumor suppressors while others are overexpressed in cancer and may therefore promote tumorigenesis. However, the role of DNA sensor pathways in tumorigenesis has not been elucidated in detail and much work remains to be done. It is currently not clear whether the ability of sensors to recognize cytosolic DNA is functionally important for cancer development, and their role, if any, in cancer immunosurveillance remains to be clarified. Understanding the role of RNA and DNA sensors in cancer may reveal new insights into the interplay between tumorigenesis, inflammation and immunosurveillance.

REFERENCES

[1] Halazonetis TD, Gorgoulis VG, Bartek J. An Oncogene-induced DNA damage model for cancer development. Science 2008;319(5868):1352–5.
[2] Gasser S, Raulet DH. The DNA damage response, immunity and cancer. Semin Cancer Biol 2006;16(5):344–7.
[3] Medzhitov R. TLR-mediated innate immune recognition. Semin Immunol 2007;19(1):1–2.

[4] Akira S, Takeda K. Toll-like receptor signalling. Nat Rev Immunol 2004;4(7):499–511.

[5] Fuertes MB, Kacha AK, Kline J, Woo S-R, Kranz DM, Murphy KM, et al. Host type I IFN signals are required for antitumor CD8+ T cell responses through CD8{alpha} + dendritic cells. J Exp Med 2011;208(10):2005–16.

[6] Dunn GP, Bruce AT, Sheehan KCF, Shankaran V, Uppaluri R, Bui JD, et al. A critical function for type I interferons in cancer immunoediting. Nat Immunol 2005;6(7):722–9.

[7] Diamond MS, Kinder M, Matsushita H, Mashayekhi M, Dunn GP, Archambault JM, et al. Type I interferon is selectively required by dendritic cells for immune rejection of tumors. J Exp Med 2011;208(10):1989–2003.

[8] Raulet DH, Guerra N. Oncogenic stress sensed by the immune system: role of natural killer cell receptors. Nat Rev Immunol 2009;9(8):568–80.

[9] Takaoka A, Wang Z, Choi MK, Yanai H, Negishi H, Ban T, et al. DAI (DLM-1/ ZBP1) is a cytosolic DNA sensor and an activator of innate immune response. Nature 2007;448(7152):501–5.

[10] Rebsamen M, Heinz LX, Meylan E, Michallet M-CEC, Schroder K, Hofmann K, et al. DAI/ZBP1 recruits RIP1 and RIP3 through RIP homotypic interaction motifs to activate NF-kappaB. EMBO Rep 2009;10(8):916–22.

[11] Schwartz T, Behlke J, Lowenhaupst K, Heinemann U, Rich A. Structure of the DLM-1 – Z-DNA complex reveals a conserved family of Z-DNA-binding proteins. Nat Struct Biol 2001;8(9):761–5.

[12] Deigendesch N, Koch-Nolte F, Rothenburg S. ZBP1 subcellular localization and association with stress granules is controlled by its Z-DNA binding domains. Nucleic Acids Res 2006;34(18):5007–20.

[13] Buchan JR, Parker R. Eukaryotic stress granules: the ins and outs of translation. Mol Cell 2009;36(6):932–41.

[14] White JP, Lloyd RE. Regulation of stress granules in virus systems. Trends Microbiol 2012;20(4):175–83.

[15] Fu Y, Comella N, Tognazzi K, Brown LF, Dvorak HF, Kocher O. Cloning of DLM-1, a novel gene that is up-regulated in activated macrophages, using RNA differential display. Gene 1999;240(1):157–63.

[16] Wang G, Christensen LA, Vasquez KM. Z-DNA-forming sequences generate large-scale deletions in mammalian cells. Proc Natl Acad Sci USA 2006;103(8):2677–82.

[17] Gariglio M, Mondini M, De Andrea M, Landolfo S. The multifaceted interferon-inducible p200 family proteins: from cell biology to human pathology. J Interferon Cytokine Res 2011;31(1):159–72.

[18] Hornung V, Ablasser A, Charrel-Dennis M, Bauernfeind F, Horvath G, Caffrey DR, et al. AIM2 recognizes cytosolic dsDNA and forms a caspase-1-activating inflammasome with ASC. Nature 2009;458(7237):514–8.

[19] Fernandes-Alnemri T, Yu J-W, Datta P, Wu J, Alnemri ES. AIM2 activates the inflammasome and cell death in response to cytoplasmic DNA. Nature 2009;458(7237):509–13.

[20] Bürckstümmer T, Baumann C, Blüml S, Dixit E, Dürnberger G, Jahn H, et al. An orthogonal proteomic-genomic screen identifies AIM2 as a cytoplasmic DNA sensor for the inflammasome. Nat Immunol 2009;10(3):266–72.

[21] Unterholzner L, Keating SE, Baran M, Horan KA, Jensen SB, Sharma S, et al. IFI16 is an innate immune sensor for intracellular DNA. Nat Immunol 2010;11(11):997–1004.

[22] Kerur N, Veettil MV, Sharma-Walia N, Bottero V, Sadagopan S, Otageri P, et al. IFI16 Acts as a nuclear pathogen sensor to induce the inflammasome in response to Kaposi sarcoma-associated herpesvirus infection. Cell Host Microbe 2011;9(5):363–75.

[23] Johnstone RW, Trapani JA. Transcription and growth regulatory functions of the HIN-200 family of proteins. Mol Cell Biol 1999;19(9):5833–8.

[24] Asefa B, Klarmann KD, Copeland NG, Gilbert DJ, Jenkins NA, Keller JR. The interferon-inducible p200 family of proteins: a perspective on their roles in cell cycle regulation and differentiation. Blood Cells Mol Dis 2004;32(1):155–67.

[25] Cresswell KS, Clarke CJP, Jackson JT, Darcy PK, Trapani JA, Johnstone RW. Biochemical and growth regulatory activities of the HIN-200 family member and putative tumor suppressor protein, AIM2. Biochem Biophys Res Commun 2005;326(2):417–24.

[26] Trapani JA, Browne KA, Dawson MJ, Ramsay RG, Eddy RL, Show TB, et al. A novel gene constitutively expressed in human lymphoid cells is inducible with interferon-gamma in myeloid cells. Immunogenetics 1992;36(6):369–76.

[27] Wei W, Clarke CJP, Somers GR, Cresswell KS, Loveland KA, Trapani JA, et al. Expression of IFI 16 in epithelial cells and lymphoid tissues. Histochem Cell Biol 2003;119(1):45–54.

[28] Luan Y, Lengyel P, Liu C-J. p204, a p200 family protein, as a multifunctional regulator of cell proliferation and differentiation. Cytokine Growth Factor Rev 2008;19(5-6):357–69.

[29] Choubey D, Lengyel P. Binding of an interferon-inducible protein (p202) to the retinoblastoma protein. J Biol Chem 1995;270(11):6134–40.

[30] Ding B, Lengyel P. p204 protein is a novel modulator of Ras activity. J Biol Chem 2008;283(9):5831–48.

[31] Duan X, Ponomareva L, Veeranki S, Panchanathan R, Dickerson E, Choubey D. Differential roles for the interferon-inducible IFI16 and AIM2 innate immune sensors for cytosolic DNA in cellular senescence of human fibroblasts. Mol Cancer Res 2011;9(5):589–602.

[32] Song LL, Alimirah F, Panchanathan R, Xin H, Choubey D. Expression of an IFN-inducible cellular senescence gene, IFI16, is up-regulated by p53. Mol Cancer Res 2008 Nov;6(11):1732–41.

[33] Raffaella R, Gioia D, De Andrea M, Cappello P, Giovarelli M, Marconi P, et al. The interferon-inducible IFI16 gene inhibits tube morphogenesis and proliferation of primary, but not HPV16 E6/E7-immortalized human endothelial cells. Exp Cell Res 2004;293(2):331–45.

[34] Di Micco R, Fumagalli M, Cicalese A, Piccinin S, Gasparini P, Luise C, et al. Oncogene-induced senescence is a DNA damage response triggered by DNA hyper-replication. Nature 2006;444(7119):638–42.

[35] Johnstone RW, Wei W, Greenway A, Trapani JA. Functional interaction between p53 and the interferon-inducible nucleoprotein IFI 16. Oncogene 2000;19(52):6033–42.

[36] Fujiuchi N. Requirement of IFI16 for the maximal activation of p53 induced by ionizing radiation. J Biol Chem 2004;279(19):20339–20344.

[37] Gugliesi F. Up-regulation of the interferon-inducible IFI16 gene by oxidative stress triggers p53 transcriptional activity in endothelial cells. J Leukocyte Biol 2005;77(5):820–9.

[38] Aglipay JA, Lee SW, Okada S, Fujiuchi N, Ohtsuka T, Kwak JC, et al. A member of the Pyrin family, IFI16, is a novel BRCA1-associated protein involved in the p53-mediated apoptosis pathway. Oncogene 2003;22(55):8931–8.

[39] Azzimonti B, Pagano M, Mondini M, De Andrea M, Valente G, Monga G, et al. Altered patterns of the interferon-inducible gene IFI16 expression in head and neck squamous cell carcinoma: immunohistochemical study including correlation with retinoblastoma protein, human papillomavirus infection and proliferation index. Histopathology 2004;45(6):560–72.

[40] De Andrea M, Gioia D, Mondini M, Azzimonti B, Renò F, Pecorari G, et al. Effects of IFI16 overexpression on the growth and doxorubicin sensitivity of head and neck squamous cell carcinoma-derived cell lines. Head Neck 2007;29(9):835–44.

[41] Alimirah F, Chen J, Davis FJ, Choubey D. IFI16 in human prostate cancer. Mol Cancer Res 2007;5(3):251–9.

[42] Liu C, Ding B, Wang H, Lengyel P. The MyoD-inducible p204 protein overcomes the inhibition of myoblast differentiation by Id proteins. Mol Cell Biol 2002;22(9):2893–905.

[43] Lengyel P, Liu CJ. The p200 family protein p204 as a modulator of cell proliferation and differentiation: a brief survey. Cell Mol Life Sci 2009;67(3):335–40.

[44] Dawson MJ, Trapani JA. IFI 16 gene encodes a nuclear protein whose expression is induced by interferons in human myeloid leukaemia cell lines. J Cell Biochem 2004;57(1):39–51.

[45] Choubey D, Walter S, Geng Y, Xin H. Cytoplasmic localization of the interferon-inducible protein that is encoded by the AIM2 (absent in melanoma) gene from the 200-gene family. FEBS Lett 2000;474(1):38–42.

[46] Roberts T, Idris A, Dunn J, Kelly G, Burnton C, Hodgson S, et al. HIN-200 proteins regulate caspase activation in response to foreign cytoplasmic DNA. Science 2009;323(5917):1057–60.

[47] Li B. p202, an interferon-inducible modulator of transcription, inhibits transcriptional activation by the p53 tumor suppressor protein, and a segment from the p53-binding protein 1 that binds to p202 overcomes this inhibition. J Biol Chem 1996;271(44):27544–27555.

[48] Abdelhaleem M. Do human RNA helicases have a role in cancer?. Biochim Biophys Acta 2004;1704(1):37–46.

[49] Linder P. Dead-box proteins: a family affair—active and passive players in RNP-remodeling. Nucleic Acids Res 2006;34(15):4168–80.

[50] Desmet CJ, Ishii KJ. Nucleic acid sensing at the interface between innate and adaptive immunity in vaccination. Nat Rev Immunol 2012;12(7):479–91.

[51] Zhang Z, Kim T, Bao M, Facchinetti V, Jung SY, Ghaffari AA, et al. DDX1, DDX21, and DHX36 helicases form a complex with the adaptor molecule TRIF to sense dsRNA in dendritic cells. Immunity 2011;34(6):866–78.

[52] Kim T, Pazhoor S, Bao M, Zhang Z, Hanabuchi S, Facchinetti V, et al. Aspartate-glutamate-alanine-histidine box motif (DEAH)/RNA helicase A helicases sense microbial DNA in human plasmacytoid dendritic cells. Proc Natl Acad Sci USA 2010;107(34):15181–15186.

[53] Abdelhaleem M, Maltais L, Wain H. The human DDX and DHX gene families of putative RNA helicases. Genomics 2003;81(6):618–22.

[54] Reddy TR, Tang H, Xu W, Wong-Staal F. Sam68, RNA helicase A and Tap cooperate in the post-transcriptional regulation of human immunodeficiency virus and type D retroviral mRNA. Oncogene 2000;19(32):3570–5.

[55] Anderson SF, Schlegel BP, Nakajima T, Wolpin ES, Parvin JD. BRCA1 protein is linked to the RNA polymerase II holoenzyme complex via RNA helicase A. Nature Genet 1998;19(3):254–6.

[56] Ohno M, Shimura Y. A human RNA helicase-like protein, HRH1, facilitates nuclear export of spliced mRNA by releasing the RNA from the spliceosome. Genes Dev 1996;10(8):997–1007.

[57] Li J, Tang H, Mullen TM, Westberg C, Reddy TR, Rose DW, et al. A role for RNA helicase A in post-transcriptional regulation of HIV type 1. Proc Natl Acad Sci USA 1999;96(2):709–14.

[58] Zhang S, Herrmann C, Grosse F. Pre-mRNA and mRNA binding of human nuclear DNA helicase II (RNA helicase A). J Cell Sci 1999;112(Pt 7):1055–64.

[59] Tang H, Gaietta GM, Fischer WH, Ellisman MH, Wong-Staal F. A cellular cofactor for the constitutive transport element of type D retrovirus. Science 1997;276(5317):1412–5.

[60] Tang H, Wong-Staal F. Specific interaction between RNA helicase A and Tap, two cellular proteins that bind to the constitutive transport element of type D retrovirus. J Biol Chem 2000;275(42):32694–32700.

[61] Yedavalli VSRK, Neuveut C, Chi Y-H, Kleiman L, Jeang K-T. Requirement of DDX3 DEAD box RNA helicase for HIV-1 Rev-RRE export function. Cell 2004;119(3):381–92.

[62] Tang H, McDonald D, Middlesworth T, Hope TJ, Wong-Staal F. The carboxyl terminus of RNA helicase A contains a bidirectional nuclear transport domain. Mol Cell Biol 1999;19(5):3540–50.

[63] Fujii R, Okamoto M, Aratani S, Oishi T, Ohshima T, Taira K, et al. A role of RNA helicase A in cis-acting transactivation response element-mediated transcriptional regulation of human immunodeficiency virus type 1. J Biol Chem 2001;276:5445–51.

[64] Jain A, Bacolla A, Chakraborty P, Grosse F, Vasquez KM. Human DHX9 helicase unwinds triple-helical DNA structures. Biochemistry 2010;49(33):6992–9.

[65] Chakraborty P, Grosse F. WRN helicase unwinds Okazaki fragment-like hybrids in a reaction stimulated by the human DHX9 helicase. Nucleic Acids Res 2010;38(14):4722–30.

[66] Wei X, Pacyna-Gengelbach M, Schlüns K, An Q, Gao Y, Cheng S, et al. Analysis of the RNA helicase A gene in human lung cancer. Oncol Rep 2004;11(1):253–8.

[67] Myöhänen S, Baylin SB. Sequence-specific DNA binding activity of RNA helicase A to the p16INK4a promoter. J Biol Chem 2001;276(2):1634–42.

[68] Aratani S, Fujii R, Fujita H, Fukamizu A, Nakajima T. Aromatic residues are required for RNA helicase A mediated transactivation. Int J Mol Med 2003;12(2):175–80.

[69] Schlegel BP, Starita LM, Parvin JD. Overexpression of a protein fragment of RNA helicase A causes inhibition of endogenous BRCA1 function and defects in ploidy and cytokinesis in mammary epithelial cells. Oncogene 2003;22(7):983–91.

[70] Tran H, Schilling M, Wirbelauer C, Hess D, Nagamine Y. Facilitation of mRNA deadenylation and decay by the exosome-bound, DExH protein RHAU. Mol Cell 2004;13(1):101–11.

[71] Iwamoto F, Stadler M, Chalupníková K, Oakeley E, Nagamine Y. Transcription-dependent nucleolar cap localization and possible nuclear function of DExH RNA helicase RHAU. Exp Cell Res 2008;314(6):1378–91.

[72] Chalupníková K, Lattmann S, Selak N, Iwamoto F, Fujiki Y, Nagamine Y. Recruitment of the RNA helicase RHAU to stress granules via a unique RNA-binding domain. J Biol Chem 2008;283(50):35186–35198.

[73] Vaughn JP, Creacy SD, Routh ED, Joyner-Butt C, Jenkins GS, Pauli S, et al. The DEXH protein product of the DHX36 gene is the major source of tetramolecular quadruplex G4-DNA resolving activity in HeLa cell lysates. J Biol Chem 2005;280(46):38117–38120.

[74] Lattmann S, Giri B, Vaughn JP, Akman SA, Nagamine Y. Role of the amino terminal RHAU-specific motif in the recognition and resolution of guanine quadruplex-RNA by the DEAH-box RNA helicase RHAU. Nucleic Acids Res 2010;38(18):6219–33.

[75] Maizels N. Dynamic roles for G4 DNA in the biology of eukaryotic cells. Nature Struct Molec Biol 2006;13(12):1055–9.

[76] Godbout R, Packer M, Katyal S, Bléoo S. Cloning and expression analysis of the chicken DEAD box gene DDX1. Biochim Biophys Acta 2002;1574(1):63–71.

[77] Chen H-C, Lin W-C, Tsay Y-G, Lee S-C, Chang C-J. An RNA helicase DDX1, interacting with poly(A) RNA and heterogeneous nuclear ribonucleoprotein K. J Biol Chem 2002;277(43):40403–40409.

[78] Bléoo S, Sun X, Hendzel MJ, Rowe JM, Packer M, Godbout R. Association of human DEAD box protein DDX1 with a cleavage stimulation factor involved in 3′-end processing of pre-MRNA. Mol Biol Cell 2001;12(10):3046–59.

[79] Godbout R, Squire J. Amplification of a DEAD box protein gene in retinoblastoma cell lines. Proc Natl Acad Sci USA 1993;90(16):7578–82.

[80] Godbout R, Packer M, Bie W. Overexpression of a DEAD box protein (DDX1) in neuroblastoma and retinoblastoma cell lines. J Biol Chem 1998;273(33):21161–21168.

[81] Squire JA, Thorner PS, Weitzman S, Maggi JD, Dirks P, Doyle J, et al. Co-amplification of MYCN and a DEAD box gene (DDX1) in primary neuroblastoma. Oncogene 1995;10(7):1417–22.

[82] Fuller-Pace FV. DExD/H box RNA helicases: multifunctional proteins with important roles in transcriptional regulation. Nucleic Acids Res 2006;34(15):4206–15.

[83] Henning D. Silencing of RNA helicase II/Gu inhibits mammalian ribosomal RNA production. J Biol Chem 2003;278(52):52307–52314.

[84] Yang H, Henning D, Valdez BC. Functional interaction between RNA helicase II/Gu(alpha) and ribosomal protein L4. FEBS J 2005;272(15):3788–802.

[85] Yang H. Down-regulation of RNA helicase II/Gu results in the depletion of 18 and 28 S rRNAs in Xenopus oocyte. J Biol Chem 2003;278(40):38847–38859.

[86] Zhang Z, Yuan B, Bao M, Lu N, Kim T, Liu Y-J. The helicase DDX41 senses intracellular DNA mediated by the adaptor STING in dendritic cells. Nat Immunol 2011;12(10):1–8.

[87] Irion U, Leptin M, Siller K, Fuerstenberg S, Cai Y, Doe CQ, et al. Abstrakt, a DEAD box protein, regulates Insc levels and asymmetric division of neural and mesodermal progenitors. Curr Biol 2004;14(2):138–44.

[88] Irion U, Leptin M. Developmental and cell biological functions of the Drosophila DEAD-box protein abstrakt. Curr Biol 1999;9(23):1373–81.

[89] Schmucker D, Vorbrüggen G, Yeghiayan P, Fan HQ, Jäckle H, Gaul U. The Drosophila gene abstrakt, required for visual system development, encodes a putative RNA helicase of the DEAD box protein family. Mech Dev 2000;91(1–2):189–96.

[90] Reed AL, Yamazaki H, Kaufman JD, Rubinstein Y, Murphy B, Johnson AC. Molecular cloning and characterization of a transcription regulator with homology to GC-binding factor. J Biol Chem 1998;273(34):21594–21602.

[91] Yang P, An H, Liu X, Wen M, Zheng Y, Rui Y, et al. The cytosolic nucleic acid sensor LRRFIP1 mediates the production of type I interferon via a [beta]-catenin-dependent pathway. Nat Immunol 2010;11(6):487–94.

[92] Clevers H. Wnt/beta-catenin signaling in development and disease. Cell 2006;127(3):469–80.

[93] Kypta RM, Waxman J. Wnt/β-catenin signalling in prostate cancer. Nat Rev Urol 2012;9:418–28.

[94] Städeli R, Hoffmans R, Basler K. Transcription under the control of nuclear arm/β-catenin. Current Biol 2006;16(10):R378–85.

[95] Michaelson JS, Leder P. Beta-catenin is a downstream effector of Wnt-mediated tumorigenesis in the mammary gland. Oncogene 2001;20(37):5093–9.

[96] Ohtsuka H, Oikawa M, Ariake K, Rikiyama T, Motoi F, Katayose Y, et al. GC-binding factor 2 interacts with dishevelled and regulates Wnt signaling pathways in human carcinoma cell lines. Int J Cancer 2011;129(7):1599–610.

[97] Suriano AR, Sanford AN, Kim N, Oh M, Kennedy S, Henderson MJ, et al. GCF2/LRRFIP1 represses tumor necrosis factor alpha expression. Mol Cell Biol 2005;25(20):9073–81.

[98] Shibutani M, Lazarovici P, Johnson AC, Katagiri Y, Guroff G. Transcriptional downregulation of epidermal growth factor receptors by nerve growth factor treatment of PC12 cells. J Biol Chem 1998;273(12):6878–84.

[99] Rikiyama T, Curtis J, Oikawa M, Zimonjic DB, Popescu N, Murphy BA, et al. GCF2: expression and molecular analysis of repression. Biochim Biophys Acta 2003;1629(1-3):15–25.

[100] Ariake K, Ohtsuka H, Motoi F, Douchi D, Oikawa M, Rikiyama T, et al. GCF2/ LRRFIP1 promotes colorectal cancer metastasis and liver invasion through integrin-dependent RhoA activation. Cancer Lett 2012;325(1):99–107.

[101] Sjöblom T, Jones S, Wood LD, Parsons DW, Lin J, Barber TD, et al. The consensus coding sequences of human breast and colorectal cancers. Science 2006;314(5797):268–74.

[102] Li Y, Li W, Yang Y, Lu Y, He C, Hu G, et al. MicroRNA-21 targets LRRFIP1 and contributes to VM-26 resistance in glioblastoma multiforme. Brain Res 2009;1286:13–18.

[103] Shen D-W, Pouliot LM, Gillet J-P, Ma W, Johnson AC, Hall MD, et al. The transcription factor GCF2 is an upstream repressor of the small GTPAse RhoA, regulating membrane protein trafficking, sensitivity to doxorubicin, and resistance to cisplatin. Mol Pharm 2012;9(6):1822–33.

[104] Liu F, Gu J. Retinoic acid inducible gene-I, more than a virus sensor. Protein Cell 2011;2(5):351–7.

[105] Hornung V, Ellegast J, Kim S, Brzózka K, Jung A, Kato H, et al. 5′-Triphosphate RNA is the ligand for RIG-I. Science 2006;314(5801):994–7.

[106] Rehwinkel J, Tan CP, Goubau D, Schulz O, Pichlmair A, Bier K, et al. RIG-I detects viral genomic RNA during negative-strand RNA virus infection. Cell 2010;140(3):397–408.

[107] Chiu Y-H, MacMillan JB, Chen ZJ. RNA polymerase III detects cytosolic DNA and induces type I interferons through the RIG-I pathway. Cell 2009;138(3):576–91.

[108] Su Z-Z, Sarkar D, Emdad L, Barral PM, Fisher PB. Central role of interferon regulatory factor-1 (IRF-1) in controlling retinoic acid inducible gene-I (RIG-I) expression. J Cell Physiol 2007;213(2):502–10.

[109] Besch R, Poeck H, Hohenauer T, Senft D, Häcker G, Berking C, et al. Proapoptotic signaling induced by RIG-I and MDA-5 results in type I interferon-independent apoptosis in human melanoma cells. J Clin Invest 2009;119(8):2399–411.

[110] Liu TX, Zhang JW, Tao J, Zhang RB, Zhang QH, Zhao CJ, et al. Gene expression networks underlying retinoic acid-induced differentiation of acute promyelocytic leukemia cells. Blood 2000;96(4):1496–504.

[111] Tsai F-M, Shyu R-Y, Jiang S-Y. RIG1 inhibits the Ras/mitogen-activated protein kinase pathway by suppressing the activation of Ras. Cell Signal 2006;18(3):349–58.

[112] Sims GP, Rowe DC, Rietdijk ST, Herbst R, Coyle AJ. HMGB1 and RAGE in inflammation and cancer. Annu Rev Immunol 2010;28:367–88.

[113] Lange SS, Vasquez KM. HMGB1: the jack-of-all-trades protein is a master DNA repair mechanic. Mol Carcinog 2009;48(7):571–80.

[114] van Beijnum JR, Petersen K, Griffioen AW. Tumor endothelium is characterized by a matrix remodeling signature. Front Biosci (Schol Ed) 2009;1:216–25.

[115] Sparvero LJ, Asafu-Adjei D, Kang R, Tang D, Amin N, Im J, et al. RAGE (Receptor for Advanced Glycation Endproducts), RAGE ligands, and their role in cancer and inflammation. J Transl Med 2009;7:17.

[116] Ellerman JE, Brown CK, de Vera M, Zeh HJ, Billiar T, Rubartelli A, et al. Masquerader: high mobility group box-1 and cancer. Clin Cancer Res 2007;13(10):2836–48.

[117] Brezniceanu M-L, Völp K, Bösser S, Solbach C, Lichter P, Joos S, et al. HMGB1 inhibits cell death in yeast and mammalian cells and is abundantly expressed in human breast carcinoma. FASEB J 2003;17(10):1295–7.

[118] Chung HW, Lee S-G, Kim H, Hong DJ, Chung JB, Stroncek D, et al. Serum high mobility group box-1 (HMGB1) is closely associated with the clinical and pathologic features of gastric cancer. J Transl Med 2009;7:38.

[119] Schlueter C, Weber H, Meyer B, Rogalla P, Röser K, Hauke S, et al. Angiogenetic signaling through hypoxia: HMGB1: an angiogenetic switch molecule. Am J Pathol 2005;166(4):1259–63.

[120] Taguchi A, Blood DC, del Toro G, Canet A, Lee DC, Qu W, et al. Blockade of RAGE-amphoterin signalling suppresses tumour growth and metastases. Nature 2000;405(6784):354–60.

[121] Apetoh L, Ghiringhelli F, Tesniere A, Obeid M, Ortiz C, Criollo A, et al. Toll-like receptor 4-dependent contribution of the immune system to anticancer chemotherapy and radiotherapy. Nat Med 2007;13(9):1050–9.

[122] Morio T, Kim H. Ku, Artemis, and ataxia-telangiectasia-mutated: signalling networks in DNA damage. Int J Biochem Cell Biol 2008;40(4):598–603.

[123] Rossi ML, Ghosh AK, Bohr VA. Roles of Werner syndrome protein in protection of genome integrity. DNA Repair 2010;9(3):331–44.

[124] Zhang X, Brann TW, Zhou M, Yang J, Oguariri RM, Lidie KB, et al. Cutting edge: Ku70 is a novel cytosolic DNA sensor that induces type III rather than type I IFN. J Immunol 2011;186:4541–5.

[125] Takaoka A, Taniguchi T. Cytosolic DNA recognition for triggering innate immune responses. Adv Drug Deliv Rev 2008;60(7):847–57.

[126] Ishikawa H, Barber GN. The STING pathway and regulation of innate immune signaling in response to DNA pathogens. Cell Mol Life Sci 2010;68(7):1157–65.

[127] Ishikawa H, Barber GN. STING is an endoplasmic reticulum adaptor that facilitates innate immune signalling. Nature 2008;455(7213):674–8.

[128] Heider MR, Munson M. Exorcising the exocyst complex. Traffic 2012;13(7):898–907.

[129] Ishikawa H, Ma Z, Barber GN. STING regulates intracellular DNA-mediated, type I interferon-dependent innate immunity. Nature 2009;461(7265):788–92.

[130] Camonis JH, White MA. Ral GTPases: corrupting the exocyst in cancer cells. Trends Cell Biol 2005;15(6):327–32.

[131] Sakurai-Yageta M, Recchi C, Le Dez G, Sibarita J-B, Daviet L, Camonis J, et al. The interaction of IQGAP1 with the exocyst complex is required for tumor cell invasion downstream of Cdc42 and RhoA. J Cell Biol 2008;181(6):985–98.

[132] Moskalenko S, Henry DO, Rosse C, Mirey G, Camonis JH, White MA. The exocyst is a Ral effector complex. Nat Cell Biol 2001;4(1):66–72.

[133] Jin L, Waterman PM, Jonscher KR, Short CM, Reisdorph NA, Cambier JC. MPYS, a novel membrane tetraspanner, is associated with major histocompatibility complex class II and mediates transduction of apoptotic signals. Mol Cell Biol 2008;28(16):5014–26.

[134] Jin L, Lenz LL, Cambier JC. Cellular reactive oxygen species inhibit MPYS induction of IFNβ. PLoS ONE 2010;5(12):e15142.

[135] Heiber JF, Barber GN. Evaluation of innate immune signaling pathways in transformed cells. Methods Mol Biol 2012;797:217–38.

[136] Kawai T, Akira S. The role of pattern-recognition receptors in innate immunity: update on Toll-like receptors. Nat Immunol 2010;11(5):373–84.

[137] Salaun B, Coste I, Rissoan M-C, Lebecque SJ, Renno T. TLR3 can directly trigger apoptosis in human cancer cells. J Immunol 2006;176(8):4894–901.

[138] Han K-J, Su X, Xu L-G, Bin L-H, Zhang J, Shu H-B. Mechanisms of the TRIF-induced interferon-stimulated response element and NF-kappaB activation and apoptosis pathways. J Biol Chem 2004;279(15):15652–15661.

[139] Weber A, Kirejczyk Z, Besch R, Potthoff S, Leverkus M, Häcker G. Proapoptotic signalling through Toll-like receptor-3 involves TRIF-dependent activation of caspase-8 and is under the control of inhibitor of apoptosis proteins in melanoma cells. Cell Death Differ 2010;17(6):942–51.

[140] Feoktistova M, Geserick P, Kellert B, Dimitrova DP, Langlais C, Hupe M, et al. cIAPs block ripoptosome formation, a RIP1/caspase-8 containing intracellular cell death complex differentially regulated by cFLIP isoforms. Mol Cell 2011;43(3):449–63.

[141] Fitzgerald KA, McWhirter SM, Faia KL, Rowe DC, Latz E, Golenbock DT, et al. IKKepsilon and TBK1 are essential components of the IRF3 signaling pathway. Nat Immunol 2003;4(5):491–6.

[142] Shen RR, Hahn WC. Emerging roles for the non-canonical IKKs in cancer. Oncogene 2011;30(6):631–41.

[143] Ou Y-H, Torres M, Ram R, Formstecher E, Roland C, Cheng T, et al. TBK1 directly engages Akt/PKB survival signaling to support oncogenic transformation. Mol Cell 2011;41(4):458–70.

[144] Chien Y, Kim S, Bumeister R, Loo Y-M, Kwon SW, Johnson CL, et al. RalB GTPase-mediated activation of the IκB family kinase TBK1 couples innate immune signaling to tumor cell survival. Cell 2006;127(1):157–70.

[145] Clément J-F, Meloche S, Servant MJ. The IKK-related kinases: from innate immunity to oncogenesis. Cell Res 2008;18(9):889–99.

[146] Barbie DA, Tamayo P, Boehm JS, Kim SY, Moody SE, Dunn IF, et al. Systematic RNA interference reveals that oncogenic KRAS-driven cancers require TBK1. Nature 2009;461(7269):108–12.

[147] Chien Y, White MA. Characterization of RalB-Sec5-TBK1 function in human oncogenesis. Meth Enzymol 2008;438:321–9.

[148] Qin B, Cheng K. Silencing of the IKKε gene by siRNA inhibits invasiveness and growth of breast cancer cells. Breast Cancer Res 2010;12(5):R74.

[149] Hsu S, Kim M, Hernandez L, Grajales V, Noonan A, Anver M, et al. IKK-ε coordinates invasion and metastasis of ovarian cancer. Cancer Res 2012;72(21):5494–504.

[150] Péant B, Forest V, Trudeau V, Latour M, Mes-Masson A-M, Saad F. IκB-kinase-ε (IKKε/IKKi/IκBKε) expression and localization in prostate cancer tissues. Prostate 2011;71(10):1131–8.

[151] Taniguchi T, Ogasawara K, Takaoka A, Tanaka N. IRF family of transcription factors as regulators of host defense. Annu Rev Immunol 2001;19:623–55.

[152] Falvo JV, Parekh BS, Lin CH, Fraenkel E, Maniatis T. Assembly of a functional beta interferon enhanceosome is dependent on ATF-2-c-jun heterodimer orientation. Mol Cell Biol 2000;20(13):4814–25.

[153] Merika M, Williams AJ, Chen G, Collins T, Thanos D. Recruitment of CBP/p300 by the IFN beta enhanceosome is required for synergistic activation of transcription. Mol Cell 1998;1(2):277–87.

[154] Barber GN. Cytoplasmic DNA innate immune pathways. Immunol Rev 2011;243(1):99–108.

[155] Civas A, Génin P, Morin P, Lin R, Hiscott J. Promoter organization of the interferon-A genes differentially affects virus-induced expression and responsiveness to TBK1 and IKKepsilon. J Biol Chem 2006;281(8):4856–66.

[156] Sakaguchi S, Negishi H, Asagiri M, Nakajima C, Mizutani T, Takaoka A, et al. Essential role of IRF-3 in lipopolysaccharide-induced interferon-beta gene expression and endotoxin shock. Biochem Biophys Res Commun 2003;306(4):860–6.

[157] Sato M, Suemori H, Hata N, Asagiri M, Ogasawara K, Nakao K, et al. Distinct and essential roles of transcription factors IRF-3 and IRF-7 in response to viruses for IFN-alpha/beta gene induction. Immunity 2000;13(4):539–48.

[158] Honda K, Yanai H, Negishi H, Asagiri M, Sato M, Mizutani T, et al. IRF-7 is the master regulator of type-I interferon-dependent immune responses. Nature 2005;434(7034):772–7.

[159] Honda K, Yanai H, Takaoka A, Taniguchi T. Regulation of the type I IFN induction: a current view. Internat Immunol 2005;17(11):1367–78.

[160] Kim T-K, Lee J-S, Oh S-Y, Jin X, Choi Y-J, Lee T-H, et al. Direct transcriptional activation of promyelocytic leukemia protein by IFN regulatory factor 3 induces the p53-dependent growth inhibition of cancer cells. Cancer Res 2007;67(23):11133–40.

[161] Kim TY, Lee K-H, Chang S, Chung C, Lee H-W, Yim J, et al. Oncogenic potential of a dominant negative mutant of interferon regulatory factor 3. J Biol Chem 2003;278(17):15272–78.

[162] Roberts ZJ, Goutagny N, Perera P-Y, Kato H, Kumar H, Kawai T, et al. The chemotherapeutic agent DMXAA potently and specifically activates the TBK1–IRF-3 signaling axis. J Exp Med 2007;204(7):1559–69.

[163] Tokunaga T, Naruke Y, Shigematsu S, Kohno T, Yasui K, Ma Y, et al. Aberrant expression of interferon regulatory factor 3 in human lung cancer. Biochem Biophys Res Commun 2010;397(2):202–7.

[164] Korherr C, Gille H, Schäfer R, Koenig-Hoffmann K, Dixelius J, Egland KA, et al. Identification of proangiogenic genes and pathways by high-throughput functional genomics: TBK1 and the IRF3 pathway. Proc Natl Acad Sci USA 2006;103(11):4240–5.

[165] Bidwell BN, Slaney CY, Withana NP, Forster S, Cao Y, Loi S, et al. Silencing of Irf7 pathways in breast cancer cells promotes bone metastasis through immune escape. Nat Med 2012;18(8):1224–34.

[166] Romieu-Mourez R, Solis M, Nardin A, Goubau D, Baron-Bodo V, Lin R, et al. Distinct roles for IFN regulatory factor (IRF)-3 and IRF-7 in the activation of antitumor properties of human macrophages. Cancer Res 2006;66(21):10576–10585.

[167] Pahl HL. Activators and target genes of Rel/NF-kappaB transcription factors. Oncogene 1999;18(49):6853.

[168] Karin M, Greten FR. NF-κB: linking inflammation and immunity to cancer development and progression. Nat Rev Immunol 2005;5(10):749–59.

[169] Beroukhim R, Mermel CH, Porter D, Wei G, Raychaudhuri S, Donovan J, et al. The landscape of somatic copy-number alteration across human cancers. Nature 2010;463(7283):899–905.

[170] Sovak MA, Bellas RE, Kim DW, Zanieski GJ, Rogers AE, Traish AM, et al. Aberrant nuclear factor-kappaB/Rel expression and the pathogenesis of breast cancer. J Clin Invest 1997;100(12):2952–60.

[171] Grivennikov SI, Greten FR, Karin M. Immunity, inflammation, and cancer. Cell 2010;140(6):883–99.

[172] Luo J-L, Maeda S, Hsu L-C, Yagita H, Karin M. Inhibition of NF-kappaB in cancer cells converts inflammation- induced tumor growth mediated by TNFalpha to TRAIL-mediated tumor regression. Cancer Cell 2004;6(3):297–305.

[173] Moore RJ, Owens DM, Stamp G, Arnott C, Burke F, East N, et al. Mice deficient in tumor necrosis factor-alpha are resistant to skin carcinogenesis. Nat Med 1999;5(7):828–31.

[174] Arnott CH, Scott KA, Moore RJ, Robinson SC, Thompson RG, Balkwill FR. Expression of both TNF-alpha receptor subtypes is essential for optimal skin tumour development. Oncogene 2004;23(10):1902–10.

[175] Gilbert LA, Hemann MT. Context-specific roles for paracrine IL-6 in lymphomagenesis. Genes Dev 2012;26(15):1758–68.

[176] Grivennikov S, Karin E, Terzic J, Mucida D, Yu G-Y, Vallabhapurapu S, et al. IL-6 and Stat3 are required for survival of intestinal epithelial cells and development of colitis-associated cancer. Cancer Cell 2009;15(2):103–13.

[177] Bollrath J, Phesse TJ, Burstin von VA, Putoczki T, Bennecke M, Bateman T, et al. gp130-mediated Stat3 activation in enterocytes regulates cell survival and cell-cycle progression during colitis-associated tumorigenesis. Cancer Cell 2009;15(2):91–102.

[178] Naugler WE, Sakurai T, Kim S, Maeda S, Kim K, Elsharkawy AM, et al. Gender disparity in liver cancer due to sex differences in MyD88-dependent IL-6 production. Science 2007;317(5834):121–4.

[179] Wong VW-S, Yu J, Cheng AS-L, Wong GL-H, Chan H-Y, Chu ES-H, et al. High serum interleukin-6 level predicts future hepatocellular carcinoma development in patients with chronic hepatitis B. Int J Cancer 2009;124(12):2766–70.

[180] Dunn GP, Koebel CM, Schreiber RD. Interferons, immunity and cancer immunoediting. Nat Rev Immunol 2006;6(11):836–48.

[181] Schmidt M, Nagel S, Proba J, Thiede C, Ritter M, Waring JF, et al. Lack of interferon consensus sequence binding protein (ICSBP) transcripts in human myeloid leukemias. Blood 1998;91(1):22–9.

[182] Takaoka A, Hayakawa S, Yanai H, Stoiber D, Negishi H, Kikuchi H, et al. Integration of interferon-alpha/beta signalling to p53 responses in tumour suppression and antiviral defence. Nature 2003;424(6948):516–23.

[183] Shimada K, Crother TR, Karlin J, Dagvadorj J, Chiba N, Chen S, et al. Oxidized mitochondrial DNA activates the NLRP3 inflammasome during apoptosis. Immunity 2012;36(3):401–14.

[184] DeYoung KL, Ray ME, Su YA, Anzick SL, Johnstone RW, Trapani JA, et al. Cloning a novel member of the human interferon-inducible gene family associated with control of tumorigenicity in a model of human melanoma. Oncogene 1997;15(4):453–7.

[185] Vilaysane A, Muruve DA. The innate immune response to DNA. Sem Immunol 2009;21(4):208–14.

[186] Hornung V, Latz E. Intracellular DNA recognition. Nat Rev Immunol 2010;10(2):123–30.

[187] Takeshita F, Ishii KJ. Intracellular DNA sensors in immunity. Curr Opin Immunol 2008;20(4):383–8.

[188] Davis BK, Wen H, Ting JP-Y. The inflammasome NLRs in immunity, inflammation, and associated diseases. Annu Rev Immunol 2011;29:707–35.

[189] Landolfo S, Gariglio M, Gribaudo G, Lembo D. The Ifi 200 genes: an emerging family of IFN-inducible genes. Biochimie 1998;80(8):721–8.

[190] Jin T, Perry A, Jiang J, Smith P, Curry JA, Unterholzner L, et al. Structures of the HIN domain:DNA complexes reveal ligand binding and activation mechanisms of the AIM2 inflammasome and IFI16 receptor. Immunity 2012;36(4):561–71.

[191] Ray ME, Su YA, Meltzer PS, Trent JM. Isolation and characterization of genes associated with chromosome-6 mediated tumor suppression in human malignant melanoma. Oncogene 1996;12(12):2527–33.

[192] Michel S, Kloor M, Singh S, Gdynia G, Roth W, Knebel Doeberitz von M, et al. Coding microsatellite instability analysis in microsatellite unstable small intestinal adenocarcinomas identifies MARCKS as a common target of inactivation. Mol Carcinog 2010;49(2):175–82.

[193] Woerner SM, Kloor M, Mueller A, Rueschoff J, Friedrichs N, Buettner R, et al. Microsatellite instability of selective target genes in HNPCC-associated colon adenomas. Oncogene 2005;24(15):2525–35.

[194] Woerner SM, Benner A, Sutter C, Schiller M, Yuan YP, Keller G, et al. Pathogenesis of DNA repair-deficient cancers: a statistical meta-analysis of putative Real Common Target genes. Oncogene 2003;22(15):2226–35.

[195] Woerner SM, Kloor M, Schwitalle Y, Youmans H, Doeberitz MK, Gebert J, et al. The putative tumor suppressor AIM2 is frequently affected by different genetic alterations in microsatellite unstable colon cancers. Genes Chromosomes Cancer 2007;46(12):1080–9.

[196] Patsos G, Germann A, Gebert J, Dihlmann S. Restoration of absent in melanoma 2 (AIM2) induces G2/M cell cycle arrest and promotes invasion of colorectal cancer cells. Int J Cancer 2010;126(8):1838–49.

[197] Ding Y, Wang L, Su L-K, Frey JA, Shao R, Hunt KK, et al. Antitumor activity of IFIX, a novel interferon-inducible HIN-200 gene, in breast cancer. Oncogene 2004;23(26):4556–66.

[198] Asefa B, Dermott JM, Kaldis P, Stefanisko K, Garfinkel DJ, Keller JR. p205, A potential tumor suppressor, inhibits cell proliferation via multiple pathways of cell cycle regulation. FEBS Letters 2006 Feb;580(5):1205–14.

[199] Chen I-F. AIM2 suppresses human breast cancer cell proliferation in vitro and mammary tumor growth in a mouse model. Molec Cancer Therap 2006;5(1):1–7.

[200] Ghiringhelli F, Apetoh L, Tesniere A, Aymeric L, Ma Y, Ortiz C, et al. Activation of the NLRP3 inflammasome in dendritic cells induces IL-1beta-dependent adaptive immunity against tumors. Nat Med 2009;15(10):1170–8.

[201] Franchi L, Eigenbrod T, Muñoz-Planillo R, Nuñez G. The inflammasome: a caspase-1-activation platform that regulates immune responses and disease pathogenesis. Nat Immunol 2009;10(3):241–7.

[202] Martinon F, Mayor A, Tschopp J. The inflammasomes: guardians of the body. Annu Rev Immunol 2009;27:229–65.

[203] Zaki MH, Vogel P, Body-Malapel M, Lamkanfi M, Kanneganti T-D. IL-18 production downstream of the Nlrp3 inflammasome confers protection against colorectal tumor formation. J Immunol 2010;185(8):4912–20.

[204] McConnell BB, Vertino PM. TMS1/ASC: the cancer connection. Apoptosis 2004;9(1):5–18.

[205] Masumoto J. ASC, a novel 22-kDa protein, aggregates during apoptosis of human promyelocytic leukemia HL-60 cells. J Biol Chemi 1999;274(48):33835–33838.

[206] Ohtsuka T, Ryu H, Minamishima YA, Macip S, Sagara J, Nakayama KI, et al. ASC is a Bax adaptor and regulates the p53-Bax mitochondrial apoptosis pathway. Nat Cell Biol 2004;6(2):121–8.

[207] Ohtsuka T, Liu X-F, Koga Y, Kitajima Y, Nakafusa Y, Ha C-W, et al. Methylation-induced silencing of ASC and the effect of expressed ASC on p53-mediated chemosensitivity in colorectal cancer. Oncogene 2005;25(12):1807–11.

[208] Virmani A, Rathi A, Sugio K, Sathyanarayana UG, Toyooka S, Kischel FC, et al. Aberrant methylation of TMS1 in small cell, non small cell lung cancer and breast cancer. Int J Cancer 2003;106(2):198–204.

[209] Yokoyama T, Sagara J, Guan X, Masumoto J, Takeoka M, Komiyama Y, et al. Methylation of ASC/TMS1, a proapoptotic gene responsible for activating procaspase-1, in human colorectal cancer. Cancer Lett 2003;202(1):101–8.

[210] Conway KE, McConnell BB, Bowring CE, Donald CD, Warren ST, Vertino PM. TMS1, a novel proapoptotic caspase recruitment domain protein, is a target of methylation-induced gene silencing in human breast cancers. Cancer Res 2000;60(22):6236–42.

[211] Moriai R, Tsuji N, Kobayashi D, Yagihashi A, Namiki Y, Takahashi H, et al. A proapoptotic caspase recruitment domain protein gene, TMS1, is hypermethylated in human breast and gastric cancers. Anticancer Res 2002 Nov;22(6C):4163–8.

[212] Terasawa K, Sagae S, Toyota M, Tsukada K, Ogi K, Satoh A, et al. Epigenetic inactivation of TMS1/ASC in ovarian cancer. Clin Cancer Res 2004;10:2000–6.

[213] Tamandani DMK, Sobti RC, Shekari M, Huria A. CpG island methylation of TMS1/ASC and CASP8 genes in cervical cancer. European J Med Res 2009;14(2):71.

[214] Machida EO. Hypermethylation of ASC/TMS1 is a sputum marker for late-stage lung cancer. Cancer Res 2006;66(12):6210–8.

[215] Collard RL, Harya NS, Monzon FA, Maier CE, O'Keefe DS. Methylation of the ASC gene promoter is associated with aggressive prostate cancer. Prostate 2006;66(7):687–95.

[216] Thornberry NA, Bull HG, Calaycay JR, Chapman KT, Howard AD, Kostura MJ, et al. A novel heterodimeric cysteine protease is required for interleukin-1β processing in monocytes. Nature 1992;356(6372):768–74.

[217] Cerretti DP, Kozlosky CJ, Mosley B, Nelson N. Molecular cloning of the interleukin-1 beta converting enzyme. Science 1992;256(5053):97–100.

[218] Jee CD, Lee HS, Bae SI, Yang HK, Lee YM, Rho MS, et al. Loss of caspase-1 gene expression in human gastric carcinomas and cell lines. Int J Oncol 2005;26(5):1265.
[219] Winter RN, Kramer A, Borkowski A, Kyprianou N. Loss of caspase-1 and caspase-3 protein expression in human prostate cancer. Cancer Res 2001;61(3):1227–32.
[220] Gringhuis SI, Kaptein TM, Wevers BA, Theelen B, van der Vlist M, Boekhout T, et al. Dectin-1 is an extracellular pathogen sensor for the induction and processing of IL-1β via a noncanonical caspase-8 inflammasome. Nat Immunol 2012;13(3):246–54.
[221] Krelin Y, Voronov E, Dotan S, Elkabets M, Reich E, Fogel M, et al. Interleukin-1beta-driven inflammation promotes the development and invasiveness of chemical carcinogen-induced tumors. Cancer Res 2007;67(3):1062–71.
[222] Hagemann T, Balkwill F, Lawrence T. Inflammation and cancer: a double-edged sword. Cancer Cell 2007;12(4):300–1.
[223] Tu S, Bhagat G, Cui G, Takaishi S, Kurt-Jones EA, Rickman B, et al. Overexpression of interleukin-1beta induces gastric inflammation and cancer and mobilizes myeloid-derived suppressor cells in mice. Cancer Cell 2008;14(5):408–19.
[224] Zitvogel L, Kepp O, Galluzzi L, Kroemer G. Inflammasomes in carcinogenesis and anticancer immune responses. Nat Immunol 2012;13(4):343–51.
[225] Qin Y, Ekmekcioglu S, Liu P, Duncan LM, Lizée G, Poindexter N, et al. Constitutive aberrant endogenous interleukin-1 facilitates inflammation and growth in human melanoma. Molec Cancer Res 2011;9(11):1537–50.
[226] Okamoto M, Liu W, Luo Y, Tanaka A, Cai X, Norris DA, et al. Constitutively active inflammasome in human melanoma cells mediating autoinflammation via caspase-1 processing and secretion of interleukin-1beta. J Biol Chem 2010;285(9):6477–88.
[227] Scherbarth S, Orr FW. Intravital videomicroscopic evidence for regulation of metastasis by the hepatic microvasculature: effects of interleukin-1alpha on metastasis and the location of B16F1 melanoma cell arrest. Cancer Res 1997;57(18):4105–10.
[228] Elaraj DM, Weinreich DM, Varghese S, Puhlmann M, Hewitt SM, Carroll NM, et al. The role of interleukin 1 in growth and metastasis of human cancer xenografts. Clin Cancer Res 2006 Feb 15;12(4):1088–96.
[229] Weinreich DM, Elaraj DM, Puhlmann M, Hewitt SM, Carroll NM, Feldman ED, et al. Effect of interleukin 1 receptor antagonist gene transduction on human melanoma xenografts in nude mice. Cancer Res 2003;63(18):5957–61.
[230] Vidal-Vanaclocha F, Fantuzzi G, Mendoza L, Fuentes AM, Anasagasti MJ, Martin J, et al. IL-18 regulates IL-1beta-dependent hepatic melanoma metastasis via vascular cell adhesion molecule-1. Proc Natl Acad Sci USA 2000;97(2):734–9.
[231] Salcedo M, Diehl A, Olsson-Alheim M, Sundbäck J, van Kaer L, Karre K, et al. Altered expression of Ly49 inhibitory receptors on natural killer cells from MHC class I deficient mice. J Immunol 1997;158(7):3174–80.
[232] Salcedo R, Worschech A, Cardone M, Jones Y, Gyulai Z, Dai R-M, et al. MyD88-mediated signaling prevents development of adenocarcinomas of the colon: role of interleukin 18. J Exp Med 2010;207(8):1625–36.
[233] Cao R, Farnebo J, Kurimoto M, Cao Y. Interleukin-18 acts as an angiogenesis and tumor suppressor. FASEB J 1999;13(15):2195–202.
[234] Coughlin CM, Salhany KE, Wysocka M, Aruga E, Kurzawa H, Chang AE, et al. Interleukin-12 and interleukin-18 synergistically induce murine tumor regression which involves inhibition of angiogenesis. J Clin Invest 1998;101(6):1441–52.

[23]

CHAPTER 10

DNA Damage Responses in Atherosclerosis

Kenichi Shimada, Timothy R. Crother and Moshe Arditi

Department of Biomedical Sciences and Pediatric Infectious Diseases and Immunology, Cedars-Sinai Medical Center; Infectious and Immunologic Diseases Research Center, Cedars-Sinai Medical Center, Los Angeles, CA; David Geffen School of Medicine, University of California at Los Angeles, USA

BACKGROUND

Atherosclerosis is the developmental process of atheromatous plaques in which an artery wall thickens as a result of the accumulation of vascular smooth muscle cells (VSMCs) and inflammatory cells, together with lipid and extracellular matrix proteins [1–3]. Advanced atherosclerotic plaques also show calcification and have a necrotic core. There is increasing evidence that DNA damage in cells within the lesion plays an important role in atherogenesis.

Mammalian cells undergo 10,000 measurable DNA modification events per cell per hour [4]. DNA adduct formation, however, may increase substantially under conditions of oxidative stress [5]. In human cells, and eukaryotic cells in general, DNA is found in two cellular locations – inside the nucleus and inside the mitochondria. Nuclear DNA (nDNA) exists as chromatin during non-replicative stages of the cell cycle and is condensed into aggregate structures known as chromosomes during cell division. In either state the DNA is highly compacted and wound up around bead-like proteins called histones. Whenever a cell needs to express the genetic information encoded in its nDNA the required chromosomal region is unraveled, genes located therein are expressed, and then the region is condensed back to its resting conformation. Mitochondrial DNA (mtDNA) is located inside mitochondria organelles, exists in multiple copies, and is also tightly associated with a number of proteins to form a complex known as the nucleoid. Inside mitochondria, reactive oxygen species (ROS) or free radicals, byproducts of the constant production of adenosine triphosphate (ATP) via oxidative phosphorylation, create a highly oxidative environment that is known

Biological DNA Sensor.
DOI: http://dx.doi.org/10.1016/B978-0-12-404732-7.00010-1

231

to damage mtDNA. A critical enzyme in counteracting the toxicity of these species is superoxide dismutase, which is present in both the mitochondria and cytoplasm of eukaryotic cells. In general, DNA damage resulting from attack by oxygen free radicals includes base modifications, sugar damage, strand breaks, abasic sites, and DNA protein cross-links [6]. 8-Hydroxyguanine (8-oxoG) is one of the most abundant and stable lesions in DNA generated by oxidative stress [7], and can result in mismatch paring [8].

Given these properties of DNA damage, modification and mutation, it can be seen that DNA damage is a special problem in non-dividing or slowly dividing cells, where unrepaired damage will tend to accumulate over time. On the other hand, in rapidly dividing cells, unrepaired DNA damage that does not kill the cell by blocking replication will tend to cause replication errors leading to mutation. The great majority of mutations that are not neutral in their effect are deleterious to a cell's survival. Thus, in a population of cells comprising a tissue with replicating cells, mutant cells will tend to be lost. However, infrequent mutations that provide a survival advantage will tend to clonally expand at the expense of neighboring cells in the tissue. This advantage to the cell is disadvantageous to the whole organism, because such mutant cells can give rise to cancer. Thus, DNA damage in frequently dividing cells, because it gives rise to mutations, is a prominent cause of cancer. In contrast, DNA damage in infrequently dividing cells contributes to aging [9].

Chronic inflammation is an important atherosclerotic promoter and DNA damage may also be linked to the increased chronic inflammatory responses seen with age. Cells senesced by DNA double strand breaks produce inflammatory mediators such as IL-1, IL-6, IL-8, GM-CSF, GRO, MCP-2, MMP-1, -2, -3, -12, -13, and -14 [10–15]. Thus, DNA damage has recently begun to attract attention as a cause of chronic inflammation.

EVIDENCE OF DNA DAMAGE IN ATHEROSCLEROSIS

The pathogenesis of atherosclerosis is closely related to many molecular mechanisms, including the modulation of oxidation-sensitive signaling, the expression of cytokines and adhesion molecules, the recruitment of macrophages and T cells, and the consequences of chronic inflammatory conditions [3]. However, at present, there is increasing evidence that DNA damage in both circulating cells, and cells of the vessel wall may play a pivotal role in atherosclerosis [16].

Oxidative DNA damage and DNA adducts have been detected in the VSMCs of human abdominal aorta affected by atherosclerosis, and their levels were significantly correlated with known atherogenic risk factors, including age, number of cigarettes per day currently smoked, arterial pressure, blood cholesterol, and triglycerides [17]. High levels of lipid peroxidation-derived etheno DNA adducts have also been observed in human atherosclerotic lesions [18]. DNA strand breaks, oxidative pyrimidines, and altered purines are also significantly elevated in patients with coronary artery disease compared with controls [19]. These studies demonstrated that not only endogenous agents (oxidative stress and lipid peroxidation) are associated with disease progression in human atherosclerosis, but that exogenous agents (environmental factors) are also involved.

8-oxoG can be detected in plaque VSMCs, macrophages, and ECs in experimental animal models of atherosclerosis, but not in VSMCs of normal arteries [20]. The number of DNA strand breaks in cells derived from freshly isolated segments of the thoracic aorta was also significantly higher. Thus it appears that DNA damage is a direct correlate of the extent of atherosclerosis in animal models.

In addition to non-specific DNA damage, there is also a significantly higher incidence of a common mitochondrial DNA deletion (mtDNA4977) in atherosclerosis [21]. A recent electron microscopic study revealed that human atherosclerotic lesions have marked structural alterations of intimal cell mitochondria [22].

Telomeres consist of tandem repeats of the sequence TTAGGG at the end of eukaryotic chromosomes that prevent chromosomal fusion by cellular repair processes [23]. Telomere dysfunction is another key aspect in the pathogenesis of cardiovascular and metabolic disease [24]. Telomeres are crucial for genomic stability and integrity, preventing chromosomal end deterioration and fusion by cellular DNA repair processes. Shortened telomeres are evident in atherosclerosis, as observed in plaque VSMCs and ECs relative to the normal vessel wall, and in circulating EC progenitor cells [25]. Telomeres in leukocytes are also shorter in patients with atherosclerosis compared to control subjects [26]. Thus, in addition to specific damage to the individual bases of the DNA, damage to the structure and integrity of the chromosomes is associated with atherosclerosis.

TYPES OF DNA DAMAGE

Damage to the DNA can originate from multiple extrinsic and intrinsic sources. Extrinsic sources comprise chemicals and radiation, such as UV damage, and viruses. Intrinsic sources include spontaneous chemical reactions and ROS [27]. DNA damage in vascular cells, caused by both extrinsic toxic factors and endogenous cellular oxidative stress, may thus represent a pivotal mediator in the initial development of atherogenesis. Oxidative DNA can produce several kinds of DNA damage, e.g., oxidized bases, abasic sites and single- and double-strand breaks [28,29]. There are >100 types of oxidative base modifications in mammalian DNA [30,31]. Some oxidative modified bases block DNA replication, while others are misread by the replication machinery and lead to base substitutions in the DNA [29] (Figure 10.1).

Abasic Sites, Depurination and Depyrimidination

Abasic sites in DNA arise naturally by a hydrolytic process, which is markedly accelerated by chemical modification of DNA bases. Abasic sites result in spontaneous depurination and depyrimidination reactions [32]. Depurination involves the loss of purine bases (adenine and guanine) from DNA. In spontaneously occurring depurination reactions, the N–glycosyl bound to deoxyribose is broken by hydrolysis, leaving the DNA's sugar–phosphate chain intact, producing an abasic site. Depyrimidination is much less frequent than depurination since

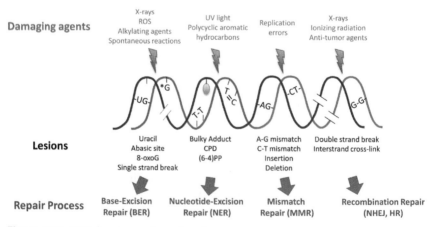

Figure 10.1 *DNA damage and repair systems.*

the N-glycosyl bound between a pyrimidine base and the deoxyribose is more stable than the corresponding bond for purine bases [33]. During DNA synthesis, dAMP is inserted preferentially opposite abasic sites in DNA, giving arise to a potentially mutagenic chain extension or frameshift error [34,35]. Damaged DNA bases are removed by DNA N-glycosylases, generating abasic sites as an intermediate during base excision repair. Abasic sites can also be produced by ROS, as well as being produced in intermediate steps of the base excision repair pathway. Inefficient or incomplete base excision repair might leave abasic sites in the DNA, resulting in permanent damage.

Deamination is the removal of an amino group from DNA bases. Almost all DNA bases undergo deamination in spontaneous reactions, with the exception of thymine – which does not have an amino group. The most common type of deamination reaction in cells results in the conversion of cytosine to an altered DNA base, uracil [36]. Most deamination products are bases that do not occur naturally in DNA (except 5-methylcytosine), and this facilitates the identification and excision of the deaminated base by a DNA glycosylase enzyme (e.g. uracil-DNA glycosylases (UDG)).

DNA Strand Breaks

DNA strand breaks are produced in intermediate events of natural reactions such as the process of V(D)J recombination during lymphocyte development, which is a kind of programmed double-strand break [37,38]. On the other hand, DNA strand breaks can be caused by oxidative DNA damage or by ionizing radiation (e.g. X-rays and gamma rays) as well as drugs like bleomycin [39,40]. These breaks in the DNA backbone can sometimes cause serious genomic instability, carcinogenesis, and cell death. Defective single-strand break repair often results in neurological diseases rather than carcinogenesis or progeria [41]. Since ROS are one of the major causes of the single-strand breaks, and the high level of oxygen consumption in the nervous system makes it more susceptible to defects in single-strand break repair, it makes sense that single-strand breaks may contribute to neurological disorders [42].

Unrepaired double stranded DNA breaks lead to genomic rearrangements, a common and serious problem for all cells and organisms. These double stranded breaks are associated in patients with some progeroid syndromes such as Werner syndrome, ataxia telangiectasia, and Fanconi's anemia [43,44].

Cyclobutane Pyrimidine Dimers

Cyclobutane pyrimidine dimers (CPD) are characterized by covalent bridging linkages between adjacent pyrimidines in the same DNA strand, and are most frequently produced in photochemical reactions. Ultraviolet light induces the formation of covalent linkages by reactions localized on the C=C double bonds [45]. Such dimers interfere with base pairing during DNA replication, leading to mutations.

MITOCHONDRIAL DNA DAMAGE

DNA damage is present in both genomic and mitochondrial DNA. The mitochondrial genome encodes proteins required for oxidative phosphorylation and ATP synthesis, and RNAs needed for mitochondrial protein translation. The mtDNA is densely packed with genes [46] and is located in the proximity of the main site of ROS generation, the inner mitochondrial membrane. Additionally, with its lack of protective histones and its limited capability for DNA repair, mtDNA is more susceptible than nuclear DNA to oxidative damage. In fact, the steady-state levels of oxidatively induced lesions observed in mtDNA can be several-fold higher than in nDNA [47–49]. The accumulation of DNA damage is thought to play an important role in aging and aging-associated diseases like atherosclerosis and Alzheimer's disease [50,51]. It was once considered to be a consequence of the absence of DNA repair in mitochondria, dubbed the mitochondria theory of aging. However, it is currently well established that some DNA repair pathways actively take place in mitochondria. Moreover, new mitochondrial DNA repair enzymes have been detected in mitochondria. Common oxidative DNA lesions in mtDNA are 8-oxoG and TG [31,52]. Three DNA glycosylases have been identified in mammalian mitochondria, OGG1, Nth1 and Neil1. The OGG1 protein is the major DNA glycosylase for the repair of the 8-oxodG lesion in DNA [53,54]. Studies with a fluorescent protein tagged construct of Ogg1 have confirmed localization of Ogg1 in both the nucleus and in the mitochondria [55,56]. To date there has been one isoform of the OGG1 protein in rodents and two isoforms in humans, α-OGG1 and β-OGG1, identified [57]. β-OGG1 appears to be localized exclusively in mitochondria [55,57] and α-OGG1 is mainly present in the nucleus [58].

NEIL might preferentially repair other oxidative lesions such as 4,6-diamino-5-formamidopyrimidine (FapyAde) and 2,6-diamino-4-hydroxy-5-formamidopyrimidine (FapyGua) [59]. Three isoforms of NEIL have been identified in mammalian tissue [60,61], NEIL1, NEIL2 and NEIL3. NEIL1 and NEIL2 have bi-functional enzymatic activity that in addition to its repair function, generates DNA strand breaks with 3'-phosphate termini [62] that are then removed by polynucleotide kinase (PNK) [63].

The substrate for NTH1 overlaps with those of with NEIL1 and NEIL2. The mammalian NTH1 is an ortholog of *E. coli* endonuclease III (Nth). It is a DNA glycosylase with an AP lyase activity that excises a wide range of pyrimidine lesions, including thymine glycol and 5-hydroxy-cytosine. It has been described to remove Fapy residues in human cells [64]. NTH1 localizes mostly in mitochondria in mice [55,65].

In mammals, two different forms of uracil-DNA glycosylase (UNG), nuclear UNG2 and mitochondrial UNG1, are encoded by the UNG gene and are generated by alternative transcription initiation sites and alternative splicing [66]. They share a common catalytic domain, and hence substrate specificity for uracil.

The final step of mitochondrial BER, the ligation of the remaining single-strand nick, is catalyzed by a DNA ligase. In humans this is accomplished by the mitochondrial ligase, DNA ligase III, a splice variant from the LIG3 gene which encodes both the nuclear and mitochondrial enzymes [67]. Nuclear and mitochondrial variants appear to differentially interact with other repair proteins such as XRCC1. XRCC1 interacts with the nuclear variant of LIG3 and stabilizes it [68], while mitochondrial DNA ligase III has been described to be independent of XRCC1 [69].

DNA polymerase inserts the correct nucleotides into the gap in the next step of the BER pathway. DNA polymerase γ (POLγ) plays a role in all replication, recombination and repair processes involving mtDNA. Interestingly, POLγ knock-in mice accelerated the mtDNA mutation rate resulted in premature aging although POLγ KO mice showed embryonic lethality [70–72].

Additionally, mitochondrial fusion and fission control the integrity of mtDNA whereby normal mitochondrial fusion promotes stability of the mitochondrial genome and excessive mitochondrial fission leads to preferential accumulation of mutated mtDNA [73,74].

DNA DAMAGE RESPONSES IN ATHEROSCLEROSIS

Various exogenous and endogenous sources can generate damaged DNA. In response to DNA damage, cells activate the appropriate DNA damage repair and checkpoint pathways or apoptosis. Lesions of single or double-stranded DNA lead to the activation of cell cycle checkpoints and a number of DNA damage repair pathways including mismatch repair (MMR), base excision repair (BER), nucleotide excision repair (NER), translesion DNA synthesis (TLS), homologous recombination (HR) and non-homologous end joining (NHEJ) [75]. Impaired repair of damaged DNA can lead to the accumulation of mutations and genomic instability. In addition, defective repair of damaged DNA can lead to aging as well as predisposing to various genetic diseases including cancer and immunodeficiency [9,76] (Table 10.1).

8-Oxoguanine DNA glycosylase (OGG1) did have altered expression levels between atherosclerotic plaques of the carotid artery and control artery compared to controls [77]. Interestingly, OGG1$^{-/-}$ mice have increased TNF-α production on a cholesterol rich diet [78]. Furthermore, OGG1 transgenic mice increased OGG1 activity by 115% in mitochondria and by 28% in nuclei, and decreased cardiac fibrosis after transaortic constriction (TAC) [79].

Statins are known for their ability to lower cholesterol levels thus providing anti-atherosclerotic benefits. More recently, a mechanism was found by which statins accelerated DNA repair via Hdm2 phosphorylation [80], NBS1 stabilization, and phosphorylation of H2AX and the mouse DNA repair enzyme ATM [81]. Consequently, statins attenuate DNA damage, cell senescence and telomere shortening in VSMCs and may thereby promote plaque stability in atherosclerosis [80]. Pernice et al. observed a reduction in chromosomal damage in culture lymphocytes from end-stage renal disease patients with different degrees of carotid artery atherosclerosis treated with simvastatin, compared to untreated cells [82]. These findings support the concept that the anti-atherogenic action of statins may therefore be partly ascribed to their ability to provide protection against genomic damage.

MMR has been extensively investigated in molecular medicine. Microsatellite instability (MSI) is the condition of genetic hypermutability that results from a dysfunctional MMR system. In other words, MSI is the phenotypic evidence that MMR isn't functioning normally. DNA MMR corrects errors that spontaneously occur during DNA replication like single base mismatches or short insertions and deletions. According to 50 human autopsy cases of atherosclerosis, loss of heterozygosity was observed in several MMR genes, including hMSH2, hPMS1, and hMLH1 [83]. These alterations in MMR genes indicate the presence of decreased fidelity in DNA repair in atherosclerotic tissues.

Table 10.1 DNA damage, repair pathways, and atherosclerotic associations

	DNA Damage Targets, Repair Function	Atherosclerosic Associations	References
Base excision repair (BER)			
OGG1	8-oxoG opposite G	OGG1 expression was not changed in human atherosclerotic plaques. OGG1 null mice have increased TNF-α on a cholesterol rich diet. OGG1 TG mice show decrease in TAC-induced cardiac fibrosis.	[77,78,79]
APE1 (Ref-1)	AP endonuclease	Expression was increased in human atherosclerotic plaque. APE1 in heterozygous mice (APE1$^{+/-}$) causes cardiovascular diseases such as hypertension. Overexpression of APE1 promotes endothelial cell survival in response to hypoxia and TNF-α.	[77,104,105]
XRCC1	LIG3 accessory factor	XRCC1 gene polymorphism is associated with the risk of coronary artery disease and elevated levels of chromosomal damage in CAD patients.	[129]
POLB	BER in nuclear DNA	oxLDL downregulates POLB in monocytes.	[103]
Nucleotide excision repair (NER)			
ERCC1 (XPG)	5′ incision DNA binding subunit	Eccr1d/− mice accelerated aorta senescence, hypertension.	[130]
ERCC2 (XPD)	5′ to 3′ DNA helicase	Endothelium–dependent vasodilator dysfunction was increased in XpdTTD mice. XPD gene polymorphism is associated with the risk of coronary artery disease and elevated levels of chromosomal damage in CAD patients.	[129,130]
Mismatch repair (MMR)			
MSH2	Mismatch (MSH2–MSH6) and loop (MSH2– MSH3) recognition	Loss of heterozygosity occurs in atherosclerosis.	[83]
MLH1/PMS1	MutL homologs, forming heterodimer	Loss of heterozygosity occurs in atherosclerosis.	[83]

(Continued)

Table 10.1 (Continued)

	DNA Damage Targets, Repair Function	Atherosclerosic Associations	References
Non-homologous end joining (NHEJ)			
DNA-PK (PRKDC)	Nuclear DNA-dependent serine/threonine protein kinase	The expression was increased in human atherosclerotic plaque.	[77]
Homologous recombination (HR)			
ATM	Ataxia telangiectasia	Atm$^{+/-}$Apoe$^{-/-}$ mice promote atherosclerosis and metabolic syndrome.	[87]
NBS1 (NBN)	Mutated in Nijmegen breakage syndrome	STATIN activates NBS1 stability and DNA repair.	[80]
Other conserved DNA damage response genes			
TP53 (p53)	Regulation of the cell cycle	Expression and phosphorylation were increased in atherosclerotic plaque. p53$^{-/-}$ Apoe$^{-/-}$ mice developed accelerated aortic atherosclerosis.	[89,90]
SIRT1	Class III histone deacetylase (HDAC)	Endothelial expression of SIRT1 decreases atherosclerosis in Apoe null mice. Heart-specific SIRT1 overexpression mice have reduced atherosclerosis. SIRT1 gene polymorphism is associated with cholesterol metabolism and coronary artery calcification in hemodialysis patients.	[99,100,101,102]
SIRT6	Deacetylase and mono–ADP ribosyltransferase enzyme	SIRT6 TG mice accumulated significantly less visceral fat, LDL-cholesterol, and triglycerides when fed HFD.	[131]
PARP1	Protects strand interruptions	PARP1 expression was elevated in atherosclerotic plaques. PARP inhibitor reduced plaque development. PARP1 deletion reduced plaque formation in Apoe KO mice.	[77,92,93,94]

Ataxia-telangiectasia (A-T) mutated (ATM) protein plays a role as an important signaling molecule to repair DNA damage, in particular for DNA double strand breaks [84]. ATM is located within the nucleus of most cells as dimers or larger multimers with the kinase domain of each molecule being blocked by others. In response to double strand breaks, each ATM molecule phosphorylates the other on ser [19,81] leading to monomerization and activation of ATM [85]. The active ATM proteins are then recruited to double strand break sites, an event which is facilitated by the MRN complex, serving as a platform for further enzymatic reactions as well as initiating phosphorylation events leading to regulation of specific cell cycle checkpoints. The main target for ATM in this pathway is the tumor suppressor gene p53, which it phosphorylates on ser [15], enhancing its transcriptional activity [86]. Mercer et al. demonstrated that the ATM haploinsufficiency generates defects in cell proliferation, apoptosis and mitochondrial dysfunction in VSMCs and macrophages [87]. This failure of DNA repair resulted in ketosis, hyperlipidemia and increased fat storage, promoting atherosclerosis and metabolic syndrome [87]. Furthermore, ATM is a sensor of ROS. Guo et al. showed that in the presence of high levels of oxidative stress, ATM is oxidized and Ox-ATM induced activation of the MRN complex, even in the absence of DNA double strand breaks [88].

p53 has been described as 'the guardian of the genome' because of its role in conserving stability by preventing genome mutation. It has been observed that p53 expression is increased in advanced human plaques compared with normal vessels, where it may induce growth arrest and apoptosis [89]. Adenoviral expression of p53 induced plaque rupture and apoptosis in $Apoe^{-/-}$ mice [90]. Confusingly, $p53^{-/-}/Apoe^{-/-}$ mice displayed comparatively enhanced aortic plaques but not branchiocephallic plaques [89,91]. This study also found that endogenous p53 promoted macrophage apoptosis, but protected VSMCs and stromal cells against apoptosis, highlighting the complexity of its role in atherogenesis.

The activation of base excision repair (BER) key protein, poly (ADT-ribose) polymerase 1 (PARP1), appears to be involved in the vascular dysfunction associated with circulatory shock, myocardial ischemia reperfusion injury, heart failure, hypertension, diabetes, and cardiovascular aging. Furthermore, elevated levels of PARP1 were found in human atherosclerotic plaques [77]. Interestingly, in an $Apoe^{-/-}$ mouse model of atherosclerosis fed on a high-fat diet, PARP inhibition improved endothelial function [92]. However, pharmacological inhibition of PARP induced endothelial dysfunction. According to genetic evidence, *Parp1* deletion resulted in a reduction in atherosclerotic plaque development although $Parp1^{-/-}Apoe^{-/-}$ mice displayed higher total cholesterol levels than $Parp1^{+/+}Apoe^{-/-}$ mice [93]. Perhaps, endogenous PARP1 is required for atherosclerosis due to an increase in adhesion molecules on endothelial cells rather than through DNA damage repair [93]. PARP1 can localize to the mitochondria and is associated with mitochondrial DNA integrity although it is predominantly a nuclear enzyme [94].

SIRT stands for sirtuin (silent mating type information regulation 2 homolog), and SIRT1 is an enzyme (class III histone deacetylase) and may have mono-ADP-ribosyltransferase activity to regulate cellular stress responses and metabolism [95]. SIRT1 transcription is under the control of at least two negative feedback loops that keep its induction tightly regulated under conditions of oxidative stress. SIRT1 promoter can be activated by E2F1 and HIC1 during cellular stress, and SIRT1 represses p53-dependent apoptosis in response to DNA damage and oxidative stress and promotes cell survival under cellular stress, induced by etoposide treatment or irradiation [96]. SIRT1 regulates APE1 endonuclease in the BER pathway [97]. SIRT1 is highly expressed in endothelial cells and controls their angiogenic function [98]: it is involved in vascular growth of cultured endothelium, in the formation of the vascular network of the developing zebrafish, and even in ischemia-induced neovascularization in the adult mouse. Human atherosclerotic plaques have reduced SIRT1 expression and increased DNA damage in VSMCs [99]. Endothelial-specific overexpression of SIRT1 decreases atherosclerosis in *Apoe* null mice by improving vascular function [100]. VSMC specific SIRT1 deletion resulted in increase DNA damage, apoptosis, medial degeneration, and plaque vulnerability. In addition, SIRT1 suppresses NF-κB-mediated endothelial tissue factor activation and coagulation [101], and inhibits scavenger receptor LOX-1, resulting in reduced uptake of ox-LDL, thus preventing macrophage foam cell formation [102]. These findings demonstrate clearly the atheroprotective effects of SIRT1.

DNA polymerase inserts the correct nucleotides in the generated gap after BER activity. It has been reported that DNA β-polymerase is down-regulated by ox-LDL in monocytes [103], thus there is the potential for DNA damage repair dysregulation in atherosclerosis.

The mammalian AP-endonuclease APE1/Ref-1 is a ubiquitous and remarkably multifunctional protein. It plays a central role in the base excision repair (BER) pathway for damaged bases and DNA single-strand breaks induced by reactive oxygen species (ROS) and alkylating agents, and also in repairing apurinic/apyrimidinic (AP) sites generated spontaneously or after excision of oxidized and alkylated bases by DNA glycosylases. Although APE1 null mice are embryonic lethal, partial loss of APE1 in the heterozygous mice (APE1$^{+/-}$) causes cardiovascular diseases such as hypertension [104]. APE1$^{+/-}$ mice have markedly elevated blood pressure, clearly indicating APE1's important role in maintaining blood pressure. This phenotype is accompanied by decreased basal nitric oxide (NO) production in arteries from APE1$^{+/-}$ mice [104]. However, expression of endothelial nitric oxide synthase (eNOS) cannot be a major determinant of the hypertensive phenotype of APE1$^{+/-}$ mice, because eNOS levels are actually increased in the vessels of these mice, despite their impaired ability to generate nitric oxide. The redox activator function of APE1 was shown to be involved in the upregulation of H-Ras expression and activation of AKT, which potently activates eNOS [104]. Therefore, it is critical to emphasize that APE1 may coordinate multiple nuclear events that impact vascular reactivity and blood pressure control through multiple mechanisms, in addition to its effects on NO generation. APE expression was associated with hypoxia–induced apoptosis in HUVEC and CPACE cells [105], which were significantly inhibited by overexpression of APE1. Deletion of the redox-sensitive domain of APE1 (aa 1–127 residues) abolished the anti-apoptotic function of APE1 in both hypoxia and tumor necrosis factor α (TNF-α) induced endothelial cell death [105]. This study suggests that upregulation of APE1 promotes endothelial cell survival in response to hypoxia and TNF-α through NF-κB-independent and NF-κB-dependent signaling cascades.

DNA DAMAGE SENSING AND INFLAMMATION IN ATHEROSCLEROSIS

Oxidative DNA damage, associated with a concomitant activation of DNA repair enzymes, increases significantly in cells present in the atherosclerotic plaques but not in those found in adjacent normal tissue or normal arteries [20].

ROS are generated by enzymatic systems of vascular wall cells, including NAD(P)H oxidase, xanthine oxidase and nitric oxide synthase. However, of the endogenous sources of oxidative stress, mitochondria are the main cellular site of ROS generation, and the mitochondrial function declines with aging due to a progressive accumulation of oxidative damage [106].

Importantly, several studies have shown that both oxidative mitochondrial DNA damage and progressive respiratory chain dysfunction are associated with atherosclerosis in human studies and animal models of oxidative stress [107]. These accumulating data suggest that mitochondria play a significant role in the regulation of cardiovascular cell function. In addition, mitochondrial ROS generation might contribute to telomere shortening [106].

Although numerous studies have linked DNA damage with atherosclerosis, it is poorly understood how DNA damage leads to proatherogenesis. Loss of DNA integrity might be associated with apoptosis. In cells where the DNA damage is too much to repair, or cells that are driven to proliferate, loss of DNA integrity induces apoptosis to remove genetically damaged cells. In fact, apoptosis is evident in endothelial cells, macrophages and VSMCs in atherosclerotic plaques although the pathophysiological significance of apoptosis in atherosclerosis remains unclear. Endothelial cell death is associated with both atherogenesis and plaque erosion [108], whereas VSMC death may promote thinning of the fibrous cap and plaque rupture [109]. Cell death can be induced by ROS and oxidized lipids in vascular cells, whereas p53 expression in an experimental neointima can induce apoptosis and promote plaque rupture [90].

Another proatherogenesis mechanism induced by oxidized DNA damage has been described recently. As mentioned above, mitochondria are the central hub by which ROS is generated. Emerging data have demonstrated that the inflammasome (a multi-protein complex (of which there are several varieties) that activates caspase-1) plays a significant role in atherogenesis [110]. Of the various inflammasomes, the nucleotide binding domain [111] and leucine-rich repeat (NLR) pyrin domain containing 3 (NLRP3) inflammasome is the most studied and has been directly linked to atherogenesis. After inflammasome activation, the active caspase-1 can cleave pro-IL-1β, releasing mature IL-1β, which is then secreted by the cell. It is well described that IL-1 plays a proatherogenic role [112] and it has been proposed to be secreted from apoptotic macrophages and monocytes [113]. In addition to IL-1β, the inflammasome can also activate IL-18 and IL-1α, both of which have been linked to atherosclerosis [110,114,115]. Elevated blood cholesterol is highly associated with development of atherosclerosis and cholesterol crystals can activate the NLRP3 inflammasome, promoting IL-1 mediated inflammation [110,116]. In addition to cholesterol crystals, oxLDL, oxycholesterols (e.g. 7-ketocholesterol and 25-OH cholesterol), hypoxia, apolipoprotein L6, Fas–FasL, advanced glycation end-products, and cigarette smoke are thought of as atherosclerotic danger stimuli (Figure 10.2). Angiotensin II (Ang II) was shown to be an important risk factor for accelerated atherosclerosis [117]. Angiotensin II may also contribute as an danger signal in atherosclerosis since it increases ROS and promotes oxidative DNA damage, promoting senescence through telomeric and nontelomeric DNA damage [118]. Although the specifics of NLRP3 inflammasome activation remain unclear, mitochondrial dysfunction and mitochondrial ROS generation are directly linked to NLRP3 inflammasome activation [119–121]. During mitochondrial dysfunction and apoptosis, mitochondrial DNA is oxidized and released into the cytosol where it binds to NLRP3. This results in NLRP3 inflammasome activation and IL-1β secretion. Interestingly, the aforementioned oxidized base lesion, 8-oxoguanosine, is induced in mtDNA after NLRP3 activating stimuli and was present in the mtDNA that bound to NLRP3. Furthermore, mtDNA that contains 8-oxoguanosine lesions preferentially activates the NLRP3 inflammasome, compared with non-oxidized mtDNA. Moreover, 8-OH-dG alone can competitively block oxidized mtDNA from binding and activating NLRP3 inflammasome, thereby blocking IL-1β release. This suggests that oxidized mtDNA induced activation of NLRP3 from cells undergoing

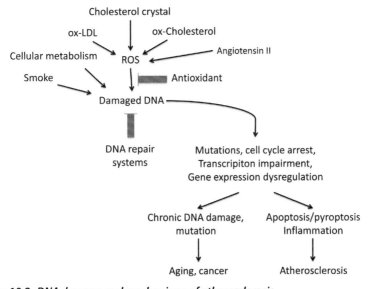

Figure 10.2 *DNA damage and mechanisms of atherosclerosis.*

apoptosis (or mitochondrial dysfunction) may be a key step in inducing inflammation and accelerating atherosclerosis. A question remains whether oxidized mtDNA binds to the NLRP3 protein directly or whether an unknown cofactor is required (although *in vitro* studies suggest that direct binding is possible). The potential importance of mitochondrial dysfunction and NLRP3 inflammasome activation has been reinforced by the observations that the crucial role of the cellular movement of calcium (albeit with extensive use of pharmacological inhibitors) [122], which is commonly associated with mitochondrial damage prior to apoptosis, is required for the assembly and activation of the NLRP3 inflammasome.

In addition to mtDNA directly activating the NLRP3 inflammasome inside macrophages, circulating mtDNA released from cellular injury may play a role as a danger signal and induce systemic inflammation [123]. It is thought that Toll-like receptor 9 (TLR9) and the absent in melanoma 2 (AIM2) inflammasome, both of which can recognize dsDNA, may be responsible for this inflammation, although questions remain as to how the released mt dsDNA can be stable and cause endocrine systemic inflammation in a DNase rich environment *in vivo*. TLR9 expression is also increased in atherosclerotic plaques upon oxidative stress, however direct involvement of TLR9 in atherosclerosis is still unclear [124]. AIM2, a member of the Ifi202/IFI16 family, senses cytosolic dsDNA and activates the inflammasome [125–127]. However, there has not been any direct evidence to indicate the involvement of the AIM2 inflammasome in atherosclerosis yet, although these proteins are concomitantly upregulated with p53 DNA damage repair upon IFN-γ stimulation [128].

CONCLUSION

Loss of mitochondrial function, a key element of atherogenesis, is caused by an increase in ROS, resulting in accumulation of mitochondrial DNA damage, respiratory chain deficiencies, apoptosis and cell death, and inflammation, favoring plaque formation and instability.

Growing evidence supports the notion that cell metabolism, oxidative stress, DNA damage, DNA damage responses, apoptosis and inflammation are all linked in the regulation of atherogenesis. These pathways all merge together in a complex system that requires further study and may provide unique and effective targets for new therapies and prevention of atherosclerosis.

ACKNOWLEDGMENT

This work was supported by NIH Grants HL66436 and AI1072726 to M. Arditi.

REFERENCES

[1] Ross R. Atherosclerosis – an inflammatory disease. N Engl J Med 1999;340(2):115–26.
[2] Hansson GK, Klareskog L. Pulling down the plug on atherosclerosis: cooling down the inflammasome. Nat Med 2011;17(7):790–1.
[3] Libby P, Ridker PM, Hansson GK. Progress and challenges in translating the biology of atherosclerosis. Nature 2011;473(7347):317–25.
[4] Billen D. Spontaneous DNA damage and its significance for the 'negligible dose' controversy in radiation protection. Radiat Res 1990;124(2):242–5.
[5] Wang D, Kreutzer DA, Essigmann JM. Mutagenicity and repair of oxidative DNA damage: insights from studies using defined lesions. Mutat Res 1998;400(1–2):99–115.
[6] Albertine KH, Soulier MF, Wang Z, et al. Fas and fas ligand are up-regulated in pulmonary edema fluid and lung tissue of patients with acute lung injury and the acute respiratory distress syndrome. Am J Pathol 2002;161(5):1783–96.
[7] Fortini P, Pascucci B, Parlanti E, D'Errico M, Simonelli V, Dogliotti E. 8-Oxoguanine DNA damage: at the crossroad of alternative repair pathways. Mutat Res 2003;531(1–2):127–39.
[8] Cheng KC, Cahill DS, Kasai H, Nishimura S, Loeb LA. 8-Hydroxyguanine, an abundant form of oxidative DNA damage, causes G→T and A→C substitutions. J Biol Chem 1992;267(1):166–72.
[9] Best BP. Nuclear DNA damage as a direct cause of aging. Rejuvenation Res 2009;12(3):199–208.
[10] Acosta JC, O'Loghlen A, Banito A, et al. Chemokine signaling via the CXCR2 receptor reinforces senescence. Cell 2008;133(6):1006–18.
[11] Kuilman T, Michaloglou C, Vredeveld LC, et al. Oncogene-induced senescence relayed by an interleukin-dependent inflammatory network. Cell 2008;133(6):1019–31.
[12] Kuilman T, Michaloglou C, Mooi WJ, Peeper DS. The essence of senescence. Genes Dev 2010;24(22):2463–79.
[13] Freund A, Orjalo AV, Desprez PY, Campisi J. Inflammatory networks during cellular senescence: causes and consequences. Trends Mol Med 2010;16(5):238–46.
[14] Rodier F, Coppé JP, Patil CK, et al. Persistent DNA damage signalling triggers senescence-associated inflammatory cytokine secretion. Nat Cell Biol 2009;11(8):973–9.
[15] Rodier F, Campisi J. Four faces of cellular senescence. J Cell Biol 2011;192(4):547–56.
[16] Mahmoudi M, Mercer J, Bennett M. DNA damage and repair in atherosclerosis. Cardiovasc Res 2006;71(2):259–68.
[17] De Flora S, Izzotti A, Walsh D, Degan P, Petrilli GL, Lewtas J. Molecular epidemiology of atherosclerosis. FASEB J 1997;11(12):1021–31.
[18] Nair J, De Flora S, Izzotti A, Bartsch H. Lipid peroxidation-derived etheno-DNA adducts in human atherosclerotic lesions. Mutat Res 2007;621(1-2):95–105.
[19] Botto N, Masetti S, Petrozzi L, et al. Elevated levels of oxidative DNA damage in patients with coronary artery disease. Coron Artery Dis 2002;13(5):269–74.
[20] Martinet W, Knaapen MW, De Meyer GR, Herman AG, Kockx MM. Oxidative DNA damage and repair in experimental atherosclerosis are reversed by dietary lipid lowering. Circ Res 2001;88(7):733–9.
[21] Botto N, Berti S, Manfredi S, et al. Detection of mtDNA with 4977 bp deletion in blood cells and atherosclerotic lesions of patients with coronary artery disease. Mutat Res 2005;570(1):81–8.

[22] Chistiakov DA, Sobenin IA, Bobryshev YV, Orekhov AN. Mitochondrial dysfunction and mitochondrial DNA mutations in atherosclerotic complications in diabetes. World J Cardiol 2012;4(5):148–56.

[23] Blasco MA. Telomeres and human disease: ageing, cancer and beyond. Nat Rev Genet 2005;6(8):611–22.

[24] Wang JC, Bennett M. Aging and atherosclerosis: mechanisms, functional consequences, and potential therapeutics for cellular senescence. Circ Res 2012;111(2):245–59.

[25] Matthews C, Gorenne I, Scott S, et al. Vascular smooth muscle cells undergo telomere-based senescence in human atherosclerosis: effects of telomerase and oxidative stress. Circ Res 2006;99(2):156–64.

[26] Saliques S, Teyssier JR, Vergely C, et al. Circulating leukocyte telomere length and oxidative stress: a new target for statin therapy. Atherosclerosis 2011;219(2):753–60.

[27] Chen JH, Hales CN, Ozanne SE. DNA damage, cellular senescence and organismal ageing: causal or correlative? Nucleic Acids Res 2007;35(22):7417–28.

[28] Cervelli T, Borghini A, Galli A, Andreassi MG. DNA damage and repair in atherosclerosis: current insights and future perspectives. Int J Mol Sci 2012;13(12):16929–44.

[29] Cooke MS, Evans MD, Dizdaroglu M, Lunec J. Oxidative DNA damage: mechanisms, mutation, and disease. FASEB J 2003;17(10):1195–214.

[30] Croteau DL, Bohr VA. Repair of oxidative damage to nuclear and mitochondrial DNA in mammalian cells. J Biol Chem 1997;272(41):25409–12.

[31] Dizdaroglu M. Chemical determination of free radical-induced damage to DNA. Free Radic Biol Med 1991;10(3–4):225–42.

[32] Nakamura J, Swenberg JA. Endogenous apurinic/apyrimidinic sites in genomic DNA of mammalian tissues. Cancer Res 1999;59(11):2522–6.

[33] Lindahl T. Instability and decay of the primary structure of DNA. Nature 1993;362(6422):709–15.

[34] Strauss BS. The 'A rule' of mutagen specificity: a consequence of DNA polymerase bypass of non-instructional lesions? Bioessays 1991;13(2):79–84.

[35] Randall SK, Eritja R, Kaplan BE, Petruska J, Goodman MF. Nucleotide insertion kinetics opposite abasic lesions in DNA. J Biol Chem 1987;262(14):6864–70.

[36] Krokan HE, Drabløs F, Slupphaug G. Uracil in DNA – occurrence, consequences and repair. Oncogene 2002;21(58):8935–48.

[37] Taccioli GE, Gottlieb TM, Blunt T, et al. Ku80: product of the XRCC5 gene and its role in DNA repair and V(D)J recombination. Science 1994;265(5177):1442–5.

[38] Walker JR, Corpina RA, Goldberg J. Structure of the Ku heterodimer bound to DNA and its implications for double-strand break repair. Nature 2001;412(6847):607–14.

[39] Sharma S. Age-related nonhomologous end joining activity in rat neurons. Brain Res Bull 2007;73(1–3):48–54.

[40] Povirk LF, Wübter W, Köhnlein W, Hutchinson F. DNA double-strand breaks and alkali-labile bonds produced by bleomycin. Nucleic Acids Res 1977;4(10):3573–80.

[41] Caldecott KW. Single-strand break repair and genetic disease. Nat Rev Genet 2008;9(8):619–31.

[42] Caldecott KW. DNA damage responses and neurological disease. Preface. DNA Repair (Amst) 2008;7(7):1009.

[43] Fishel ML, Vasko MR, Kelley MR. DNA repair in neurons: so if they don't divide what's to repair? Mutat Res 2007;614(1–2):24–36.

[44] McKinnon PJ, Caldecott KW. DNA strand break repair and human genetic disease. Annu Rev Genomics Hum Genet 2007;8:37–55.

[45] Yaar M, Gilchrest BA. Photoageing: mechanism, prevention and therapy. Br J Dermatol 2007;157(5):874–87.

[46] Kasamatsu H, Robberson DL, Vinograd J. A novel closed-circular mitochondrial DNA with properties of a replicating intermediate. Proc Natl Acad Sci USA 1971;68(9):2252–7.

[47] Barja G, Herrero A. Oxidative damage to mitochondrial DNA is inversely related to maximum life span in the heart and brain of mammals. FASEB J 2000;14(2):312–8.

[48] Hamilton ML, Guo Z, Fuller CD, et al. A reliable assessment of 8-oxo-2-deoxy-guanosine levels in nuclear and mitochondrial DNA using the sodium iodide method to isolate DNA. Nucleic Acids Res 2001;29(10):2117–26.

[49] Richter C, Park JW, Ames BN. Normal oxidative damage to mitochondrial and nuclear DNA is extensive. Proc Natl Acad Sci USA 1988;85(17):6465–7.

[50] Wallace DC. A mitochondrial paradigm of metabolic and degenerative diseases, aging, and cancer: a dawn for evolutionary medicine. Annu Rev Genet 2005;39:359–407.

[51] Yu E, Mercer J, Bennett M. Mitochondria in vascular disease. Cardiovasc Res 2012;95(2):173–82.

[52] Ames BN. Endogenous oxidative DNA damage, aging, and cancer. Free Radic Res Commun 1989;7(3-6):121–8.

[53] Zharkov DO, Rosenquist TA, Gerchman SE, Grollman AP. Substrate specificity and reaction mechanism of murine 8-oxoguanine-DNA glycosylase. J Biol Chem 2000;275(37):28607–17.

[54] Bjorås M, Luna L, Johnsen B, et al. Opposite base-dependent reactions of a human base excision repair enzyme on DNA containing 7,8-dihydro-8-oxoguanine and abasic sites. EMBO J 1997;16(20):6314–22.

[55] Takao M, Aburatani H, Kobayashi K, Yasui A. Mitochondrial targeting of human DNA glycosylases for repair of oxidative DNA damage. Nucleic Acids Res 1998;26(12):2917–22.

[56] Kang D, Nishida J, Iyama A, et al. Intracellular localization of 8-oxo-dGTPase in human cells, with special reference to the role of the enzyme in mitochondria. J Biol Chem 1995;270(24):14659–65.

[57] Nishioka K, Ohtsubo T, Oda H, et al. Expression and differential intracellular localization of two major forms of human 8-oxoguanine DNA glycosylase encoded by alternatively spliced OGG1 mRNAs. Mol Biol Cell 1999;10(5):1637–52.

[58] Hashiguchi K, Stuart JA, de Souza-Pinto NC, Bohr VA. The C-terminal alphaO helix of human Ogg1 is essential for 8-oxoguanine DNA glycosylase activity: the mitochondrial beta-Ogg1 lacks this domain and does not have glycosylase activity. Nucleic Acids Res 2004;32(18):5596–608.

[59] Chan MK, Ocampo-Hafalla MT, Vartanian V, et al. Targeted deletion of the genes encoding NTH1 and NEIL1 DNA N-glycosylases reveals the existence of novel carcinogenic oxidative damage to DNA. DNA Repair (Amst) 2009;8(7):786–94.

[60] Hazra TK, Kow YW, Hatahet Z, et al. Identification and characterization of a novel human DNA glycosylase for repair of cytosine-derived lesions. J Biol Chem 2002;277(34):30417–20.

[61] Hazra TK, Izumi T, Boldogh I, et al. Identification and characterization of a human DNA glycosylase for repair of modified bases in oxidatively damaged DNA. Proc Natl Acad Sci USA 2002;99(6):3523–8.

[62] Dou H, Mitra S, Hazra TK. Repair of oxidized bases in DNA bubble structures by human DNA glycosylases NEIL1 and NEIL2. J Biol Chem 2003;278(50):49679–84.

[63] Wiederhold L, Leppard JB, Kedar P, et al. AP endonuclease-independent DNA base excision repair in human cells. Mol Cell 2004;15(2):209–20.

[64] Luna L, Bjørås M, Hoff E, Rognes T, Seeberg E. Cell-cycle regulation, intracellular sorting and induced overexpression of the human NTH1 DNA glycosylase involved in removal of formamidopyrimidine residues from DNA. Mutat Res 2000;460(2):95–104.

[65] Ikeda S, Kohmoto T, Tabata R, Seki Y. Differential intracellular localization of the human and mouse endonuclease III homologs and analysis of the sorting signals. DNA Repair (Amst) 2002;1(10):847–54.

[66] Nilsen H, Otterlei M, Haug T, et al. Nuclear and mitochondrial uracil-DNA glycosylases are generated by alternative splicing and transcription from different positions in the UNG gene. Nucleic Acids Res 1997;25(4):750–5.

[67] Lakshmipathy U, Campbell C. The human DNA ligase III gene encodes nuclear and mitochondrial proteins. Mol Cell Biol 1999;19(5):3869–76.

[68] Thompson LH, West MG. XRCC1 keeps DNA from getting stranded. Mutat Res 2000;459(1):1–18.

[69] Lakshmipathy U, Campbell C. Mitochondrial DNA ligase III function is independent of Xrcc1. Nucleic Acids Res 2000;28(20):3880–6.

[70] Trifunovic A, Wredenberg A, Falkenberg M, et al. Premature ageing in mice expressing defective mitochondrial DNA polymerase. Nature 2004;429(6990):417–23.

[71] Hance N, Ekstrand MI, Trifunovic A. Mitochondrial DNA polymerase gamma is essential for mammalian embryogenesis. Hum Mol Genet 2005;14(13):1775–83.

[72] Kujoth GC, Hiona A, Pugh TD, et al. Mitochondrial DNA mutations, oxidative stress, and apoptosis in mammalian aging. Science 2005;309(5733):481–4.

[73] Chen H, Vermulst M, Wang YE, et al. Mitochondrial fusion is required for mtDNA stability in skeletal muscle and tolerance of mtDNA mutations. Cell 2010;141(2):280–9.

[74] Bess AS, Crocker TL, Ryde IT, Meyer JN. Mitochondrial dynamics and autophagy aid in removal of persistent mitochondrial DNA damage in Caenorhabditis elegans. Nucleic Acids Res 2012;40(16):7916–31.

[75] Bohgaki T, Bohgaki M, Hakem R. DNA double-strand break signaling and human disorders. Genome Integr 2010;1(1):15.

[76] Gennery AR, Cant AJ, Jeggo PA. Immunodeficiency associated with DNA repair defects. Clin Exp Immunol 2000;121(1):1–7.

[77] Martinet W, Knaapen MW, De Meyer GR, Herman AG, Kockx MM. Elevated levels of oxidative DNA damage and DNA repair enzymes in human atherosclerotic plaques. Circulation 2002;106(8):927–32.

[78] Wu M, Audet A, Cusic J, Seeger D, Cochran R, Ghribi O. Broad DNA repair responses in neural injury are associated with activation of the IL-6 pathway in cholesterol-fed rabbits. J Neurochem 2009;111(4):1011–21.

[79] Wang J, Wang Q, Watson LJ, Jones SP, Epstein PN. Cardiac overexpression of 8-oxoguanine DNA glycosylase 1 protects mitochondrial DNA and reduces cardiac fibrosis following transaortic constriction. Am J Physiol Heart Circ Physiol 2011;301(5):H2073–80.

[80] Mahmoudi M, Gorenne I, Mercer J, Figg N, Littlewood T, Bennett M. Statins use a novel Nijmegen breakage syndrome-1-dependent pathway to accelerate DNA repair in vascular smooth muscle cells. Circ Res 2008;103(7):717–25.

[81] Kang J, Ferguson D, Song H, et al. Functional interaction of H2AX, NBS1, and p53 in ATM-dependent DNA damage responses and tumor suppression. Mol Cell Biol 2005;25(2):661–70.

[82] Pernice F, Floccari F, Caccamo C, et al. Chromosomal damage and atherosclerosis. A protective effect from simvastatin. Eur J Pharmacol 2006;532(3):223–9.

[83] Flouris GA, Arvanitis DA, Parissis JT, Arvanitis DL, Spandidos DA. Loss of heterozygosity in DNA mismatch repair genes in human atherosclerotic plaques. Mol Cell Biol Res Commun 2000;4(1):62–5.

[84] Dar ME, Winters TA, Jorgensen TJ. Identification of defective illegitimate recombinational repair of oxidatively-induced DNA double-strand breaks in ataxia-telangiectasia cells. Mutat Res 1997;384(3):169–79.

[85] Bakkenist CJ, Kastan MB. DNA damage activates ATM through intermolecular autophosphorylation and dimer dissociation. Nature 2003;421(6922):499–506.

[86] Cheng Q, Chen L, Li Z, Lane WS, Chen J. ATM activates p53 by regulating MDM2 oligomerization and E3 processivity. EMBO J 2009;28(24):3857–67.

[87] Mercer JR, Cheng KK, Figg N, et al. DNA damage links mitochondrial dysfunction to atherosclerosis and the metabolic syndrome. Circ Res 2010;107(8):1021–31.

[88] Blackmore HL, Piekarz AV, Fernandez-Twinn DS, et al. Poor maternal nutrition programmes a pro-atherosclerotic phenotype in ApoE$^{-/-}$ mice. Clin Sci (Lond) 2012;123(4):251–7.

[89] Mercer J, Figg N, Stoneman V, Braganza D, Bennett MR. Endogenous p53 protects vascular smooth muscle cells from apoptosis and reduces atherosclerosis in ApoE knockout mice. Circ Res 2005;96(6):667–74.

[90] von der Thüsen JH, van Vlijmen BJ, Hoeben RC, et al. Induction of atherosclerotic plaque rupture in apolipoprotein E$^{-/-}$ mice after adenovirus-mediated transfer of p53. Circulation 2002;105(17):2064–70.

[91] Guevara NV, Kim HS, Antonova EI, Chan L. The absence of p53 accelerates atherosclerosis by increasing cell proliferation in vivo. Nat Med 1999;5(3):335–9.

[92] Benkö R, Pacher P, Vaslin A, Kollai M, Szabó C. Restoration of the endothelial function in the aortic rings of apolipoprotein E deficient mice by pharmacological inhibition of the nuclear enzyme poly(ADP-ribose) polymerase. Life Sci 2004;75(10):1255–61.

[93] von Lukowicz T, Hassa PO, Lohmann C, et al. PARP1 is required for adhesion molecule expression in atherogenesis. Cardiovasc Res 2008;78(1):158–66.

[94] Rossi MN, Carbone M, Mostocotto C, et al. Mitochondrial localization of PARP-1 requires interaction with mitofilin and is involved in the maintenance of mitochondrial DNA integrity. J Biol Chem 2009;284(46):31616–24.

[95] Michan S, Sinclair D. Sirtuins in mammals: insights into their biological function. Biochem J 2007;404(1):1–13.

[96] Dehennaut V, Leprince D. Implication of HIC1 (Hypermethylated In Cancer 1) in the DNA damage response. Bull Cancer 2009;96(11):E66–72.

[97] Yamamori T, DeRicco J, Naqvi A, et al. SIRT1 deacetylates APE1 and regulates cellular base excision repair. Nucleic Acids Res 2010;38(3):832–45.

[98] Potente M, Ghaeni L, Baldessari D, et al. SIRT1 controls endothelial angiogenic functions during vascular growth. Genes Dev 2007;21(20):2644–58.

[99] Gorenne I, Kumar S, Gray K, et al. Vascular smooth muscle cell sirtuin 1 protects against DNA damage and inhibits atherosclerosis. Circulation 2013;127(3):386–96.

[100] Zhang QJ, Wang Z, Chen HZ, et al. Endothelium-specific overexpression of class III deacetylase SIRT1 decreases atherosclerosis in apolipoprotein E-deficient mice. Cardiovasc Res 2008;80(2):191–9.

[101] Breitenstein A, Stein S, Holy EW, et al. Sirt1 inhibition promotes in vivo arterial thrombosis and tissue factor expression in stimulated cells. Cardiovasc Res 2011;89(2):464–72.

[102] Stein S, Lohmann C, Schäfer N, et al. SIRT1 decreases Lox-1-mediated foam cell formation in atherogenesis. Eur Heart J 2010;31(18):2301–9.

[103] Chen KH, Srivastava DK, Singhal RK, Jacob S, Ahmed AE, Wilson SH. Modulation of base excision repair by low density lipoprotein, oxidized low density lipoprotein and antioxidants in mouse monocytes. Carcinogenesis 2000;21(5):1017–22.

[104] Jeon BH, Gupta G, Park YC, et al. Apurinic/apyrimidinic endonuclease 1 regulates endothelial NO production and vascular tone. Circ Res 2004;95(9):902–10.

[105] Hall JL, Wang X, Adamson V, Zhao Y, Gibbons GH. Overexpression of Ref-1 inhibits hypoxia and tumor necrosis factor-induced endothelial cell apoptosis through nuclear factor-kappab-independent and -dependent pathways. Circ Res 2001;88(12):1247–53.

[106] Passos JF, Saretzki G, von Zglinicki T. DNA damage in telomeres and mitochondria during cellular senescence: is there a connection? Nucleic Acids Res 2007;35(22):7505–13.

[107] Madamanchi NR, Runge MS. Mitochondrial dysfunction in atherosclerosis. Circ Res 2007;100(4):460–73.

[108] Stoneman VE, Bennett MR. Role of apoptosis in atherosclerosis and its therapeutic implications. Clin Sci (Lond) 2004;107(4):343–54.

[109] Geng YJ, Libby P. Evidence for apoptosis in advanced human atheroma. Colocalization with interleukin-1 beta-converting enzyme. Am J Pathol 1995;147(2):251–66.

[110] Duewell P, Kono H, Rayner KJ, et al. NLRP3 inflammasomes are required for atherogenesis and activated by cholesterol crystals. Nature 2010;464(7293):1357–61.

[111] Kirii H, Niwa T, Yamada Y, et al. Lack of interleukin-1beta decreases the severity of atherosclerosis in ApoE-deficient mice. Arterioscler Thromb Vasc Biol 2003;23(4):656–60.

[112] Isoda K, Sawada S, Ishigami N, et al. Lack of interleukin-1 receptor antagonist modulates plaque composition in apolipoprotein E-deficient mice. Arterioscler Thromb Vasc Biol 2004;24(6):1068–73.

[113] Hogquist KA, Nett MA, Unanue ER, Chaplin DD. Interleukin 1 is processed and released during apoptosis. Proc Natl Acad Sci USA 1991;88(19):8485–9.

[114] Sutterwala FS, Ogura Y, Szczepanik M, et al. Critical role for NALP3/CIAS1/ Cryopyrin in innate and adaptive immunity through its regulation of caspase-1. Immunity 2006;24(3):317–27.

[115] Beasley D, Cooper AL. Constitutive expression of interleukin-1alpha precursor promotes human vascular smooth muscle cell proliferation. Am J Physiol 1999;276(3 Pt 2):H901–12.

[116] Rajamäki K, Lappalainen J, Oörni K, et al. Cholesterol crystals activate the NLRP3 inflammasome in human macrophages: a novel link between cholesterol metabolism and inflammation. PLoS One 2010;5(7):e11765.

[117] Keidar S. Angiotensin LDL peroxidation and atherosclerosis. Life Sci 1998;63(1):1–11.

[118] Herbert KE, Mistry Y, Hastings R, Poolman T, Niklason L, Williams B. Angiotensin II-mediated oxidative DNA damage accelerates cellular senescence in cultured human vascular smooth muscle cells via telomere-dependent and independent pathways. Circ Res 2008;102(2):201–8.

[119] Zhou R, Yazdi AS, Menu P, Tschopp J. A role for mitochondria in NLRP3 inflammasome activation. Nature 2011;469(7329):221–5.

[120] Nakahira K, Haspel JA, Rathinam VA, et al. Autophagy proteins regulate innate immune responses by inhibiting the release of mitochondrial DNA mediated by the NALP3 inflammasome. Nat Immunol 2011;12(3):222–30.

[121] Shimada K, Crother TR, Karlin J, et al. Oxidized mitochondrial DNA activates the NLRP3 inflammasome during apoptosis. Immunity 2012;36(3):401–14.

[122] Murakami T, Ockinger J, Yu J, et al. Critical role for calcium mobilization in activation of the NLRP3 inflammasome. Proc Natl Acad Sci U S A 2012;109(28):11282–7.

[123] Zhang Q, Raoof M, Chen Y, et al. Circulating mitochondrial DAMPs cause inflammatory responses to injury. Nature 2010;464(7285):104–7.

[124] Ding Z, Liu S, Wang X, Khaidakov M, Dai Y, Mehta JL. Oxidant stress in mitochondrial DNA damage, autophagy and inflammation in atherosclerosis. Sci Rep 2013;3:1077.

[125] Hornung V, Ablasser A, Charrel-Dennis M, et al. AIM2 recognizes cytosolic dsDNA and forms a caspase-1-activating inflammasome with ASC. Nature 2009;458(7237):514–8.

[126] Fernandes-Alnemri T, Yu JW, Datta P, Wu J, Alnemri ES. AIM2 activates the inflammasome and cell death in response to cytoplasmic DNA. Nature 2009;458(7237):509–13.

[127] Bürckstümmer T, Baumann C, Blüml S, et al. An orthogonal proteomic-genomic screen identifies AIM2 as a cytoplasmic DNA sensor for the inflammasome. Nat Immunol 2009;10(3):266–72.

[128] Kim KS, Kang KW, Seu YB, Baek SH, Kim JR. Interferon-gamma induces cellular senescence through p53-dependent DNA damage signaling in human endothelial cells. Mech Ageing Dev 2009;130(3):179–88.

[129] Guven M, Guven GS, Oz E, et al. DNA repair gene XRCC1 and XPD polymorphisms and their association with coronary artery disease risks and micronucleus frequency. Heart Vessels 2007;22(6):355–60.

[130] Durik M, Kavousi M, van der Pluijm I, et al. Nucleotide excision DNA repair is associated with age-related vascular dysfunction. Circulation 2012;126(4):468–78.

[131] Kanfi Y, Peshti V, Gil R, et al. SIRT6 protects against pathological damage caused by diet-induced obesity. Aging Cell 2010;9(2):162–73.

The DNA Sensing Pathway and Its Impact on Vaccinology

CHAPTER 11

DNA Vaccine: Does it Target the Double Stranded-DNA Sensing Pathway?[*]

Cevayir Coban[1], Miyuki Tozuka[2], Nao Jounai[2], Kouji Kobiyama[2], Fumihiko Takeshita[2], Choon Kit Tang[1,3] and Ken J. Ishii[2,3]

[1]Laboratory of Malaria Immunology, W PI Immunology Frontier Research Center (IFReC), Osaka University, Osaka, Japan
[2]Laboratory of Adjuvant Innovation, National Institute of Biomedical Innovation, Osaka, Japan
[3]Laboratory of Vaccine Science, IFReC, Osaka University, Osaka, Japan

INTRODUCTION

Vaccines are expected to give lifetime protection from infectious diseases, ideally with a single immunization. Modern understanding of vaccination has led to vaccines which protect not only from infectious diseases, but also non–infectious disorders, such as allergic diseases, autoimmune diseases and cancer. During the 1990s, DNA vaccination was introduced as a byproduct of DNA-based gene therapy, so that it could be successfully administered into skin or muscle to induce both cellular and humoral immune responses to many viral or non-viral antigens including hepatitis B, malaria, HIV and cancer [1–4]. The major advantage of DNA vaccines was their ability to elicit CD8[+] T cell responses along with Th1-biased antibody production in a very simple way, with administration of plasmid DNA and encoding antigen. All the safety concerns were addressed by the FDA, and the DNA vaccines were found to be safe and well-tolerated in humans [5]. At present, although several DNA vaccines have been approved for veterinary medicine since 2005 [6,7], the promise of DNA vaccines in humans has been hampered by low immunogenicity. To date, various approaches have been proposed to solve this problem (detailed strategies to improve DNA vaccines can be found in recent review articles [8–10]).

---

[*]Abbreviations used: B-DNA, B form right-handed helical structure deoxyribonucleic acid; STING, stimulator of interferon genes; TBK1, TANK-binding kinase 1; IFN, interferon; IRF, interferon regulatory factor; CTL, cytotoxic T lymphocytes.

Biological DNA Sensor.
DOI: http://dx.doi.org/10.1016/B978-0-12-404732-7.00011-3

In this chapter we will introduce DNA vaccines and the current knowledge on their mechanism of action. We will also discuss recent developments to improve immunogenicity of DNA vaccines for future use in humans.

DNA VACCINES: PROOF OF CONCEPT

A DNA vaccine is composed of a bacterial plasmid which expresses the protein of interest (an antigen) under the control of a mammalian promoter to enable it to function in the transfected mammalian cells. Once the plasmid DNA is administered *in vivo*, the encoded protein is expressed in the host cells. Since the *in vivo* uptake of plasmid DNA by host cells (most likely stromal cells such as muscle cells) is usually not efficient, various transfection modalities such as electroporation have been introduced [8,11]. Overall, proteins successfully expressed by DNA plasmids can be processed as peptides; they bind to MHC Class I or Class II molecules and are presented by antigen presenting cells (APCs) such as dendritic cells (DCs) for cellular immune activation. Moreover, expressed proteins (either secreted or processed, via $CD4^+$ T cell engagement) can also activate B cells for antibody production. Thus, this unique ability of DNA vaccines to introduce antigen to the host immune system occurs by eliciting strong Th1 type $CD4^+$ T cells and $CD8^+$ cytotoxic T cells which act against pathogens. Table 11.1 summarizes the uniqueness and advantages of DNA vaccines over protein vaccines.

DNA VACCINES: MOUSE VS. MAN

More than 40 human clinical trials have been conducted by using DNA vaccines against infectious diseases, as well as autoimmune diseases (reviewed in refs [10,12]). The outcomes of many of these clinical trials

Table 11.1 Unique characteristics of DNA vaccines over protein vaccines

Advantages
- Proper folding of targeted protein
- Cost effectiveness of plasmid purification over protein purification
- Easy manufacturing and storage on a large scale

Safety profile
- Higher safety profile due to being non-live and non-replicating
- No evidence of genome integration
- No anti-DNA antibody production even after multiple immunizations

were disappointing. The main obstacle was the generation of lower immunogenicity in humans compared with other mammals. On the other hand, safety issues (i.e. integration of plasmid DNA into genomic DNA, risk of autoimmunity or tolerance to antigen) that were addressed in these studies have concluded DNA vaccination to be a safe form of immunization.

The reasons for the lower immune responses induced by DNA vaccines in humans are currently unknown. However, successful demonstrations have been achieved in veterinary applications and DNA vaccine products have been licensed for use in animals since 2005. For example, DNA vaccines have been produced to target canine melanoma, West Nile virus in horses and fish hematopoietic necrosis viruses [6,7,13]. Hence, these successful veterinary applications have indicated that selecting an appropriate immunogenic antigen that is specific for the target disease might be the key for success in human application. Overall, with the improvements in antigen design and advancements in delivery methods, great improvement in DNA vaccine immunogenicity has been observed over the last 20 years, and recently there has been renewed interest in these vaccines [12].

Moreover, studies involving the investigation of DNA-sensing machinery have also added new incentive to improve DNA vaccine immunogenicity in humans. It was found that if DNA is present in the cytoplasm of mammalian cells, it can trigger host immune responses, but the unknown sensor that detects cytosolic DNA is poorly understood. Hence, lower expression levels of certain components of the DNA-sensing machinery or their effects on DNA delivery and processing could be behind the lower immunogenicity seen in humans [14].

WHAT IMMUNOLOGICAL RESPONSES ARE TRIGGERED AFTER DNA VACCINATION?

It is evident from several sets of experimental data that DNA vaccine administration directly transfects stromal/tissue resident cells (such as muscle cells) or APCs such as DCs. There are two components in DNA vaccination which should be considered for the generation of immune responses: (1) the DNA plasmid itself; and (2) the antigens encoded by these DNA plasmids and their expression profile.

The first component, the transfected DNA plasmid used in DNA vaccines, has been shown to be immunogenic and could induce a strong type I IFN response [15]. Hence, the DNA plasmid may be engaged with an

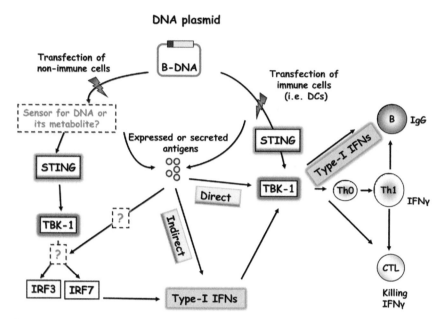

Figure 11.1 *DNA vaccine by electroporation utilizes the STING–TBK1 pathway for adaptive immune responses.* When introduced by electroporation into cells (immune or non-immune cells), the transfected DNA plasmid or its metabolites are engaged with an unknown sensor and induce the production of type I interferons (IFNs) via the STING–TBK1 complex. The antigens encoded by plasmid DNA are also expressed. These antigens may be directly (in immune cells) or indirectly (from non-immune cells) processed and presented to T cells, collectively leading to the induction of antigen-specific B cells, CD4+ T cells and CD8+ T cells that are possibly promoted by type I IFNs controlled by STING–TBK1.

unidentified cytosolic DNA sensor and induces the activation of TRAF-family-member-associated NF-κB activator (TANK) binding kinase 1 (TBK1) and IκB kinase-ε (IKKε) through the stimulation of IFN genes (STING), leading to the activation of interferon-regulatory factor 3 (IRF3) and resulting in the production of type I IFNs (Figure 11.1) [15–17].

The second component, the encoded antigen, may be found in professional APCs such as DCs in one of two ways: (i) DNA plasmids could be taken up and the antigen directly expressed and processed in DCs leading to MHC class I presentation to CD8 T cells; or (ii) it may be indirectly captured by DCs after release from transfected stromal (i.e. muscle) cells and then cross-presented to CD8+ T cells or presented to naive CD4+ T cells on MHC class II molecules (Figure 11.1). Importantly,

TBK1-dependent type I IFN expression by stromal cells and DCs is likely to be important for promoting the cross-presentation activity of DCs, and subsequent differentiation of T cells and the induction of B cell responses. Further studies have suggested that DNA vaccination requires TBK1 activation in hematopoietic cells (presumably DCs); however, TBK1 activity in non-hematopoietic cells (presumably stromal cells) is only essential for $CD8^+$ T cell activation. In contrast, the activation of antigen-specific $CD4^+$ T cells requires TBK1 activity in both the hematopoietic and non-hematopoietic compartments. Altogether, direct presentation as well as cross-presentation of processed antigens is equally essential for the generation of adaptive immune responses to DNA vaccines that STING–TBK1-dependent type I IFN is required to promote in these responses [14,16,18].

CYTOSOLIC DOUBLE STRANDED (DS)-DNA SENSING BY IMMUNE SYSTEM

The host immune system recognizes several microbes and/or their products by intracellular or membrane localized pattern-recognition receptors (PRRs); this recognition activates complex downstream signaling molecules/pathways such as Toll-like receptors (TLR) to defend against pathogens [19,20]. How PRRs detect DNA on the cell surface and in endosomes has been characterized extensively in the case of TLR-mediated signaling pathways. TLR9 was one of the first receptors found to play an important role in the recognition of foreign (i.e. bacterial origin) DNA and its triggering immune responses [21]. Particularly, foreign DNA can bind to TLR9 in the endosome and stimulate the production of type I IFNs in DCs. Subsequently it was found that not only foreign DNA, but also dsDNA derived from host cells, if introduced into the cytosol (mainly in the B form, a right-handed helical structure), could induce immune responses [15]. This was quite a surprising finding, because so-called 'inert' host genetic material, once introduced into the cytosol by transfection, could activate fibroblasts, macrophages and DCs to produce robust amounts of type I IFNs, with an effect independent of the CpG motifs and TLR9, but completely dependent on TBK1. TBK1 is a non-canonical IκB kinase which comprises another inducible IκB kinase (IKK-i, also known as IKK-ε), and these kinases directly phosphorylate IRF3 and IRF7.

Mechanisms for how cytosolic DNA is sensed and detected have been proposed by several groups and various proteins were identified. However, given that the metabolites of nucleic acids could also have significant effects on the innate immune system [22], how DNA and/or its metabolites are recognized will need further investigation to reach a universal consensus. Nevertheless, various cytosolic DNA-recognition molecules were identified upstream of TBK1 that could be important in understanding the mechanisms of DNA vaccines:

1. **DAI:** DNA-dependent activator of IFN-regulatory factors (DAI), also called DLM-1 or Z-DNA binding protein 1 (ZBP1), was found to bind to dsDNA and enhanced its association with the IRF3 and the TBK1, resulting in the DNA-mediated activation of innate immune responses [23]. However, the role of DAI is very cell type specific [24]; its role in innate signaling in humans remains to be further clarified, as DAI-deficient mice showed normal type I IFN responses to dsDNA [16].

2. **STING:** It was found that, upon various differentially originated cytoplasmic dsDNA stimulations, an endoplasmic reticulum (ER)-localized molecule STING (stimulator of interferon genes, also called MITA, ERIS or TMEM173) relocalizes with TBK1 from the endoplasmic reticulum to the perinuclear vesicles and activates IRF3/IRF7 to stimulate type I IFN production, as well as activation of NF-κB. However, how STING regulates NF-κB is not well understood [25]. Important recent progress has revealed that STING is also a sensor of bacterial secondary messenger molecules (i.e. c-di-GMP or c-di-AMP) [26]. This raises the important possibility that cytosolic dsDNA stimulation might produce secondary products such as c-di-GMP/c-di-AMP which might be recognized by STING.

3. **IFI16:** Interferon-gamma inducible protein 16 (also called p204) is a member of the PYHIN protein family that contains a pyrin domain and two DNA-binding HIN domains. Studies have suggested that IFI16 depletion by RNA interference reduces the induction of type I IFN by synthetic DNA which might be dependent on STING [27].

4. **HMGB:** In response to cytosolic DNA, high mobility group box (HMGB) proteins showed defects in IFN production, suggesting that HMBGs are required for the regulation of STING/TBK1 pathways, although the mechanisms are unclear [28].

5. **TRIM56:** Another molecule TRIM56 (interferon-inducible tripartite-motif 56), an interferon-inducible E3 ubiquitin ligase, was recently identified as a modulator of STING dimerization upstream of TBK1 to confer dsDNA-mediated type I IFN responses [29].

6. **DDX41:** More recently, DDX41, a member of the DEXDc family of helicases, was found to sense intracellular DNA in myeloid dendritic cells (mDCs) and to be responsible for the type I IFN and cytokine responses to DNA and DNA viruses [26]. DNA sensing capability of DDX41 was found to be dependent on STING: DDX41 and DNA together with STING co-localized in the cytosol. Moreover, not only DNA from viruses, but also dsDNA from bacteria (small bacterial nucleic acids called cyclic dinucleotides) together with cyclic di-GMP and cyclic di-AMP were found to bind to central DEAD-box domain (DEADc), albeit not directly to STING [30,31]. However, further studies are needed to understand the interactions between DDX41, dsDNA, cyclic dinucleotides and STING [32].

DOES PLASMID DNA ACT IN A COMMON IMMUNOLOGICAL WAY OF DNA SENSING?

This was particularly interesting because previous studies have shown minimal involvement of TLR9 in the activity of plasmid DNA *in vivo* [18,33,34]. It was shown later that the optimal DNA vaccine immunogenicity (antigen-specific T and B cell induction) required type I IFNs with clear evidence of no activity found in IFNαR2-deficient mice [16]. TLR signaling did not play a role (MyD88/TRIF-double deficient mice showed normal responses to DNA vaccines), but TBK1 was the major player in the DNA vaccine immunogenicity [16]. Additional studies using interferon-beta promoter stimulator 1 (IPS-1)-deficient mice also suggested that possible RNA generation is less likely to act as a TBK1-dependent adjuvant, because IPS-1, which is an adaptor for retinoic acid-inducible gene I (RIG-I) and melanoma differentiation-associated gene 5 (MDA5) deficiency had no role on the immune responses generated by DNA vaccination [16]. Importantly, studies with STING-deficient mice confirmed that STING was an essential molecule for plasmid DNA immunogenicity [17]. Collectively, these studies strongly suggested that STING and TBK1 are the essential components of the DNA recognition machinery, and that a double-stranded structure could be an essential adjuvant element for DNA vaccine-induced immunogenicity.

Our group has focused on the role of several dsDNA-signaling molecules in the immunogenicity of DNA vaccines. Table 11.2 summarizes the current knowledge on the dsDNA-sensing machinery and its known

Table 11.2   Known signaling molecules in the dsDNA-induced immune responses with their role in the mechanism of DNA vaccines

Knockout Mice	DNA Vaccine					Reference
	Antigen	Route	Ab Responses	CD4$^+$ T cells	CD8$^+$ T cells	
ZBP-1 (DAI)	LacZ	i.m e.p.	→	→	→	[16]
TRIM56	?	?	?	?	?	[29]
IFI16	?	?	?	?	?	[27]
DDX41	?	?	?	?	?	[30]
STING	OVA	i.m e.p.	→	→	→	[17]
TBK1	LacZ NP	i.m e.p.	→	→	→	[16]
IRF3	OVA-Luc	i.m.	→	→	→	[35]
IRF7	?	?	?	?	?	?
IFNαR2	LacZ	i.m. e.p.	→	→	→	[16]

impact on the mechanism of DNA vaccination. Although ZBP-1 (DAI) and its function as an intracellular DNA sensor is a controversial issue, DNA vaccine immunization of ZBP-1(DAI)-deficient mice nevertheless revealed a minimal role of ZBP-1 in innate or adaptive immune responses to DNA vaccine *in vivo* [16]. Currently no data are available for the possible involvement of IFI16, TRIM56 or DDX41 in the mechanism of DNA vaccine; this should be explored in future by using available mice deficient for these molecules.

In the downstream signaling of the STING–TBK1 complex, however, IRF3-deficient mice elicited strong antigen-specific humoral responses after DNA vaccinations, while CD4$^+$ and CD8$^+$ T cell responses (including the production of Th1, Th2 and Th17 cytokines) were severely impaired [35]. It should be noted however that this previous study used only the intramuscular (i.m.) route for the immunizations; electroporation (e.p.) of plasmid DNA might have resulted in intact T cell responses, which needs further confirmation. Nevertheless, if IRF3 is not involved in the DNA vaccine-mediated immune responses, one might hypothesize that other downstream molecules of TBK1 (i.e. IRF7) could be leading the production of antibodies (Table 11.2). Supporting this hypothesis, our preliminary data using IRF7-deficient mice have implied that the requirement of type I IFN for the immunogenicity of DNA vaccine might be affected not only by the DNA plasmid backbone, but also other factors such as the features of antigens expressed from the DNA plasmid backbone (Tozuka et al., unpublished observations).

DNA SENSING MACHINERY: IS THERE A HOPE TO IMPROVE DNA VACCINE IMMUNOGENICITY IN HUMANS?

These discoveries in which cytosolic plasmid DNA-induced immuno-genicity was attributed to its adjuvant properties (mediated via STING and TBK1 kinase) have led researchers to evaluate whether such immuno-genicity could be improved by incorporating overexpression of these signaling molecules as an external adjuvant to DNA vaccines. The initial pioneer studies have shown that although dispensable for the immune signaling of DNA vaccines, adapter molecules such as the myeloid differ-entiation primary response gene (MyD88) or Toll/IL-1 receptor (TIR)-domain-containing adaptor inducing interferon-β (TRIF), if incorporated into DNA vaccines as dual-promoter plasmids and overexpressed, greatly enhance humoral as well as cellular responses to a given antigen [36]. A similar approach was applied for RIG-I and MDA5. Although IPS-1-deficiency has played no role in the immunogenicity of DNA vaccines [16], upstream of IPS-1, RIG-I recognizes cytoplasmic double-stranded RNA (dsRNA) of viral or non-viral origin. Some types of syn-thetic dsDNA sequences (i.e. Poly(dA:dT) (synthetic AT-rich dsDNA, but not other types of DNA including plasmid DNA)), if transfected, were reported to transcribe dsRNA containing 5′-triphosphate ends that has the ability to induce type I IFNs via the RIG-I pathway by involv-ing DNA-dependent RNA polymerase III [37]. Based on this, Luke et al. have generated DNA vaccines co-expressing antigen and potent RIG-I ligands from RNA polymerase III promoters as the RIG-I agonist [38]. The resultant vectors potently induced type I IFN production in cell culture and increased influenza-specific serum antibody binding avid-ity of DNA vaccines in mice. In addition, a similar approach in chick-ens using the co-administration of influenza-HA plasmid and plasmid DNA for MDA5 expression resulted in enhanced immunogenicity as well as protection against a lethal H5N1 challenge infection [39]. Similarly, although ZBP1/DAI had no role in the immunogenicity of DNA vac-cines, the genetic adjuvanticity of DAI has been shown [40]. Another study has evaluated the usage of transcription factors of the interferon regulatory factor (IRF) family (IRF-1, IRF-3, and IRF-7) as adjuvants to DNA vaccines. Plasmid DNA encoding IRF-1, but not IRF-3 or IRF-7, enhanced antigen-specific immune responses [41]. We have also tested the effect of TBK1-encoding plasmids on DNA vaccine immunogenic-ity against *Plasmodium* antigens, and found that the DNA vaccine plasmid cocktail may improve, at least, anti-malarial antigen humoral immune responses [18]. However, one should note that, although TBK1 (or in some extent IRF3) was shown to have a critical role for DNA vaccine-induced immune responses (Table 11.2), their minimal or limited role as an external adjuvant to plasmid DNA raises the possibility that either antigens expressed from the DNA plasmid backbone, or plasmid DNA metabolites, especially after strong transfection by electroporation, might be having a crucial role in the immune response and its signaling pathways.

CAN DNA PLASMID METABOLITES BE ADJUVANT?

In earlier studies, it was shown that the synthetic non-coding dsDNA could be added to protein-based vaccines to improve vaccine efficacy [42]. This implied that the immunostimulatory properties of dsDNA are capable of improving vaccine efficacy, namely by being an adjuvant. However, it may also suggest that other byproducts resulting from DNA metabolism could be involved in elevating the immunogenicity of the vaccine [22], although we do not currently have clear evidence as to whether *in vivo* transfection of plasmid DNA could release other metabolites that are capable of inducing type I IFN production.

Cyclic-di-GMP (c-di-GMP) is a small nucleotide with a cyclic conformation made up of two guanine bases that are coupled by ribose and phosphate groups. It is mainly found in bacteria and is used as a second messenger during the signal transduction processes. When c-di-GMP was transfected into mammalian cells, it was found to be strongly immunogenic and to induce high levels of type I IFN [43]. Subsequent studies found that c-di-GMP utilizes the same signaling pathway as dsDNA in inducing type I IFN. C-di-GMP can directly bind to DDX41 and STING to activate downstream TBK1-IRF3 signaling for the induction of type I IFN responses [5,30,31]. For this immunogenic property, c-di-GMP was studied and successfully proven to be an effective adjuvant for vaccines targeting the mucosal route [44,45]. A mammalian equivalent of this bacterial second messenger, cyclic-GMP-AMP (cGAMP), was recently discovered by Wu et al. [46]. Both the mammalian and bacterial second messengers share similar immunogenic properties. cGAMP induces strong type I IFN responses when introduced into cells and it targets STING for the activation of IRF3 dimerization. In mammalian cells, cGAMP was found to be produced when dsDNA is transfected into cells or during DNA virus infection. The mechanism by which cGAMP is produced within the cells was elucidated by the same group of researchers. They showed that cGAMP is generated by a cytoplasmic molecule, cyclic GMP-AMP synthetase (cGAS), upon detection of dsDNA in the cytoplasm [47]. cGAS was found to function upstream of STING and was not responsive to RNA stimulation.

The above studies have suggested that other compounds target the DNA sensing pathway, and have a capability for the induction of type I IFN responses, possibly functioning as adjuvants. For this reason, our laboratory has recently investigated the application of 5, 6-dimethylxanthenone-4-acetic acid (DMXAA) as an adjuvant [49]. DMXAA is a synthetic small molecule compound and a potent type I IFN inducer that resembles viral infections and dsDNA in the inflammatory signaling events it triggers [48]. Furthermore, DMXAA utilizes the TBK1–IRF3 signaling pathway without the involvement of TLRs or RNA helicases for its mechanism of type I IFN induction. Our recent studies have suggested that DMXAA could act as a vaccine adjuvant via its unique property as a soluble innate immune activator and that its adjuvant effect is directly dependent on the IRF3-mediated production of type I IFNs.

These studies have highlighted the fact that the type I IFN response induced by the cytosolic DNA sensing pathway is a potent one, and that it is capable of adjuvanting vaccine responses, at least in mouse models. This has no doubt brought us back to the same question of why DNA vaccines are only efficacious in mouse models but not humans. The solution to this question can only be achieved by a greater understanding of the cellular DNA sensing mechanism, and knowledge of the difference between mouse and human in sensing DNA.

CONCLUSIONS

Ongoing research on the investigation of the DNA sensors in immunology has been improving our understanding of how DNA vaccines might function. Current knowledge indicates that DNA vaccines might be inducing innate and adaptive immune responses in three different, if not connected, ways:

1. The transfected DNA plasmid itself may be engaged with an unidentified cytosolic DNA receptor and induces the activation of the TBK1–STING pathway, resulting in the production of type I IFNs which finally act as an intrinsic built-in adjuvant.

2. The DNA plasmid-encoded antigen could be expressed in both stromal cells as well as DCs and may lead to the activation/involvement of the STING–TBK1 pathway, in direct presentation and cross-presentation of antigens which are collectively important for the DNA vaccine-induced adaptive immune responses.

3. *In vivo* transfection (i.e. by electroporation) of plasmid DNA may release metabolites such as cyclic-di-GMP which may induce the production of type I IFNs through STING–TBK1 signaling and work as adjuvant to DNA vaccines.

These ongoing studies may collectively open new research areas for further improvement of DNA vaccines for human use.

ACKNOWLEDGMENTS

These studies were supported by a Health and Labour Sciences Research Grant 'Adjuvant Database Project' of the Japanese Ministry of Health, Labour and Welfare and Grant-in-Aid for Scientific Research (KAKENHI) from the Japanese Ministry of Education, Culture, Sports, Science and Technology (MEXT) and the Japan Science and Technology Agency (JST).

REFERENCES

[1] Tang DC, DeVit M, Johnston SA. Genetic immunization is a simple method for eliciting an immune response. Nature 1992;356(6365):152–4.

[2] Ulmer JB, Donnelly JJ, Parker SE, Rhodes GH, Felgner PL, Dwarki VJ, et al. Heterologous protection against influenza by injection of DNA encoding a viral protein. Science 1993;259(5102):1745–9.

[3] Fynan EF, Webster RG, Fuller DH, Haynes JR, Santoro JC, Robinson HL. DNA vaccines: protective immunizations by parenteral, mucosal, and gene-gun inoculations. Proc Natl Acad Sci USA 1993;90(24):11478–82.

[4] Gurunathan S, Klinman DM, Seder RA. DNA vaccines: immunology, application, and optimization. Ann Rev Immunol 2000;18:927–74.

[5] Smith HA, Klinman DM. The regulation of DNA vaccines. Curr Opinion Biotechnol 2001;12(3):299–303.

[6] Davis BS, Chang GJ, Cropp B, Roehrig JT, Martin DA, Mitchell CJ, et al. West Nile virus recombinant DNA vaccine protects mouse and horse from virus challenge and expresses in vitro a noninfectious recombinant antigen that can be used in enzyme-linked immunosorbent assays. J Virol 2001;75(9):4040–7.

[7] Bergman PJ, McKnight J, Novosad A, Charney S, Farrelly J, Craft D, et al. Long-term survival of dogs with advanced malignant melanoma after DNA vaccination with xenogeneic human tyrosinase: a phase I trial. Clin Cancer Res J Am Assoc Cancer Res 2003;9(4):1284–90.

[8] Sardesai NY, Weiner DB. Electroporation delivery of DNA vaccines: prospects for success. Curr Opin Immunol 2011;23(3):421–9.

[9] Li L, Saade F, Petrovsky N. The future of human DNA vaccines. J Biotechnol 2012;162(2–3):171–82.

[10] Liu MA. DNA vaccines: an historical perspective and view to the future. Immunol Rev 2011;239(1):62–84.

[11] Gothelf A, Gehl J. What you always needed to know about electroporation based DNA vaccines. Hum Vaccines Immunotherapeut 2012;8:11.

[12] Ferraro B, Morrow MP, Hutnick NA, Shin TH, Lucke CE, Weiner DB. Clinical applications of DNA vaccines: current progress. Clin Infect Dis Publ Infect Dis Soc Am 2011;53(3):296–302.

[13] Corbeil S, Lapatra SE, Anderson ED, Jones J, Vincent B, Hsu YL, et al. Evaluation of the protective immunogenicity of the N, P, M, NV and G proteins of infectious hematopoietic necrosis virus in rainbow trout *Oncorhynchus mykiss* using DNA vaccines. Dis Aquat Org 1999;39(1):29–36.

[14] Desmet CJ, Ishii KJ. Nucleic acid sensing at the interface between innate and adaptive immunity in vaccination. Nat Rev Immunol 2012;12(7):479–91.

[15] Ishii KJ, Coban C, Kato H, Takahashi K, Torii Y, Takeshita F, et al. A Toll-like receptor-independent antiviral response induced by double-stranded B-form DNA. Nat Immunol 2006;7(1):40–8.

[16] Ishii KJ, Kawagoe T, Koyama S, Matsui K, Kumar H, Kawai T, et al. TANK-binding kinase-1 delineates innate and adaptive immune responses to DNA vaccines. Nature 2008;451(7179):725–9.

[17] Ishikawa H, Ma Z, Barber GN. STING regulates intracellular DNA-mediated, type I interferon-dependent innate immunity. Nature 2009;461(7265):788–92.

[18] Coban C, Kobiyama K, Aoshi T, Takeshita F, Horii T, Akira S, et al. Novel strategies to improve DNA vaccine immunogenicity. Current Gene Ther 2011;11(6):479–84.

[19] Ishii KJ, Koyama S, Nakagawa A, Coban C, Akira S. Host innate immune receptors and beyond: making sense of microbial infections. Cell Host Microbe 2008;3(6):352–63.

[20] Kawai T, Akira S. Toll-like receptors and their crosstalk with other innate receptors in infection and immunity. Immunity 2011;34(5):637–50.

[21] Hemmi H, Takeuchi O, Kawai T, Kaisho T, Sato S, Sanjo H, et al. A Toll-like receptor recognizes bacterial DNA. Nature 2000;408(6813):740–5.

[22] Ishii KJ, Akira S. Potential link between the immune system and metabolism of nucleic acids. Curr Opin Immunol 2008;20(5):524–9.

[23] Kawamura A, Iacovidou M, Takaoka A, Soll CE, Blumenstein M. A polyacetylene compound from herbal medicine regulates genes associated with thrombosis in endothelial cells. Bioorganic Med Chem Lett 2007;17(24):6879–82.

[24] Lippmann J, Rothenburg S, Deigendesch N, Eitel J, Meixenberger K, van Laak V, et al. IFNbeta responses induced by intracellular bacteria or cytosolic DNA in different human cells do not require ZBP1 (DLM-1/DAI). Cell Microbiol 2008;10(12):2579–88.

[25] Barber GN. Cytoplasmic DNA innate immune pathways. Immunol Rev 2011;243(1):99–108.

[26] Burdette DL, Monroe KM, Sotelo-Troha K, Iwig JS, Eckert B, Hyodo M, et al. STING is a direct innate immune sensor of cyclic di-GMP. Nature 2011;478(7370):515–8.

[27] Unterholzner L, Keating SE, Baran M, Horan KA, Jensen SB, Sharma S, et al. IFI16 is an innate immune sensor for intracellular DNA. Nature Immunol 2010;11(11):997–1004.

[28] Yanai H, Ban T, Wang Z, Choi MK, Kawamura T, Negishi H, et al. HMGB proteins function as universal sentinels for nucleic-acid-mediated innate immune responses. Nature 2009;462(7269):99–103.

[29] Tsuchida T, Zou J, Saitoh T, Kumar H, Abe T, Matsuura Y, et al. The ubiquitin ligase TRIM56 regulates innate immune responses to intracellular double-stranded DNA. Immunity 2010;33(5):765–76.

[30] Parvatiyar K, Zhang Z, Teles RM, Ouyang S, Jiang Y, Iyer SS, et al. The helicase DDX41 recognizes the bacterial secondary messengers cyclic di-GMP and cyclic di-AMP to activate a type I interferon immune response. Nat Immunol 2012;13(12):1155–61.

[31] Ouyang S, Song X, Wang Y, Ru H, Shaw N, Jiang Y, et al. Structural analysis of the STING adaptor protein reveals a hydrophobic dimer interface and mode of cyclic di-GMP binding. Immunity 2012;36(6):1073–86.

[32] Bowie AG. Innate sensing of bacterial cyclic dinucleotides: more than just STING. Nat Immunol 2012;13(12):1137–9.

[33] Babiuk S, Mookherjee N, Pontarollo R, Griebel P, van Drunen Littel-van den Hurk S, Hecker R, et al. TLR9$^{-/-}$ and TLR9$^{+/+}$ mice display similar immune responses to a DNA vaccine. Immunology 2004;113(1):114–20.

[34] Rottembourg D, Filippi CM, Bresson D, Ehrhardt K, Estes EA, Oldham JE, et al. Essential role for TLR9 in prime but not prime-boost plasmid DNA vaccination to activate dendritic cells and protect from lethal viral infection. J Immunol 2010;184(12):7100–7.

[35] Shirota H, Petrenko L, Hattori T, Klinman DM. Contribution of IRF-3 mediated IFNbeta production to DNA vaccine dependent cellular immune responses. Vaccine 2009;27(15):2144–9.

[36] Takeshita F, Tanaka T, Matsuda T, Tozuka M, Kobiyama K, Saha S, et al. Toll-like receptor adaptor molecules enhance DNA-raised adaptive immune responses against influenza and tumors through activation of innate immunity. J Virol 2006;80(13):6218–24.

[37] Chiu YH, Macmillan JB, Chen ZJ. RNA polymerase III detects cytosolic DNA and induces type I interferons through the RIG-I pathway. Cell 2009;138(3):576–91.

[38] Luke JM, Simon GG, Soderholm J, Errett JS, August JT, Gale Jr. M, et al. Coexpressed RIG-I agonist enhances humoral immune response to influenza virus DNA vaccine. J Virol 2011;85(3):1370–83.

[39] Liniger M, Summerfield A, Ruggli N. MDA5 can be exploited as efficacious genetic adjuvant for DNA vaccination against lethal H5N1 influenza virus infection in chickens. PloS One 2012;7(12):e49952.

[40] Lladser A, Mougiakakos D, Tufvesson H, Ligtenberg MA, Quest AF, Kiessling R, et al. DAI (DLM-1/ZBP1) as a genetic adjuvant for DNA vaccines that promotes effective antitumor CTL immunity. Mol Ther J Am Soc Gene Ther 2011;19(3):594–601.

[41] Castaldello A, Sgarbanti M, Marsili G, Brocca-Cofano E, Remoli AL, Caputo A, et al. Interferon regulatory factor-1 acts as a powerful adjuvant in tat DNA based vaccination. J Cell Physiol 2010;224(3):702–9.

[42] Ishii KJ, Suzuki K, Coban C, Takeshita F, Itoh Y, Matoba H, et al. Genomic DNA released by dying cells induces the maturation of APCs. J Immunol 2001;167(5):2602–7.

[43] McWhirter SM, Barbalat R, Monroe KM, Fontana MF, Hyodo M, Joncker NT, et al. A host type I interferon response is induced by cytosolic sensing of the bacterial second messenger cyclic-di-GMP. J Exp Med 2009;206(9):1899–911.

[44] Chen W, Kuolee R, Yan H. The potential of 3′,5′-cyclic diguanylic acid (c-di-GMP) as an effective vaccine adjuvant. Vaccine 2010;28(18):3080–5.

[45] Pedersen GK, Ebensen T, Gjeraker IH, Svindland S, Bredholt G, Guzman CA, et al. Evaluation of the sublingual route for administration of influenza H5N1 virosomes in combination with the bacterial second messenger c-di-GMP. PloS One 2011;6(11):e26973.

[46] Wu J, Sun L, Chen X, Du F, Shi H, Chen C, et al. Cyclic GMP-AMP is an endogenous second messenger in innate immune signaling by cytosolic DNA. Science 2013;339(6121):826–30.

[47] Sun L, Wu J, Du F, Chen X, Chen ZJ. Cyclic GMP-AMP synthase is a cytosolic DNA sensor that activates the type I interferon pathway. Science 2013;339(6121):786–91.

[48] Roberts ZJ, Goutagny N, Perera PY, Kato H, Kumar H, Kawai T, et al. The chemotherapeutic agent DMXAA potently and specifically activates the TBK1-IRF-3 signaling axis. J Exp Med 2007;204(7):1559–69.

[49] Tang CK, Aoshi T, Jounai N, Ito J, Ohata K, Kobiyama K, et al. The chemotherapeutic agent DMXAA as a unique IRF3-dependent type-2 vaccine adjuvant. PloS One 2013;8(3):e60038.

CHAPTER 12

Adjuvants Targeting the DNA Sensing Pathways – Alum Based Adjuvants

Christophe J. Desmet
Laboratory of Cellular and Molecular Immunology, Interdisciplinary Cluster of Applied Genoproteomics (GIGA) – Research Center and Faculty of Veterinary Medicine, University of Liège, Liège, Belgium

UNDERSTANDING THE MECHANISMS OF ACTION OF ALUM THROUGH VACCINE IMMUNOLOGY

The potential of aluminum-based adjuvants (which will be referred to as alum hereafter) was discovered more than eight decades ago, following the observation by Alexander T. Glenny and his coworkers that adsorption of diphtheria toxoid onto aluminum potassium sulfate led to higher antibody titers than injection of antigen alone [1]. Nowadays, alum, most often in the form of aluminum hydroxide or aluminum phosphate, is by far the most widely used vaccine adjuvant in human medicine worldwide. Alum is indeed used in many common vaccines such as the DTP (diphtheria–tetanus–pertussis), *Haemophilus influenzae* B, hepatitis A virus, hepatitis B virus, human papillomavirus, and pneumococcal vaccines [2].

In spite of decades of successful use however, the precise mechanisms at the origin of the adjuvant activity of alum remain elusive. It is actually only recently that interest in the modes of action of adjuvants has been reignited, along with the realization that efficient vaccines against problematic diseases such as HIV or malaria would likely not be obtained using empirical approaches. This revival coincides with the emergence of the field of vaccine immunology, which merges the previously rather mutually hermetic disciplines of vaccinology and immunology. Vaccine immunology notably aims at understanding the cellular and molecular immunological mechanisms at play in vaccination strategies, with the objective to harness these mechanisms for the development of novel, rationally engineered vaccines [3]. In this line of thought, the elucidation of the mechanisms of action of alum could provide invaluable information

Biological DNA Sensor.
DOI: http://dx.doi.org/10.1016/B978-0-12-404732-7.00012-5

for the improvement of the efficacy and safety of many current vaccines, as well as potential hints for the development of novel, rationally designed vaccines and adjuvants.

As a preliminary cautionary note, it has to be said that the generic term 'alum' is a semantic simplification, which covers a small variety of compounds with some distinct physico-chemical properties [2,4]. Nevertheless, these different types of alum tend to be considered as immunological equivalents in the field of vaccine immunology. Indeed, no major differences in terms of mechanisms of immunological activity have been reported so far in studies that compared different types of alum. Thus, while there is no arguing that differences between alum types or formulations may have some qualitative or quantitative impact on adaptive immune responses [5], it is likely that all alum types rely on similar core immunological mechanisms as the basis of their adjuvant activity. Consequently, for the sake of clarity and simplicity, the term 'alum' will encompass all aluminum-based adjuvants indistinctly in this chapter.

Alum is More Than a Delivery System or a Retention Adjuvant

Like other adjuvants currently in use, alum was developed on largely empirical bases. It is quite relevant to note in this regard that the major adjuvants licensed for marketed human vaccines so far, i.e. alum and the squalene-based emulsions MF59 and AS03, were intended and are still largely considered as 'delivery systems' [6]. The main mechanism of action of such adjuvants has indeed long been thought to be limited to improving the bioavailability of antigens to the immune system. This premise actually stayed in line with the mechanistic rationale behind the introduction of alum as an adjuvant. Glenny and colleagues indeed thought that antigen adsorption onto insoluble particles could enhance the immune response by allowing the antigen to persist for longer periods of time at the site of injection and by making it more readily available to cells of the immune system. Providing a basis to this assumption, early experiments in animal models showed that naïve animals could be immunized by the transfer of ground-up macroscopic alum depots recovered from immunized donor animals [7]. Furthermore, antigens were shown to persist longer when administered adsorbed on alum compared to antigen alone [8,9]. Finally, it has been reported that alum may facilitate antigen uptake and presentation by antigen-presenting cells *in vitro* [10,11]. The 'depot effect' has thus long been considered as the main explanation for the adjuvant properties of alum. Yet, it has a number of significant shortcomings. Several initial and recent studies for instance found no major positive impact of antigen adsorption or macroscopic depot formation and persistence on the immunogenicity of alum-containing antigen preparations in animal models [9,12–15].

The 'depot theory' in its simplest form is also intuitively unsatisfactory in the light of our current understanding of the induction of immunity to non-self antigens. Indeed, if detection of non-self epitopes by T cell and B cell receptors (the so-called 'signal 1') is an essential step in the initiation of adaptive immune responses, this signal alone does not result in efficient immune activation. Co-stimulatory signals (signals 2 and 3) are also required that are induced in response to immunogenic stimulation. Thus, it may be predicted that alum is more than a mere retention adjuvant and that it must provide the immune system with activating signals, either directly or indirectly.

RECENT IMMUNOLOGICAL INSIGHTS INTO THE MECHANISMS OF ACTION OF ALUM

The hallmark of alum adjuvant activity has long been known to be the induction of robust antibody responses. In contrast, alum does not induce marked cytotoxic responses, although this type of response may not be completely absent [16–19]. Cytotoxic T cell differentiation could even be enhanced by combining alum with monophosphoryl lipid A, an immunostimulatory Toll-like receptor-4 ligand [19], as is done in the clinically approved adjuvant AS04.

Alum boosts antibody production by providing T cell-dependent help to B cell responses. The characterization of the functional profile of alum-elicited auxiliary CD4$^+$ T helper (T$_H$) cell responses revealed a significant bias toward type 2 (T$_H$2) responses [20,21], which favor the production of antibodies of the IgG1 and IgE isotypes. The long-standing and as yet not definitively answered question has ever since been how alum induces T$_H$2 responses. It may be noted here that discovering the bottom-line answer to this question is not trivial, since T$_H$2 responses remain probably the least understood type of auxiliary T cell responses in terms of mechanisms of induction and biological function [22–24]. In the following sections, we review the recent insights into the cellular responses triggered by alum, and how these lead to the activation of functionally specialized T$_H$2 responses.

Innate Immune Cellular Responses to Alum

The best testimony that the action of alum is not limited to a 'depot effect' and that it is recognized by the immune system is the complex inflammatory response that takes place following its injection. Indeed, alum induces the rapid influx of inflammatory cells at sites of injection, as observed following intramuscular administration as in most clinical vaccinations, as well as following intraperitoneal injection as is often done in mouse studies. This influx consists mostly of monocytes, dendritic cells, eosinophils, and neutrophils [18,25–27]. It coincides with the establishment of an environment with molecular signatures of type 2 inflammation, characterized by the upregulation of proinflammatory cytokines such as interleukin (IL)-1β and IL-6, of chemotactic factors such as CXCL1 and CCL2, and of type 2 cytokines and chemokines such as IL-5, IL-13 and eotaxin [18,28].

The establishment of the inflammatory context in response to alum implicates a variety of immune and possibly structural cell types. Resident mast cells and macrophages seem important by their contribution to the production of many, but not all, of the proinflammatory factors induced following alum injection [18]. Both resident macrophages and mast cells are responsible for the recruitment of eosinophils to the site of injection, in the case of mast cells through the production of IL-5 and histamine [18]. Provided they are primed by endotoxin, resident macrophages also respond to alum by producing prostaglandin E_2 (PGE$_2$) [29]. Supporting a potential role of this phenomenon in some aspects of the response to alum, mice deficient for PGE synthase were shown to develop normal IgG1, but reduced IgE responses compared to their wild-type counterparts following immunization with alum and antigen [29]. The cellular mechanisms underlying this observation were not investigated. This latter observation notwithstanding, resident macrophages do not appear to have a major impact on the adjuvant activity of alum, since their depletion using clodronate-loaded liposomes prior to injection of alum and antigen does not impact on the ensuing adaptive immune responses [18]. Similarly, transgenic mice lacking mast cells display adaptive immune responses indistinguishable from those of control mice to alum-adjuvanted antigens [18]. However, IgE production has not yet been investigated in these contexts.

Eosinophils are potent producers of the prototypical type 2 cytokine IL-4, and may thus participate in the establishment of a local type 2 inflammatory environment [18,25]. Yet, different mouse strains harboring defective eosinophil differentiation exhibit normal T cell responses and antibody production in response to immunization with alum and antigen [18]. Thus, eosinophils are unlikely to significantly impact on the adjuvant activity of alum.

Neutrophils have been proposed to influence adaptive immune responses [30]. Somewhat counterintuitively, depletion of neutrophils using depleting antibodies, or neutropeny in transgenic mouse models, results in increased T and B cell responses to alum-adjuvanted antigen [31]. These effects were attributed to competition for the antigen between neutrophils and antigen-presenting cells, as well as to a possibly more direct influence of lymph node-infiltrating neutrophils on the cross-talk between antigen-presenting cells and T cells [31].

Conventional (also called myeloid) dendritic cells actually appear as the most important innate immune cells mediating the effects of alum. Conventional dendritic cells are the immune system's professional antigen-presenting cells [32,33]. As such, they are considered a crucial cellular interface between the innate and adaptive immune system. Conventional dendritic cells indeed are particularly well equipped for the uptake of antigens and their processing into major histocompatibility class I (MHC I) and MHC II molecules for presentation to $CD8^+$ and $CD4^+$ T cells, respectively (i.e. for providing 'signal 1'). Following activation, they also express high levels of co-stimulatory molecules, which are crucial for providing 'signal 2' for T cell activation [34]. They also indirectly promote antibody responses by providing $CD4^+$ T cell help to B cells [35]. Alum induces the recruitment of conventional dendritic cells to the site of injection [25,27,36] and significantly enhances the uptake of adsorbed antigens [25,37]. This process is accompanied by the upregulation of co-stimulatory surface molecules such as CD86 and CD40 on dendritic cells *in vivo* [25]. The role of conventional dendritic cells in the adjuvant activity of alum was studied in mice expressing the human diphtheria toxin receptor (DTR) under the dependence of the murine CD11c promoter (CD11c-DTR mice). In these mice, DTR expression is restricted to CD11c-expressing cells, most of which are dendritic cells. Administration of diphtheria toxin to these mice leads to the rapid depletion of conventional dendritic cells at the site of injection and in the lymph nodes draining these sites, while leaving dendritic cell populations in other organs intact [38]. Bart Lambrecht's group showed that dendritic cell depletion in CD11c-DTR mice drastically reduces antigen-specific T cell proliferation and antibody production following immunization with alum and antigen [25]. Interestingly, by using different administration routes of diphtheria toxin, the same authors could additionally show that dendritic cells that migrate from the injection site to the lymph nodes, but not lymph node-resident dendritic cells that acquire antigen through afferent lymph, are responsible for the induction of T cell responses [25]. This study thus established a crucial role of conventional dendritic cells recruited to the site of injection in the adjuvant activity of alum.

Conventional dendritic cells are not a homogenous population in peripheral tissues. They may indeed originate from committed circulating precursors called pre-DCs, or differentiate from circulating inflammatory monocytes [39–41]. Of note, the discrimination between the respective dendritic cell progenies of these precursors may be uneasy, since the expression of their distinctive surface markers is thought to be dynamic. Inflammatory monocytes are recruited in important numbers to sites of alum injection [25,27,36]. These cells display a high antigen uptake activity, which is increased by the presence of alum [25]. They subsequently migrate from the injection site to the draining lymph nodes, a process that coincides with their upregulation of typical dendritic cell markers. Supporting the idea that these cells act as antigen-presenting cells in alum-adjuvanted immunization, transfer of bone marrow-derived inflammatory monocytes induces antigen-specific T cell proliferation, which is otherwise severely impaired, in diphtheria toxin-treated CD11c-DTR mice. These observations led to the conclusion that inflammatory monocyte-derived dendritic cells are the major antigen-presenting cells in alum-adjuvanted immunization. However, the effect of the transfer of inflammatory monocytes on antibody production was not studied. Furthermore, both antigen-loaded conventional dendritic cells and inflammatory monocytes isolated from the lymph nodes of mice immunized with alum and antigen are able to activate antigen-specific T cell proliferation *ex vivo* [25]. Later studies rather suggested that inflammatory monocytes only mediate specific components of the adaptive response to alum (see below) [27,36]. It is thus likely that multiple subtypes of dendritic cells participate in the induction of T cell responses to alum. Identifying these different cell types along with their contribution to the distinct aspect of the response to alum-adjuvanted immunization will require further investigation.

T$_H$2 Cell Responses and the Induction of Antibody Responses in Alum Immunization

As outlined above, the migration of conventional dendritic cells from the injection site to the draining lymph nodes is an important link between the innate response at the injection site and the adaptive T cell response in the draining lymph nodes following alum-adjuvanted immunization. In the lymph nodes, activated antigen-specific T_H cells interact with cognate B cells at the border of the T cell- and B cell-zones. This interaction initiates isotype switching and leads to the formation of germinal centers in B cell follicles, in which B cells undergo somatic hypermutation and are selected for high affinity toward the antigen. This process is followed by the differentiation of memory B cells or plasma cells [42].

It has long been considered that the T_H2 response initiated following alum-adjuvanted immunization is functionally monolithic. This view has however become an oversimplification. Indeed, naïve T_H cells may acquire functionally distinct phenotypes with specialized functions following activation. Next to 'canonical' effector T_H cells, T_H cells have been identified that are specialized in providing help to B cells for the development of germinal centers within B cell follicles. These specialized cells were consequently named T follicular helper (T_{FH}) cells [43–45]. T_{FH} cells are essential contributors to the development of T cell-dependent antibody responses. To mediate these functions, T_{FH} cells express high levels of co-stimulatory molecules such as CD40 ligand that promote B cell proliferation and survival, as well as high levels of the cytokine IL-21, which enhances B cell differentiation and antibody production [45]. Canonical T_H cells on the other hand are mainly responsible for mediating other functions, such as effector functions in peripheral tissues [46]. The relationship between T_{FH} cells and canonical T_H cells is rather complex, and recent evidence tends to support that T_{FH} cells may arise from naïve T cells committed toward pre-T_{FH} cells, as well as from canonical effector T_H cells in case of persistent antigen exposure [45]. T_{FH} cells are known to produce lower levels of signature T_H cytokines than classical T_H cells. It has however been shown that T_{FH} cells may express IL-4 [46].

This general partition of T_H cell responses applies to alum-adjuvanted T cell responses. Indeed, naïve T_H cells from mice immunized with alum and antigen were shown to differentiate into canonical lymph node-resident T_H cells, canonical effector T cells that exit the lymph nodes, and T_{FH} cells [43]. Of note, whereas alum induces T_{FH} cell responses, it is outclassed in this regard by other experimental adjuvants. Recent studies indeed suggest that alum induces globally fewer T_{FH} cells than other experimental adjuvants, at least in part because of a lower contribution of these cells to the overall T_H cell response [43,47]. This notably results in alum generating comparably lower IgG1 titers than some other experimental adjuvants. Consequently, a better understanding of the mechanisms that determine the ratios of T_{FH} and other T_H cell subsets following alum-adjuvanted immunization may provide important clues for the development of more efficient alum-adjuvanted vaccines. A proposed explanation is that alum could allow the activation of T_H cells with lower T cell receptor affinity. T_{FH} cells were indeed proposed to preferentially differentiate from high-affinity T_H cells [43].

TRIGGERING OF INNATE IMMUNE SIGNALING FOLLOWING ALUM-ADJUVANTED IMMUNIZATION

From the evidence presented above, it may be concluded that the adjuvant activity of alum lies originally in its ability to activate innate immune responses. Much uncertainty however remains concerning the precise molecular mechanisms by which alum activates innate immunity.

Alum Does Not Fit in the Classical Framework of Immune Potentiator Adjuvants

With research intensifying in the field of vaccine immunology, a common theme emerged that helps to satisfactorily model the molecular mechanisms of action of many efficient vaccines, as well as of the fast-developing class of adjuvants referred to as immune potentiators. These vaccines and adjuvants would mostly fall within the explanatory frame of the 'foreign model' developed by Charles Janeway [48]. Charles Janeway anticipated that the induction of adaptive immune responses against pathogens requires not only antigen recognition by the adaptive immune system, but also the sensing of 'stranger' signals associated with the pathogen [49,50]. He termed these signals pathogen-associated molecular patterns (PAMPs), and proposed that they are detected by germline-encoded receptors of the innate immune system, which he named pattern-recognition receptors (PRRs). Translated to vaccination, this model predicts that PRR activation may be the initial trigger of innate immunity, which would determine the eventual outcome of the adaptive response to a given vaccine in terms of magnitude, orientation and duration [3,51,52].

Proving Charles Janeway's assumption right, the last decade has witnessed the identification of a large variety of PRRs [53,54]. These receptors are structurally varied, with representatives mostly encountered in four protein (super)families: the Toll-like receptors (TLRs) [55,56], the RIG-I-related receptors (RLRs) [57], the NOD-like receptors (NLRs) [58] and the C-type lectin receptors (CLRs) [59]. Each PRR may recognize particular PAMPs and hence provide the innate immune system with some level of specificity in the recognition of different classes of pathogens. The compartmentalization of PRRs within cells and their distinct distribution among different cell types provides another level of functional specificity [60–62]. The PAMPs that PRRs recognize belong to two main classes of molecules: molecular structures associated with microbial envelopes (such as bacterial endotoxin, flagellin and lipoproteins) and pathogen nucleic acids. These molecules are recognized as foreign material based on different discriminative factors, such as their absence in eukaryotic cells in the case of microbial envelope components, or their sequence, structure, molecular modifications, and localization in the case of nucleic acids [63,64].

The signals generated by the activation of PRRs converge on antigen-presenting cells such as dendritic cells. Antigen-presenting cells integrate the signals generated by the sensing of PAMPs through their own PRRs with the signals originating from secondary mediators released by PAMP-activated bystander cells (Figure 12.1). This process critically influences the orientation of T cell responses, and hence, the general outcome of the adaptive immune response.

Recent research on the immunology of vaccines has given center stage to PAMP-mediated PRR activation as an important mechanism underlying the activity of a majority of clinical and experimental vaccines and immune potentiator adjuvants. The most straightforward examples regarding immune potentiator adjuvants are vaccines in which PAMP-containing adjuvants are formulated in antigen preparations to boost their immunogenicity. This strategy is used in recent attempts at developing novel, rationally designed human vaccines [51]. Of note, as illustrated in other chapters of this book, many of these attempts rely on agonists of nucleic acid-sensing PRRs [52]. It is also increasingly recognized that many non-adjuvanted vaccines rely on PAMPs as 'built-in' immunological adjuvants [52]. In the case of live-attenuated vaccines for instance, PAMPs derived from the attenuated pathogen trigger the activation of the innate immune system. A good example is the live-attenuated yellow fever vaccine YF-17D, one of the most efficient antiviral vaccines ever developed. Evidence indicates that YF-17D activates dendritic cells through the concomitant stimulation of several TLRs and possibly of certain RLRs, which results in the induction of CD8$^+$ T cells and a mixed T_H1- and T_H2-type immune response [65–67]. Another example, documented in more detail in other chapters of this book, is DNA vaccines, in which the DNA

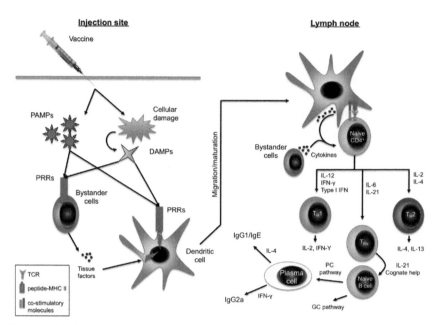

Figure 12.1 *Induction of adaptive immune responses to vaccines through pattern recognition receptor-mediated antigen-presenting cell activation.* Vaccines may contain pathogen-associated molecular patterns (PAMPs) or induce the local release of damage-associated molecular patterns (DAMPs). These PAMPs and DAMPs are detected directly by pattern recognition receptors (PRRs) expressed by antigen-presenting cells such as dendritic cells, leading to their activation, maturation and migration to the lymph nodes. PRR-mediated recognition of PAMPs and DAMPs by bystander cells may also induce the release of tissue factors that impact on the dendritic cell response. In the lymph nodes, the activated dendritic cells may present antigen to CD4$^+$ T cells, provide them with co-stimulatory signals, and stimulate their differentiation by providing a favorable cytokinic milieu. Bystander cells may provide some cytokines such as IL-4 and type I IFN. Depending on the cytokinic milieu, CD4$^+$ T cells may differentiate into various effector T helper subtypes (only T$_H$1 and T$_H$2 cells are depicted here). T$_H$ cells may also become T follicular helper cells (T$_{FH}$) and help in the activation of cognate B cells toward the plasma cell (PC) pathway or the germinal center (GC) pathway, and favor B cell isotype switching in function of their cytokinic expression profile. (MHC II: major histocompatibility complex class II).

backbone of the antigen–encoding plasmid activates innate immune DNA sensors [52,68].

To date, most documented examples of PRR–mediated activation of immunity by vaccines and/or adjuvants thus implicate PAMP–induced activation of innate immunity. May alum fit in this conceptual framework? The discovery of the first TLRs and their associated signaling pathways naturally led to the testing of their potential involvement in the adjuvant activity of alum. These studies compared the responses of control mice and mice deficient for one or both of the two critical adaptor molecules downstream of TLRs, namely myeloid differentiation primary-response gene 88 (MyD88) and TIR-domain-containing adaptor protein inducing IFNβ (TRIF) [56]. In a first study, MyD88 ablation did not have any impact on the production of IgG1 in mice immunized with alum and antigen [69]. In a second study, Gavin and collaborators studied antibody responses in mice deficient for both MyD88 and TRIF and observed no significant differences between their antibody responses and those of their wild-type counterparts [70]. TLR signaling thus does not seem to have a major role in the adjuvant activity of alum.

The 'Dangerous' Side of Alum Activity

It may be suspected that alum does not act through PAMP recognition, since it would be difficult to identify any structural mimics of pathogen-associated molecules in such compounds. Yet, as we will see below, PRRs indeed may mediate the adjuvant activity of alum even though this requires considering an alternative to the 'stranger model'.

The 'Danger' Model

Shortly after Charles Janeway proposed his 'stranger theory', a concurrent model was proposed by Polly Matzinger, which is often referred to as the 'danger model' [71]. In this model, the activation of adaptive immunity would result from the detection of endogenous 'danger' signals indicative of damage to host cells or tissues. These signals are often collectively termed damage-associated molecular patterns (DAMPs). Like in Charles Janeway's case, further studies proved Polly Matzinger's concept right and a growing list of molecules recognized as DAMPs has been reported in recent years. DAMPs generically are altered or misplaced self-molecules [72–74]. Examples of DAMPs include cleaved matrix proteins (such as low-molecular-weight hyaluronan), liberated intracellular proteins (such as heat-shock proteins, histones, high-mobility group box proteins and certain cytokines such as IL-1α and IL-33) and extracellular self-nucleic acids (see also other chapters of this book on this subject).

A majority of DAMPs are recognized as such by PRRs [72–74], so that there is some overlap of PAMP and DAMP recognition for some of these receptors. But the 'stranger' and 'danger' theories certainly have more to share. It may indeed be envisioned that in many instances, detection of both PAMPs and DAMPs cooperates or synergizes to activate innate and adaptive immune responses tailored to the type of immunogenic stimuli encountered. By extending the above concepts to vaccination, it may be expected that in some vaccines and adjuvants, immunopotentiation may result at least partly from the release of DAMPs by host cells (Figure 12.1). We have seen above that alum does not appear to fit in the 'stranger model', but the situation is proving rather different when considering alum in light of the 'danger model'.

Alum Induces Signs of Cell Damage In Vivo

Alum is not usually considered as a cytotoxic adjuvant. It is indeed well-tolerated in a vast majority of cases and does not usually induce clinically visible inflammatory reactions in vaccinated subjects. However, this does not forcefully mean that alum is devoid of any cytotoxic activity. A few earlier studies actually reported some level of cytotoxicity of alum *in vivo* [75–77]. In an elegant more recent study, Philippa Marrack's group investigated the composition of alum nodules *in vivo* [14]. It is well known that alum quickly forms nodules at sites of injection and it has long been thought that these nodules play a role in the adjuvant activity of alum. Yet, the main conclusion of Philippa Marrack's team was that nodule formation is dispensable for the adjuvant activity of alum [14]. Interestingly, in addition to alum, nodules were shown to contain large quantities of fibrin derived from thrombin-cleaved fibrinogen. Fibrin deposition is often observed in inflammatory conditions such as foreign body reactions [78]. Alum nodules were furthermore shown to also contain histones and host DNA indicative of the presence of extracellular chromatin within alum nodules. Thus, such studies compellingly indicate that alum treatment leads to the release of intracellular molecules at sites of injection, some of which could potentially be recognized as DAMPs.

Uric Acid as a Potential Major DAMP Released in Response to Alum

The notion that alum may induce the release of DAMPs is further strongly supported by the observation that alum induces the accumulation of uric acid at sites of injection [14,25,79]. Uric acid was one of the first identified DAMPs, when it was shown to greatly contribute to the immunopotentiating effects of dead lysed cells on adaptive immune responses to co-administered antigens [80]. Uric acid is released upon cell injury or death, and is also rapidly produced due to the degradation of cellular nucleic acids following cell death. When accumulating in extracellular fluids, uric acid may form crystals of monosodium urate, which may act as potent DAMPs and activate a variety of innate immune signaling pathways [81].

In their interesting study, Bart Lambrecht's group proposed that alum mediates its effects on adaptive T cell responses by the induction of uric acid accumulation at sites of injection [25]. This notion was supported by the observation that pre-treatment of mice with uricase at sites of alum injection significantly reduces antigen-specific T cell proliferative responses [25]. The same group later reported that uricase treatment also reduces IgG1 and IgE antibody responses to alum-adjuvanted antigens [79]. They also supported the view that alum and monosodium urate crystals display similar adjuvant activity in adaptive immune responses *in vivo* [79]. The precise molecular mechanisms of action of uric acid in alum activity have not been formally identified yet. It remains unknown for instance whether uric acid precipitates into monosodium urate crystals, its major form as a DAMP, at sites of alum injection.

The Inflammasome Puzzle

Almost at the same time as the aforementioned study on uric acid, a number of reports showed that alum induces the activation of inflammasomes containing the cytosolic receptor NLRP3 (NLR family, pyrin domain containing 3), a member of the NLR family of PRRs, in macrophages primed with endotoxin [36,82–85]. Inflammasomes are multiproteic assemblages that recruit and activate caspase-1, leading to the cleavage of caspase-1 substrates such as the precursors of the IL1-related proinflammatory cytokines IL-1β and IL-18 into bioactive cytokines [58]. This finding was in line with the previously established capacity of alum to induce the production of mature IL-1β from endotoxin-primed macrophages [10].

The mechanisms through which alum activates the NLRP3 inflammasome have been proposed to involve the phagocytosis of alum particles by macrophages. Ensuing rupture of the endosomal membrane would release endosomal enzymes such as cathepsin B into the cytoplasm and lead to inflammasome activation [84]. Of note, monosodium urate crystals are also potent inducers of the activation of the NLRP3 inflammasome [86]. Nevertheless, alum-induced NLRP3 inflammasome activation does not seem to require uric acid, at least *in vitro* [82,87].

If it is well established that alum triggers the activation of the NLRP3 inflammasome in endotoxin-primed macrophages, the actual impact of this pathway on adaptive immune responses remains controversial. Indeed, some initial studies reported that mice deficient for the genes coding for NLRP3 or caspase-1 develop globally reduced antigen-specific antibody responses to alum-adjuvanted antigens [82,85], whereas others only observed effects on certain antibody isotypes [36], or no effect at all [18,27,83]. There is no current experimental explanation for these discrepancies. In any case, a role for the NLRP3 inflammasome in the adjuvant activity of alum is currently difficult to reconcile with the earlier observation that MyD88 signaling is dispensable for the adjuvant activity of alum [69,70], since MyD88 is also involved in the transduction of the signal from the receptors to IL-1β and IL-18.

DNA RELEASE AND SIGNALING IN THE ADJUVANT ACTIVITY OF ALUM

As mentioned above, the administration of alum *in vivo* results in the release of cellular contents in the extracellular milieu. Among these molecules is host cellular DNA, which has been shown to act as a DAMP in a number of pathophysiological contexts. The next section will first provide a synoptic overview of the contexts in which host DNA has been identified as a DAMP and of the established and potential signaling pathways of self-DNA sensing in these contexts. We will next address the potential role of host DNA sensing in the adjuvant activity of alum.

Self-DNA as a DAMP
In Vivo Evidence of Self-DNA Acting as a DAMP
Probably the first hint that self-DNA may act as a proinflammatory DAMP came from studies in systemic lupus erythematosus (SLE) patients. SLE is a systemic autoimmune disease that affects multiple organs, including the skin, joints, kidneys, lungs and nervous system. In SLE patients, anti-nuclear antibodies are detected that are directed against entire nucleosomes as well as against naked DNA. It was therefore suspected that increased release or deficient clearance of DNA from dying host cells might underlie the pathogenesis of the disease. In support of this assumption, genetic ablation of the gene coding for DNAse I, the DNAse mainly responsible for degradation of extracellular DNA, recapitulated key features of SLE in mice [88]. Inactivating mutations in the genes coding for DNAse I or its relative DNAse I-like 3 were later found in some SLE patients [89,90] and reduced DNAse I activity in the serum was found to be a feature of a subset of patients with SLE [91]. Giving further support to the concept of self-DNA being a potential DAMP, genetic deficiency

in DNAse II, which degrades endocytosed DNA from apoptotic cells in macrophages, was shown to lead to *in utero* death of deficient animals due to lethal anemia [92,93]. This lethality depends on the overexpression of type I interferons (IFNs) and is rescued by combined deficiency in the type I IFN receptor [93]. Rescued mice nevertheless develop autoimmune polyarthritis later in life, a process that is driven by the overexpression of tumor necrosis factor-α [94]. Self-genomic DNA released from dying cells was also shown to form complexes with the antimicrobial peptide LL37, which is overexpressed by skin cells in psoriasis, and to thereby activate plasmacytoid dendritic cells [95]. Through this process, self-DNA release in the skin might contribute to the breakdown of self-tolerance in psoriasis [95]. Altogether, the aforementioned studies thus suggest that immune recognition of self-DNA participates in the induction of autoimmunity [96].

The relevance of self-DNA sensing in physiopathology may extend beyond autoimmune disorders. Such roles have so far mostly been proposed for host mitochondrial DNA. A first study showed that (oxidized) mitochondrial DNA causes arthritis upon intra-articular injection in mice and accumulates in arthritic lesions in human patients [97]. Mitochondrial DNA is also a released proinflammatory signal following trauma-induced cell death, so that circulating mitochondrial DNA has been suggested to play a role in systemic inflammatory response syndrome [98]. Finally, in line with these findings, in mice conditionally deficient for the gene coding for DNAse II in cardiomyocytes and exposed to cardiac pressure overload, mitochondrial DNA that escapes autophagosomal degradation induces cell-autonomous inflammatory responses in cardiomyocytes [99]. This inflammatory response may lead to myocarditis and heart dysfunction. A non-autoimmune situation in which self-genomic DNA appears to act as a DAMP is acetaminophen-induced liver injury. Acetaminophen hepatotoxicity is a leading cause of acute liver failure, in which the initial toxic metabolic injury leads to the release of DAMPs from necrotic and apoptotic hepatocytes and to an ensuing inflammatory response responsible for further liver damage [100]. Self-DNA released from dying hepatocytes is thought to play a major role in this secondary inflammatory response through the activation of sinusoidal endothelial cells to produce proinflammatory cytokines [100,101].

The aforementioned studies thus suggest that self-DNA from dying or damaged host cells is an important danger signal in clinically relevant pathophysiological situations. However, the mechanisms of self-DNA sensing have not been fully identified *in vivo* in most of these contexts. Because eukaryotic mitochondria derive from ancient prokaryotic endosymbionts, mitochondrial DNA exhibits structural features of prokaryotic genomes, such as hypomethylated CpG motifs. Mitochondrial DNA is therefore often presumed to activate mainly TLR9, the major receptor for extracellular hypomethylated CpG-containing DNA [62,102]. In support of this assumption, mitochondrial DNA was shown to activate TLR9 in neutrophils *in vitro* [98], and TLR9-deficient mice are partially protected from pressure overload–induced cardiomyopathy [99]. Nevertheless, oxidized mitochondrial DNA may also activate the NLRP3 inflammasome through direct binding to NLRP3 during macrophage apoptosis [103,104], which suggests that additional PRRs could also play a role in the response to mitochondrial DNA as a DAMP.

Mammalian genomic DNA is known to comprise few unmethylated CpG motifs in comparison with bacterial genomes. Yet, unmethylated CpG motifs in eukaryotic genomic DNA may activate TLR9 [105,106]. On the other hand, it also has to be noted that unmethylated CpG motifs may favor, but do not seem to be an absolute requirement for, TLR9 activation by DNA, which would rather depend on the recognition of the phosphodiester 2'deoxyribose backbone of DNA [107]. Furthermore, host genomic DNA is altered during cell death, through cleavage, dysregulated methylation and oxidation, which may favor its immunostimulatory potential [101,108]. There is experimental evidence that TLR9 may recognize self-genomic DNA in certain *in vivo* contexts. For instance, TLR9-deficient mice display reduced mortality following acetaminophen-induced liver injury, thanks to reduced inflammatory damage to liver cells [101]. Different mechanisms leading to TLR9 activation following receptor-mediated endocytosis of self-DNA (naked, in immune complexes or associated with antimicrobial peptides) have been reported *in vitro*, which may take place in SLE and psoriasis and contribute to disease pathogenesis [96,109]. The role of TLR9 in the context of SLE however remains unclear, since contradictory findings were reported in mouse models, with one report showing a protective effect, and another an aggravating effect of TLR9 deficiency on SLE parameters [110,111].

Candidate Receptors for Self-DNA

As appears from the aforementioned studies, TLR9 is the only DNA-sensing PRR that has been implicated so far in the response to self-DNA as a DAMP *in vivo*. Other pathways are however most likely to intervene as well in certain contexts. It is for instance known that aberrant type I IFN production in DNAse II-deficient mice is the result of TLR-independent activation of the transcription factors interferon response factor (IRF)-3 and -7 [92,112]. *In vitro*, the list of PRRs known to respond to double-stranded DNA, and hence possibly to self-genomic DNA, is not restricted to TLR9 (Table 12.1) [52].

PRRs able to detect DNA in *in vitro* settings have been identified mainly in three PRR groups: the TLRs, the DExD/H-box helicase superfamily (which includes the RLRs) and the AIM2 (absent in melanoma-2)-like receptors (ALRs). Within the TLR family, TLR9 as mentioned above has been identified as an intracellular receptor to endocytosed extracellular DNA [62,102]. Some members of the DExD/H-box helicase superfamily, namely DDX60, DHX9, DHX36 and DDX41, were recently proposed to sense DNA in the cytoplasm of dendritic cells and other cell types [113–115]. ALRs are a newly proposed group of DNA-sensing PRRs that comprises two members of the pyrin and HIN domain-containing protein family (PYHIN family): absent in melanoma 2 (AIM2) and IFNγ-inducible protein 16 (IFI16) [116–120]. Finally, two other putative cytosolic DNA sensors have also been proposed that belong to other protein families: ZBP1 (Z-DNA-binding protein 1; also known as DAI and DLM1) [121,122] and LRRFIP1 (leucine-rich repeat flightless-interacting protein 1) [123].

These DNA-sensing PRRs signal through a limited number of conserved signaling pathways and, depending on the activated pathways, trigger the expression of type I IFNs, of IL-1-related cytokines and of other proinflammatory cytokines (Figure 12.2). The contribution of these PRRs to innate immune responses to self-DNA *in vivo* has not been investigated yet.

Alum Induces Self-DNA Release

In line with the reported local cytotoxic effects of alum [75–77], several independent studies that analyzed sites of alum injection reported significant concentrations of extracellular double-stranded DNA at these sites [5,14,27,124]. Alum induces DNA release irrespective of the injection site (intraperitoneal or intramuscular) or of the type of alum used [5,27]. DNA release was shown to happen fast, rapidly reaching a plateau 3 hours after alum injection, and remained stable for at least 24 hours [27]. This might indicate that a balance is established between extracellular DNA accumulation in response to alum and its degradation by extracellular

Table 12.1 Proposed DNA-sensing pattern recognition receptors, localization and evidence

PRR	Localization	Evidence	References
TLR9	Endolysosomal compartment	Recognizes DNA from viruses, bacteria, protozoan parasites, as well as native and altered self-DNA (genomic and mitochondrial).	62
RIG-I	Cytoplasm	Recognizes RNA transcribed by RNA polymerase III from viral or bacterial A/T-rich DNA (in human cells only).	151,152
DDX60	Cytoplasm	Binds to dsDNA *in vitro*. Silencing reduces type I IFN responses in VSV-infected cells.	114
DHX9	Cytoplasm	Binds to CpG-B ODN. Silencing reduces proinflammatory cytokine expression in HSV-infected cells.	113
DHX36	Cytoplasm	Binds to CpG-A ODN. Silencing reduces type I IFN expression in HSV-infected cells.	113
DDX41	Cytoplasm	Binds to dsDNA *in vitro*. Silencing reduces type I IFN production following dsDNA-transfection or *L. monocytogenes*, HSV1 or adenovirus infection.	115
AIM2	Cytoplasm	Binds to dsDNA *in vitro*. Silencing inhibits the secretion of mature IL-1β in response to dsDNA and to vaccinia virus infection.	116–119
IFI16	Cytoplasm/nucleus	Binds to viral dsDNA *in vitro*. Silencing reduces type I IFN expression in response to dsDNA transfection and HSV1 infection.	120
ZBP1	Cytoplasm	Binds to dsDNA in the cell cytoplasm. Silencing reduces type I IFN and proinflammatory cytokine production in response to dsDNA transfection.	121,122
LRRFIP1	Cytoplasm	Binds to dsDNA in the cell cytoplasm. Silencing reduces type I IFN production following *L. monocytogenes* or VSV.	123

CpG ODN: CpG oligodeoxinucleotide, ds: double-stranded, HSV: herpes simplex virus, *L. monocytogenes*: *Listeria monocytogenes*, VSV: vesicular stomatitis virus.

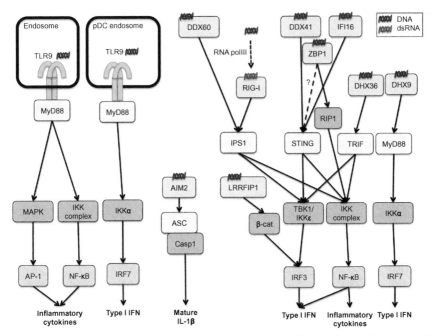

Figure 12.2 *DNA-sensing PRRs and their associated signaling pathways.* Endosomal TLR9 activates downstream signaling through the adaptor MyD88 (myeloid differentiation primary response gene 88) in the cytosol, leading to the activation of the IKK (inhibitor of NF-κB kinase) complex and MAPK (mitogen-activated protein kinases) and subsequent activation of NF-κB (nuclear factor-κB) and AP-1 (activator protein-1) transcription factors, promoting the expression of proinflammatory cytokines. In plasmacytoid dendritic cells (pDCs), TLR9 activation also leads to the expression of high levels of type I IFNs by promoting the activation of IRF7 via IKKα. The DExD/H-box helicases DDX60 and RIG-I (retinoic acid-inducible gene-I) may induce the expression of inflammatory cytokines and type I IFNs through the IPS1 (IFN-β promoter stimulator-1)-mediated activation of TBK1/IKKε and the IKK complex. In the case of RIG-I, the activation requires transcription of DNA by RNA polymerase III (RNA polIII) to produce RNA ligands. The proposed cytosolic DNA receptors DDX41, IFI16 (IFN-inducible protein 16) and possibly ZBP1 (Z-DNA binding protein 1) interact with STING (stimulator of IFN genes) to activate TBK1/IKKε and the IKK complex. ZBP1 was also shown to directly interact with RIP1 (receptor interacting protein 1) to induce NF-κB activation. LRRFIP1 (leucine rich repeat (in FLII) interacting protein 1) was suggested to potentiate IRF3 transcriptional activity through β-catenin (β-cat). In pDCs, DHX36 and DHX9 would activate TRIF-dependent and MyD88-dependent signaling, respectively. Finally, AIM2 (absent in melanoma 2) may induce inflammasome formation and caspase 1 (casp1) activation through ASC (apoptosis-associated speck-like protein containing a CARD), leading to the release of mature IL-1β. (dsRNA: double-stranded RNA, pDC: plasmacytoid dendritic cell).

DNAses. Extracellular DNA concentrations depend on the dose of alum used, but typically are in the microgram range for all concentrations of alum tested [5,27,124].

Combining the above analyses with viability assays on cells recovered from the lavage of sites of injection indicates the presence of a significant proportion of dead cells among total cells, an effect that depends on the dose of alum used [27]. Imaging using DNA–intercalating fluorescent dyes that cannot penetrate living cells also reveals that significant quantities of self-DNA are entrapped in the typical alum nodules that form at sites of intraperitoneal or intramuscular injection [14,27]. The conformation of the DNA in these depots is reminiscent of nuclei of dead lysed cells, but interestingly, linear projections of chromatin and 'blown-up' nuclei are also observed (Figure 12.3A). These latter structures are highly reminiscent of the expulsion of nuclear chromatin that happens in neutrophils activated to form extracellular traps (ETs, see below) [125]. Altogether, these studies thus indicate that alum consistently induces the release of self-DNA at sites of injection.

Mechanisms and Consequences of Self-DNA Sensing on Antigen-Specific T and B Cell Responses in Alum-Adjuvanted Immunization

Self-DNA is Involved in the Adjuvant Activity of Alum

The finding that alum induces self-DNA release being recent, only one study so far has investigated the potential involvement of self-DNA in the adjuvant properties of alum in mouse models [27]. In a first approach, alum was substituted by self-DNA as an adjuvant. Self-DNA, in concentrations similar to those measured at sites of alum injection (e.g. 2.5 µg DNA instead of 0.6 mg of alum), when mixed with an experimental antigen induces antigen-specific IgM, IgG1 and IgE antibody responses and T cell responses similar to those obtained with alum–adjuvanted antigen. This observation is consistent for intraperitoneal and intramuscular immunizations. Thus, like alum, self-DNA may act as an adjuvant that boosts T_H2-biased humoral responses.

In a second approach, local treatment with DNAse I to degrade extracellular DNA at the sites of injection was shown to reduce antigen-specific IgM, IgG1 and IgE responses, as well as T cell proliferative responses in mice immunized with alum–adjuvanted antigen (Figure 12.3B). Finally, naïve mice may be immunized by the transfer of lavage fluid from the

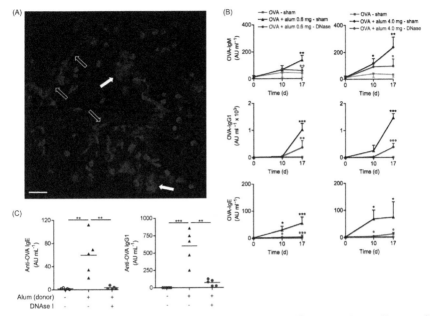

Figure 12.3 *Alum-induced self-DNA release and its effect on alum-adjuvanted humoral responses.* **(A)** Confocal microscopic imaging of extracellular DNA deposition in an alum nodule isolated from the peritoneal cavity of an immunized mouse and stained with the DNA intercalating agent 4′,6-diamidino-2-phenylindole (DAPI) (scale bar, 25 μm). DNA appears grey, empty arrows indicate linear DNA 'fibers', solid arrows indicate 'blown up' cell nuclei. **(B)** Serum titers of antigen (ovalbumin: OVA)-specific IgM (OVA-IgM), IgG1 (OVA-IgG1) and IgE (OVA-IgE) measured through time in mice that received OVA alone, or OVA and alum at the indicated dose and were boosted 10 days later with OVA. Mice received local DNAse I treatment (or sham treatment) shortly after immunization (n = 5. Error bars show means ± s.d. *, OVA versus OVA and adjuvant; °, OVA and alum versus OVA and alum followed by DNase I treatment; *,°$P<0.05$, **,°$P<0.01$, ***,°°°$P<0.001$. a.u., arbitrary unit). **(C)** *In vitro,* the acellular fraction of peritoneal lavage fluid from OVA- or OVA and alum-treated mice was mock-treated or submitted to DNase I treatment, before transfer to naive recipient mice. Recipient mice were boosted with OVA and OVA-IgG1 and OVA-IgE titers were measured (n=5. **$P<0.01$, ***$P<0.001$. a.u., arbitrary unit). *Panels* **B** *and* **C** *are reproduced with permission from Marichal et al. [27].*

site of injection of immunized donor mice, an effect that is abolished by the treatment of the fluid with DNAse I (Figure 12.3C). Altogether, these observations thus support the idea that self-DNA is an important contributor to the adjuvant effects of alum on antigen-specific T and B cell responses.

Self-DNA Triggers IRF3-Dependent Innate Immune Signaling in Alum-Adjuvanted Immunization

TLR9 may appear as a first-line candidate sensor of self-DNA in alum-adjuvanted immunization, in view of previous knowledge on the detection of self-DNA *in vivo* [126]. Yet, mice deficient for the gene coding for TLR9 display humoral and T cell responses similar to those of their wild-type counterparts in response to immunization with alum and antigen [27]. This observation is actually consistent with previous reports showing that alum adjuvant activity is independent of TLR signaling, as mentioned previously [69,70].

Activation of DNA-sensing PRRs leads notably either to the production of type I IFNs through the activation of IRF3 or to the production of mature IL-1β through the activation of inflammasomes (Figure 12.2) [52]. It was noticed that injection of antigen combined with self-DNA or alum as adjuvant induces similar production of IFNβ1 at sites of injection (Figure 12.4A, left panel), and that this response depends on IRF3. In contrast, only adjuvantation with alum, and not self-DNA, leads to enhanced production of IL-1β (Figure 12.4A, right panel). In the setting of this study, inflammasomes did not actually appear to have a major role in the effects of alum on adaptive immune responses, as evidenced by comparing the immune responses of mice deficient for the genes coding for caspase-1 or NLRP3 with wild-type counterparts [27]. These results thus suggest that self-DNA is responsible for the activation of IRF3-dependent pathways, but not for inflammasome activation following immunization with alum and antigen.

A TBK1–IRF3 Axis Promotes IgE Responses to Alum-Adjuvanted Immunization, but is Dispensable for IgG1 Production

The role of IRF3 in the adaptive immune response to alum-adjuvanted immunization was subsequently examined by comparing the responses of wild-type mice to those of mice deficient for IRF3. It was observed that IRF3-deficient mice develop significantly reduced IgE responses to alum-adjuvanted (Figure 12.4B, left panel) or self-DNA-adjuvanted antigens. Unexpectedly, however, IRF3-deficient mice develop normal antigen-specific IgG1 titers in the same conditions (Figure 12.4B, right panel). Of note, IgG2c titers are unaffected by IRF3 deficiency, suggesting that the adaptive responses are not shifted from T_H2 toward a more T_H1 profile. Similar results were obtained with different types of alum and different experimental antigens administered intraperitoneally or intramuscularly.

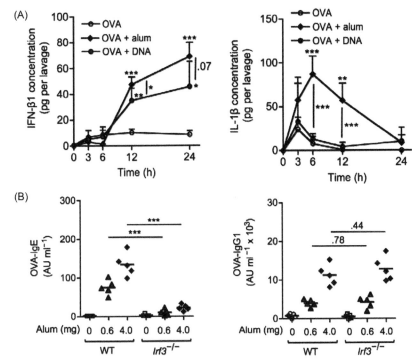

Figure 12.4 *Self-DNA activates a type I IFN response following alum-adjuvanted immunization, and IRF3 is required for IgE production.* **(A)** Quantities of IFN-β1 and IL-1β at sites of injection in mice immunized with OVA, OVA and self-DNA, or OVA and alum. **(B)** Serum titers of OVA-IgE and OVA-IgG1 in wild-type (WT) and IRF3-deficient mice immunized with OVA or OVA and alum. (*n*=5; Error bars show means ± s.d. *P<0.05, **P<0.01, ***P<0.001. a.u., arbitrary unit). *These panels are reproduced with permission from Marichal et al. [27].*

Further experiments aimed to decipher the pathway of IRF3 activation by self-DNA in alum–adjuvanted immunization. It was observed that mice deficient for TANK-binding kinase-1 (TBK-1), a major kinase upstream of IRF3 [127,128], develop reduced antigen–specific IgE responses, but normal levels of antigen–specific IgG1, in response to immunization with alum–adjuvanted antigen. These mice thus phenocopy IRF3-deficient mice, suggesting that self-DNA activates a TBK1–IRF3 axis in alum–adjuvanted immunization. Several known pathways upstream of TBK1 activation in DNA sensing were consequently tested. It was noted that mice deficient for DAI or the RLR adaptor protein MAVS (mitochondrial antiviral signaling protein) developed normal serum responses to alum and antigen. Thus, it is likely that self-DNA rather signals to IRF3 through one of the recently identified STING-dependent cytosolic DNA sensors (see below).

Altogether, the above data suggest that alum-induced host DNA release promotes IgG1 and IgE responses through distinct signaling pathways, and that IRF3 is an essential mediator of IgE production following alum immunization.

IRF3 Controls Canonical T_H2 Cell Responses to Alum-Adjuvanted Antigens

Given that IgE isotype switching by B cells is best induced by T cells expressing a T_H2 cytokine profile, the role of IRF3 in the induction of T_H2 cell responses following immunization with alum and antigen was investigated. When compared with the situation in wild-type mice, antigen-specific T cells in the draining lymph nodes of IRF3-deficient mice display reduced, but not abolished, proliferative responses in response to immunization with alum- or self-DNA-adjuvanted antigen (Figure 12.5A). Accordingly, the production of T_H2 cytokines is reduced in lymph node cells from immunized IRF3-deficient mice, although they remain higher than in naïve animals (Figure 12.5B). Thus, T_H2 responses to alum-adjuvanted immunization are partially dependent on IRF3. It is important to note that IRF3-deficient T cells are not intrinsically impaired in their ability to develop T_H2 responses, as shown in different studies [27,129]. IRF3-deficient mice furthermore may develop humoral responses fully comparable to those of wild-type animals following immunization with Freund's adjuvant [27].

Thanks to its high potential at inducing T_H2 responses, alum-adjuvanted immunization is often used in models of airway allergy [130]. Mice immunized systemically with alum and antigen, when reexposed to the antigen through inhalation, indeed develop allergic airway inflammation with many typical characteristics of human airway allergy, including airway eosinophilic inflammation, mucus overproduction and airway hyperresponsiveness [130]. Strikingly, IRF3-deficient mice immunized with alum and antigen, in contrast with their wild-type counterparts, are totally protected from airway allergy following antigen challenge of their airways (Figure 12.5C). Therefore, it may be postulated that IRF3 is implicated in the induction of canonical T_H2 cells, which are responsible for peripheral tissue inflammation [46]. Nevertheless, IRF3 deficiency only partially ablates T_H2 cell responses in lymph nodes draining immunization sites and does not impact on ensuing IgG1 responses. Therefore, IRF3-independent pathways of self-DNA detection must exist that contribute to the induction of other types of T_H cell responses involved in the promotion of IgG1 production.

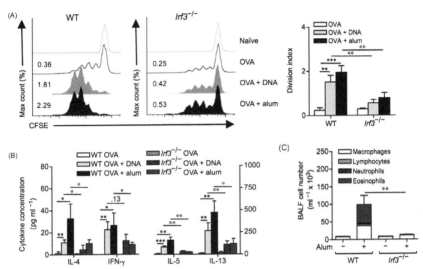

Figure 12.5 *IRF3 boosts canonical TH2 cell responses in response to self-DNA following alum-adjuvanted immunization.* **(A)** Proliferation profile (left) and division index (right) of adoptively transferred OVA-specific CD4+ OT-II cells in the lymph nodes of wild-type (WT) and IRF3-deficient mice treated with OVA, OVA and self-DNA, or OVA and alum. Inserted numbers indicate division index values. **(B)** Cytokine production upon *in vitro* restimulation with OVA of lymph node cells from WT and IRF3-deficient mice that were immunized with OVA, OVA and DNA, or OVA and alum. **(C)** Differential cell counts in the bronchoalveolar lavage fluid of WT and IRF3-deficient mice immunized with OVA and alum and challenged with aerosolized OVA to elicit airway allergy. (n=5, Error bars show means ± s.d. *, OVA versus OVA and adjuvant; °, WT OVA and adjuvant versus IRF3-deficient OVA and adjuvant; *,°$P<0.05$, **,°°$P<0.01$, ***$P<0.001$). *These panels are reproduced with permission from Marichal et al. [27].*

IRF3-Dependent Response of Inflammatory Monocytes to Self-DNA Following Alum-Adjuvanted Immunization

The role of IRF3 in innate immune responses to alum–adjuvanted immunization was investigated. At sites of alum injection, there is no significant difference in the recruitment of major inflammatory cell populations between IRF3-deficient and control mice. The inflammatory infiltrate in these mice includes eosinophils, neutrophils, inflammatory monocytes and conventional and plasmacytoid dendritic cells, as classically observed [25]. Macrophage numbers also rapidly drop at sites of immunization in IRF3-deficient mice, as expected [18,25]. Thus, there is no obvious difference in innate immune cell responses at sites of alum administration in the absence of IRF3.

The situation is different when considering the lymph nodes draining immunization sites. In IRF3-deficient mice the recruitment of inflammatory monocytes and, to a lesser extent, of conventional dendritic cells to the draining lymph nodes is reduced (Figure 12.6A). Self-DNA actually drives inflammatory monocyte recruitment to the lymph nodes. Self-DNA release and cell death at sites of injection indeed strongly correlate with the recruitment of inflammatory monocytes in draining lymph nodes, and immunization with self-DNA-adjuvanted antigen induces the recruitment of inflammatory monocytes to the lymph nodes (Figure 12.6A). More importantly, local treatment of wild-type mice with DNase I reduces this response in mice immunized with alum and antigen (Figure 12.6B).

Inflammatory monocytes may differentiate into inflammatory dendritic cells [39–41] and have been proposed to be the essential antigen-presenting cells mediating the effects of alum-adjuvanted immunization [25]. However, inflammatory monocytes isolated from the site of injection of alum-adjuvanted immunization in donor mice are unable to promote IgG1 or IgE antibody responses following adoptive transfer to naïve recipient mice [27]. This transfer nevertheless induces T_H2 cell responses in naïve mice (Figure 12.6C). It is furthermore able to fully restore antigen-specific T_H2 responses and IgE production in IRF3-deficient mice immunized with antigen and alum, while leaving IgG1 responses unaffected (Figure 12.6D). Thus, rather than the entire spectrum of adaptive immune responses promoted by alum adjuvantation, inflammatory monocytes appear to be involved only in the arm of the T_H2 response implicated in the production of IgE. It is likely that other types of antigen-presenting cells, presumably conventional dendritic cells, mediate the other aspects of the adaptive immune response to alum-adjuvanted immunization, including the induction of T_H2 cells able to promote IgG1 responses.

Inflammatory monocytes are the main producers of type I IFN at the sites of injection of alum and adjuvant, and this production is entirely dependent on IRF3 [27]. Type I IFNs however do not appear to play a role in the adaptive immune responses to alum-adjuvanted immunization, since mice deficient for the type I IFN receptor display humoral responses similar to those of wild-type mice [27]. IRF3 deficiency also does not impact on the overall maturation of inflammatory monocytes in response to alum and antigen [25,27]. However, a noticeable increase in the number of inflammatory monocytes expressing the chemokine receptor CCR7 is observed in IRF3-deficient mice immunized with alum and antigen. CCR7 is essential to the chemotactic attraction of maturing

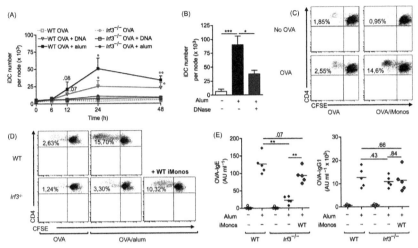

Figure 12.6 *Inflammatory monocytes are responsible for the IRF3-dependent induction of canonical T$_H$2 cell responses and IgE production in response to self-DNA sensing following alum-adjuvanted immunization.* **(A)** Recruitment of inflammatory monocyte-derived dendritic cells in the lymph nodes of wild-type (WT) and IRF3-deficient mice immunized with OVA alone, OVA and alum or OVA and self-DNA. **(B)** Effect of DNAse I treatment on the recruitment of inflammatory monocyte-derived dendritic cells in the lymph nodes of WT mice immunized with OVA and alum. **(C)** WT mice received OVA alone or OVA with inflammatory monocytes isolated from the peritoneal cavity of OVA and alum-treated WT mice (OVA/iMonos). The lymph node cells of recipient mice were labeled with CFSE and restimulated *in vitro* with or without OVA. The proliferation of OVA-specific CD4$^+$ T cells was estimated by measuring the percentage of CFSE^{low} CD4$^+$ T cells by flow cytometry (representative histograms are shown, with inserts indicating the percentage of CFSE^{low} CD4$^+$ T cells). **(D)** IRF3-deficient mice were treated i.p. with OVA and alum and received an i.p. transfer of inflammatory monocytes isolated from the peritoneal cavity of OVA and alum-treated WT mice. The lymph node cells of recipient mice were labeled with CFSE and restimulated *in vitro* with OVA. The proliferation of OVA-specific CD4$^+$ T cells was estimated by measuring the percentage of CFSE^{low} CD4$^+$ T cells by flow cytometry (inserts indicate the percentage of CFSE^{low} CD4$^+$ T cells). **(E)** OVA-IgE and OVA-IgG1 production in IRF3-deficient mice treated with OVA and alum, transferred with WT inflammatory monocytes as in (D) and boosted with OVA. As controls, WT and IRF3-deficient mice that received PBS with OVA alone or OVA and alum were used. *These panels are reproduced with permission from Marichal et al. [27].*

antigen–presenting cells to the draining lymph nodes. This observation thus may indicate that CCR7$^+$ inflammatory monocytes remain 'stuck' at sites of alum injection in case of impaired IRF3 signaling, leading to their accumulation. There is evidence that inflammatory monocyte–derived dendritic cells depend on signaling by homodimers of IL12p40 (IL12p80) to become able to respond to CCR7-activating chemokines [131,132]. Similar mechanisms seem implicated in alum-adjuvanted immunization. Indeed, immunization with antigen and alum induces the IRF3-dependent production of significant amounts of IL12p80 at sites of injection [27]. Neutralization of IL12p80 using neutralizing anti–IL12p40 antibodies reduces the recruitment of inflammatory monocytes to the draining lymph nodes and the production of antigen-specific IgE responses, but not of IgG1, in wild-type mice immunized with alum and antigen. Conversely, complementation with recombinant mouse IL12p80 restores inflammatory monocyte migration to the lymph nodes and IgE responses in IRF3-deficient mice, while leaving IgG1 production unaffected.

A MODEL FOR SELF-DNA INDUCED RESPONSES IN ALUM IMMUNIZATION

Based on data available, a model may be proposed for the induction of adaptive immune responses by the sensing of self-DNA release in alum-adjuvanted immunization (Figure 12.7). In this model, alum induces the release of self-DNA from cells dying at sites of injection. This self-DNA would in turn activate two distinct pathways, one being IRF3-dependent, the other being IRF3-independent. In the IRF3-independent pathway, self-DNA induces IgG1 production through the promotion of T_H responses endowed with high B cell helper activity. This part of the T_H2 response would thus mainly consist of T_{FH} cells. In the IRF3-dependent pathway, self-DNA stimulates the migration of antigen-loaded inflammatory monocytes to the draining lymph nodes, where they induce the differentiation of canonical T_H2 cells. T_H2 cells in turn mediate effector functions in peripheral tissues. These T_H2 cells would also promote IgE production, provided help is supplied for antibody production by the IRF3-independent pathway.

A number of significant questions remain open in this model, which will be mentioned in a 'top to bottom' discussion of the model depicted above. Potential links between self-DNA sensing and other proposed mechanisms of alum adjuvant activity will also be discussed. Of note, this discussion highlights that different mechanisms proposed so far could have important points of convergence. The possibility also exists that different mechanisms cooperate to induce full-blown adaptive immune responses to alum-adjuvanted immunization. It may be noted in this regard that elimination of self-DNA at sites of alum injection comparatively has a stronger

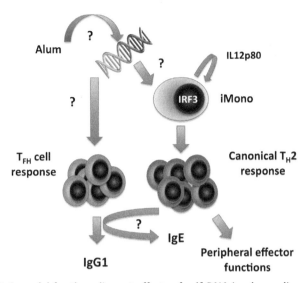

Figure 12.7 *A model for the adjuvant effects of self-DNA in alum-adjuvanted immunization.* Alum induces self-DNA release at sites of injection and activates innate immune signaling to boost adaptive immune responses. This response would comprise two components, driven by IRF3-dependent and IRF3-independent mechanisms, respectively. In the IRF3-independent pathway, DNA would induce the activation of T follicular helper (T$_{FH}$) cells helping B lymphocytes to develop into IgG1-secreting plasmocytes. In the IRF3-dependent pathway, DNA activates the migration of inflammatory monocytes (iMono) toward the draining lymph nodes. There, iMono-derived dendritic cells activate canonical T$_H$2 cells, unable to promote antibody production autonomously, but responsible for peripheral effector functions and for the production of IgE with the help of IRF3-independent T$_{FH}$ responses.

impact on IgE than on IgG1 responses. Yet, self-DNA alone may be used as an adjuvant for IgG1 and IgE responses. This therefore supports the hypothesis that the detection of self-DNA is essential to IgE responses and could cooperate with other pathways to fully induce IgG1 responses.

Alum-Induced Self-DNA Release: a Kind of ETosis?

The first and one of the most obvious questions relates to the origin of the self-DNA detected at sites of alum injection. Signs of general tissue damage and irritation have been reported [75–77], suggesting that unspecific cell damage may occur following alum administration. Yet, experimental evidence tends to support that alum-induced self-DNA release could be a more specific process. The self-DNA in the nodules that form at sites of alum injection *in vivo* is associated with significant amounts of myeloperoxydase and citrullinated histone H3 [14].

Histone H3 citrullination occurs through the deamination of arginine to citrulline. This process is mediated by the peptidyl arginine deaminase (PAD4) enzyme, which is also implicated in chromatin decondensation in the process of extracellular trap (ET) formation (ETosis). ETosis is a specific program of neutrophils and other myeloid cell types, in which the cell nucleus is decondensed and expulsed [133,134]. This process generates extracellular nets of chromatin decorated with antimicrobial proteins and granular enzymes such as myeloperoxidase. ETs have antibacterial and antiparasitic properties and bind to the pathogen surface. Thereby, they contribute to the circumvention of pathogen spread *in vivo* [133,135–137]. Interestingly, ETs have also recently been implicated in immunostimulatory activities as DAMPs [134]. ETs have for instance been proposed as important stimuli in the pathogenesis of SLE [109,138].

The presence of citrullinated histones and myeloperoxidase, along with the fiber-like appearance of chromatin in alum depots, supports the idea that alum induces a process similar to ETosis at sites of injection [14]. This hypothesis and the underlying mechanisms yet remain to be formally tested. ETosis is induced by diverse stimuli that strongly promote phagocytosis in myeloid cells [134]. Among these stimuli are monosodium urate crystals [136], which have been observed in tight association with extracellular self-DNA *in vivo* [139]. Thus, an appealing possibility could be that monosodium urate crystals forming due to local alum cytotoxic activity induce self-DNA release from recruited inflammatory cells through ETosis. On the other hand, uric acid itself is a major product of the degradation of self-nucleic acids. Thus, an opposite model could be that self-DNA released following ETosis is at the origin of the formation of monosodium urate crystals at sites of alum injection. Many similarities in their interaction with phagocytic cells have been noted between monosodium urate crystals and alum particles [87]. Although this possibility has not been investigated yet, one may envision that alum, like monosodium urate crystals, could be a direct inducer of ETosis. An intermediate possibility worth testing could be that ETosis and monosodium urate crystal formation contribute to an autoamplifying loop for the generation of immunostimulatory DAMPs at sites of alum injection. This interrelationship might also explain why alum, monosodium urate crystals and self-DNA display similar adjuvant activity on antigen-specific adaptive responses *in vivo* [27,79].

Intriguingly, a recent study showed that ETs may lower the threshold for T cell activation *in vitro* [140]. This process was shown to require T cell receptor engagement and direct contact between ET and T cells. It was however independent of TLR9. It would thus be appealing to test whether alum may induce the appearance of myeloid cells undergoing ETosis in the draining lymph nodes, and whether these ETs participate in the induction of T cell responses. It would also be interesting to investigate whether this could explain that alum is able to adjuvant the activation of T cells with seemingly lower T cell receptor affinity compared to other adjuvants [43].

What is the Identity of the IRF3-Activating Receptor for Self-DNA in Alum-Adjuvanted Immunization?

The next major question concerns the identity of the receptors for self-DNA in alum-adjuvanted immunization. As stated above, two pathways, an IRF3-dependent and an IRF3-independent pathway, are activated in response to alum-induced self-DNA release. Both pathways mediate distinct arms of the response to alum-adjuvanted immunization. The dichotomy appears to take place high upstream of the cascade of alum-adjuvanted events, since the IRF3-dependent pathway impacts on inflammatory monocyte responses, whereas the IRF3-independent pathway functions independently of this cell type.

It is likely that activation of the IRF3-dependent pathway relies on the activation of a DNA-sensing PRR. Since a role for TLR9 has already been ruled out, cytoplasmic DNA-sensing PRRs appear as next-in-line candidates [52]. A direct question would then be how extracellular self-DNA may be sensed by cytosolic receptors, which typically react to DNA that has been forced into the cytoplasm [141]. Phagocytosed alum may induce the destabilization of endosomal membranes and the subsequent release of their content into the cytoplasm, which may contribute to alum-induced NLRP3 inflammasome activation [84]. Self-DNA co-ingested with alum particles might gain access to the cytoplasm through such a mechanism. In this regard, it was recently reported that alum may facilitate the activation of cytoplasmic DNA-sensing PRRs by double-stranded DNA in reporter cells *in vitro* [124]. This would yet not explain how self-DNA alone may act as an adjuvant. One possibility could be that monosodium urate crystals derived from the degradation of self-DNA may mediate similar destabilizing effects on phagosomes [142]. It would thus be worthwhile testing whether alum treatment induces the accumulation of self-DNA in the phagosomes and cytoplasm of phago-cytic cells at sites of injection.

Most of the currently known cytoplasmic pathways to IRF3 activation in response to DNA implicate the adaptor protein STING [143–145]. It would thus be interesting to study the involvement of STING in the adjuvant effects of alum. The recently identified STING-dependent and -independent cytosolic DNA-sensing PRRs could subsequently be tested in the same context.

Another possibility is that extracellular self-DNA is sensed by an as yet unknown cell surface- or endosome-associated receptor. This hypothesis is supported by the observation that extracellular self-DNA that escapes degradation and accumulates in endosomes leads to IRF3 activation *in vivo* through TLR-independent pathways [92,112]. A possible endosomal leakage of phagocytosed self-DNA has however not been investigated in this context. Altogether, identifying a precise receptor for self-DNA in the context of alum-adjuvanted immunization will thus prove a significant task in itself.

Must There Be a DNA-Sensing Receptor for the IRF3-Independent Effects of Self-DNA in Alum-Adjuvanted Immunization?

The IRF3-independent pathway of the response to self-DNA mediates the clinically most important adjuvant effects of alum, i.e. those that promote the production of IgG1. They also are the least understood at the present time. The adjuvant activity of alum on IgG1 production is independent of the activation of IRF3, IRF7 (T. Marichal, F. Bureau & C.J. Desmet, unpublished observations) and the production of type I IFN in general [27]. Essential components of inflammasomes also appear not to be absolute requirements for the activity of alum in IgG1 responses [18,27,36,83]. Two of the central pathways downstream of DNA-sensing PRRs are thus dispensable for most of the adjuvant activity of alum. This raises the possibility that other DNA-sensing PRR-dependent pathways such as those leading to proinflammatory cytokine expression may be involved [52]. The identification of such pathways could come from studies using mice deficient for specific DNA-sensing PRRs, as proposed above for the identification of the IRF3-activating PRR in alum-adjuvanted immunization.

Another possibility not to overlook could be that there actually is no need for a specific direct receptor for self-DNA to mediate its adjuvant effect on IgG1 responses. One such scenario, as mentioned above, would be that degradation products of self–DNA such as monosodium urate crystals would mediate these adjuvant effects. There also is interesting recent evidence that alum may directly activate antigen–presenting cells without any apparent requirement for a specific receptor [87]. Indeed, alum, like monosodium urate crystals, may force lipid sorting in cell membranes through direct contact [87,146]. This results in the aggregation of immunoreceptor signaling motif (ITAM)-containing receptors, and subsequent activation of spleen tyrosine kinase (Syk)- and phosphoinositide 3-kinase (PI3K)-mediated signaling. In bone marrow–derived dendritic cells, these signaling events directly contribute to increase the binding of the cells to T lymphocytes [87]. This effect has however so far only been demonstrated *in vitro*, using cultured models of dendritic cells. Whether such a model is relevant *in vivo* in the specific types of dendritic cells encountered at sites of alum injection remains to be established. This notwithstanding, it would be worthwhile studying the potential interaction of such a mechanism with self–DNA detection *in vivo*. Indeed, both alum and self–DNA, much like self–DNA and monosodium urate crystals, form tight associations *in vivo* [14,27,139]. Discriminating between the different possible mechanisms evoked above would likely benefit from the identification of the antigen–presenting cells involved in the induction of IgG1 responses to alum–adjuvanted antigens *in vivo*.

What is the Cellular Mechanism of Induction of IgE Responses in Alum-Adjuvanted Immunization?

Even though alum-adjuvanted vaccines have no record of inducing adverse allergic reactions, it is generally preferred to avoid inducing IgE production in response to vaccination. The IRF3-dependent pathway leading to canonical T_H2 responses downstream of self–DNA sensing is responsible for the induction of IgE production in response to alum. Intriguingly though, this pathway alone is not sufficient for antibody production and requires help from IRF3-independent T_H responses. This raises the question of the exact mechanisms at play in the induction of antigen-specific IgE responses following alum-adjuvanted immunization.

The biology of IgE responses *in vivo* is poorly understood [147]. It is known that isotype switching toward IgE production in B cells requires T cell help, which provides CD40L and IL-4 stimulation. It is however unclear at present how exactly IgE$^+$ B cells differentiate. From evidence available, it seems that IgE-switched B cells may develop directly from naïve

B cells in germinal centers, or derive from germinal center-derived IgG1-switched intermediates [148,149]. A likely scenario to explain the role of IRF3-dependent T_H2 cell responses in the induction of IgE responses could be that canonical T_H2 cells at the T–B cell border provide high levels of IL-4 to B cells, as bystander cells or through cognate help, and hence initiate IgE isotype switching before the B cell enters the germinal center reaction. IRF3-independently activated T_{FH} cells would then be required to support germinal center reactions and effective production of IgE. A more direct evaluation of the phenotype and function of IRF3-dependently and IRF3-independently activated T_H2 cell responses could provide valuable indications on the respective mechanisms of induction of IgE and IgG1 responses. This knowledge might eventually prove useful in the design of alum-containing vaccines with higher T_{FH} activity and devoid of 'IgE bias'.

CONCLUSIONS AND PERSPECTIVES ON THE ROLE OF SELF-DNA SENSING IN VACCINATION

As illustrated above by the case of alum-adjuvanted immunization, the 'danger model' may theoretically help explain (some of) the mechanisms of action of current vaccines. However, DAMPs have been notably understudied in comparison with PAMPs in the context of vaccination [52]. This is likely due to the more complicated study of DAMPs compared to PAMPs, since DAMPs have to be identified after their release *in vivo*, whereas PAMPs are readily present and analyzable in immunogenic preparations. Nevertheless DAMPs, including self-DNA, are likely to be present at sites of immunization in different types of vaccinations. ETosis for instance increasingly appears as a widespread protective mechanism elicited in response to a wide variety of PAMPs and DAMPs [150]. It may thus be important to evaluate the contribution of self-DNA in types of vaccinations other than those involving alum.

A recent study brings direct support to the relevance of this proposal [124]. This study investigated the mechanisms of adjuvant activity of polyethyleneimine (PEI), an experimental adjuvant, on mucosal responses to viral subunit glycoprotein antigens. PEI represents a family of organic polycations usually used as nucleic acid transfection reagents *in vitro* and DNA vaccine delivery vehicles *in vivo*. PEI was shown to adjuvant predominantly antigen-specific IgG1 responses, consistent with an observed T_H2 bias in the adaptive response. Complexes of PEI and viral glycoproteins enhance antigen uptake by resident cells and dendritic cells *in vivo*, although they are not sufficient to activate dendritic cells directly. PEI was shown to induce the release of self-DNA at sites of injection in quantities comparable to those induced by alum. Furthermore, local treatment with DNAse I reduced PEI-adjuvanted antibody responses. This study thus establishes an important role of self-DNA sensing in the adjuvant activity of PEI. In addition, the same study indicates that IRF3-deficient mice develop reduced IgG1 titers in response to PEI-adjuvanted immunization. This is in contrast with the role of IRF3 in alum-adjuvanted immunization, in which IRF3 is not involved in the promotion of IgG1 responses [27]. Differences in the routes of administration and, hence, in the targeted antigen-presenting cells might account for these differences. PEI also was shown to facilitate the entry of DNA into the cell cytoplasm more efficiently than alum, which could result in more robust IRF3-dependent adjuvant effects in dendritic cells.

In conclusion, recent studies indicate that self-DNA sensing may be an important mediator of the effects of vaccine adjuvants. Therefore, they advocate for the further evaluation of self-DNA release as an endogenous immupotentiator signal in vaccination. In addition, different mechanisms of self-DNA sensing may be responsible for distinct components of the adaptive immune response to a vaccine, which opens perspectives for the tailoring of vaccines favoring desirable (or avoiding undesirable) types of such components.

REFERENCES

[1] Glenny AT, Pope CG, Waddington H, Wallace U. Immunological notes: XVII–XXIV. J Pathol Bacteriol 1926;29:31–40.
[2] Lindblad EB. Aluminium compounds for use in vaccines. Immunol Cell Biol 2004;82(5):497–505.
[3] Pulendran B, Ahmed R. Immunological mechanisms of vaccination. Nat Immunol 2011;131(6):509–17.
[4] Lindblad EB. Aluminium adjuvants – in retrospect and prospect. Vaccine 2004;22(27–28):3658–68.
[5] Cain DW, Sanders SE, Cunningham MM, Kelsoe G. Disparate adjuvant properties among three formulations of 'alum'. Vaccine 2012;31(4):653–60.
[6] O'Hagan DT, Ott GS, De Gregorio E, Seubert A. The mechanism of action of MF59 – An innately attractive adjuvant formulation. Vaccine 2012;30(29):4341–8.
[7] Harrison WT. Some observations on the use of alum precipitated diphtheria toxoid. Am J Public Health Nations Health 1935;25(3):298–300.
[8] Glenny A, Buttle G, Stevens M. Rate of disappearance of diphtheria toxoid injected into rabbits and guinea-pigs: toxoid precipitated with alum. J Pathol Bacteriol 1931;34:267–87.
[9] Noe SM, Green MA, HogenEsch H, Hem SL. Mechanism of immunopotentiation by aluminum-containing adjuvants elucidated by the relationship between antigen retention at the inoculation site and the immune response. Vaccine 2010;28(20):3588–94.

[10] Mannhalter JW, Neychev HO, Zlabinger GJ, Ahmad R, Eibl MM. Modulation of the human immune response by the non-toxic and non-pyrogenic adjuvant aluminium hydroxide: effect on antigen uptake and antigen presentation. Clin Exp Immunol 1985;61(1):143–51.

[11] Morefield GL, et al. Role of aluminum-containing adjuvants in antigen internalization by dendritic cells *in vitro*. Vaccine 2005;23(13):1588–95.

[12] Holt L. Developments in Diphtheria Prophylaxis. London: Wm. Heinemann; 1950.

[13] Romero Mendez IZ, Shi Y, HogenEsch H, Hem SL. Potentiation of the immune response to non-adsorbed antigens by aluminum-containing adjuvants. Vaccine 2007;25(5):825–33.

[14] Munks MW, et al. Aluminum adjuvants elicit fibrin-dependent extracellular traps *in vivo*. Blood 2010;116(24):5191–9.

[15] Hutchison S, et al. Antigen depot is not required for alum adjuvanticity. FASEB J 2012;26(3):1272–9.

[16] Rahman F, et al. Cellular and humoral immune responses induced by intradermal or intramuscular vaccination with the major hepatitis B surface antigen. Hepatology 2000;31(2):521–7.

[17] Hohn H, et al. Longitudinal analysis of the T-cell receptor (TCR)-VA and -VB repertoire in CD8+ T cells from individuals immunized with recombinant hepatitis B surface antigen. Clin Exp Immunol 2002;129(2):309–17.

[18] McKee AS, et al. Alum induces innate immune responses through macrophage and mast cell sensors, but these sensors are not required for alum to act as an adjuvant for specific immunity. J Immunol 2009;183(7):4403–14.

[19] MacLeod MK, et al. Vaccine adjuvants aluminum and monophosphoryl lipid A provide distinct signals to generate protective cytotoxic memory CD8 T cells. Proc Natl Acad Sci USA 2011;108(19):7914–9.

[20] Grun JL, Maurer PH. Different T helper cell subsets elicited in mice utilizing two different adjuvant vehicles: the role of endogenous interleukin 1 in proliferative responses. Cell Immunol 1989;121(1):134–45.

[21] Brewer JM, et al. Aluminium hydroxide adjuvant initiates strong antigen-specific Th2 responses in the absence of IL-4- or IL-13-mediated signaling. J Immunol 1999;163(12):6448–54.

[22] Palm NW, Rosenstein RK, Medzhitov R. Allergic host defences. Nature 2012;484(7395):465–72.

[23] Pulendran B, Tang H, Manicassamy S. Programming dendritic cells to induce TH2 and tolerogenic responses. Nat Immunol 2010;11(8):647–55.

[24] Pulendran B, Artis D. New paradigms in type 2 immunity. Science 2012;337(6093):431–5.

[25] Kool M, et al. Alum adjuvant boosts adaptive immunity by inducing uric acid and activating inflammatory dendritic cells. J Exp Med 2008;205(4):869–82.

[26] Calabro S, et al. Vaccine adjuvants alum and MF59 induce rapid recruitment of neutrophils and monocytes that participate in antigen transport to draining lymph nodes. Vaccine 2011;29(9):1812–23.

[27] Marichal T, et al. DNA released from dying host cells mediates aluminum adjuvant activity. Nat Med 2011;17(8):996–1002.

[28] Mosca F, et al. Molecular and cellular signatures of human vaccine adjuvants. Proc Natl Acad Sci USA 2008;105(30):10501–6.

[29] Kuroda E, et al. Silica crystals and aluminum salts regulate the production of prostaglandin in macrophages via NALP3 inflammasome-independent mechanisms. Immunity 2011;34(4):514–26.

[30] Mantovani A, Cassatella MA, Costantini C, Sb Jaillon. Neutrophils in the activation and regulation of innate and adaptive immunity. Nat Rev Immunol 2011;11(8):519–31.

[31] Yang CW, Strong BS, Miller MJ, Unanue ER. Neutrophils influence the level of antigen presentation during the immune response to protein antigens in adjuvants. J Immunol 2010;185(5):2927–34.

[32] Banchereau J, et al. Immunobiology of dendritic cells. Annu Rev Immunol 2000;18:767–811.

[33] Steinman RM. Dendritic cells *in vivo*: a key target for a new vaccine science. Immunity 2008;29(3):319–24.

[34] Diebold SS. Determination of T-cell fate by dendritic cells. Immunol Cell Biol 2008;86(5):389–97.

[35] Sornasse T, et al. Antigen-pulsed dendritic cells can efficiently induce an antibody response *in vivo*. J Exp Med 1992;175(1):15–21.

[36] Kool M, et al. Cutting edge: alum adjuvant stimulates inflammatory dendritic cells through activation of the NALP3 inflammasome. J Immunol 2008;181(6):3755–9.

[37] Ghimire TR, Benson RA, Garside P, Brewer JM. Alum increases antigen uptake, reduces antigen degradation and sustains antigen presentation by DCs *in vitro*. Immunol Lett 2012;147(1–2):55–62.

[38] van Rijt LS, et al. *In vivo* depletion of lung CD11c+ dendritic cells during allergen challenge abrogates the characteristic features of asthma. J Exp Med 2005;201(6):981–91.

[39] Naik SH, et al. Intrasplenic steady-state dendritic cell precursors that are distinct from monocytes. Nat Immunol 2006;7(6):663–71.

[40] Shortman K, Naik SH. Steady-state and inflammatory dendritic-cell development. Nat Rev Immunol 2007;7(1):19–30.

[41] Greter M, et al. GM-CSF controls nonlymphoid tissue dendritic cell homeostasis but is dispensable for the differentiation of inflammatory dendritic cells. Immunity 2012;36(6):1031–46.

[42] Goodnow CC, Vinuesa CG, Randall KL, Mackay F, Brink R. Control systems and decision making for antibody production. Nat Immunol 2010;11(8):681–8.

[43] Fazilleau N, McHeyzer-Williams LJ, Rosen H, McHeyzer-Williams MG. The function of follicular helper T cells is regulated by the strength of T cell antigen receptor binding. Nat Immunol 2009;10(4):375–84.

[44] King C. New insights into the differentiation and function of T follicular helper cells. Nat Rev Immunol 2009;9(11):757–66.

[45] Ma CS, Deenick EK, Batten M, Tangye SG. The origins, function, and regulation of T follicular helper cells. J Exp Med 2012;209(7):1241–53.

[46] Reinhardt RL, Liang H-E, Locksley RM. Cytokine-secreting follicular T cells shape the antibody repertoire. Nat Immunol 2009;10(4):385–93.

[47] Moon JJ, et al. Enhancing humoral responses to a malaria antigen with nanoparticle vaccines that expand Tfh cells and promote germinal center induction. Proc Natl Acad Sci USA 2012;109(4):1080–5.

[48] Janeway Jr. CA. Approaching the asymptote? Evolution and revolution in immunology. Cold Spring Harb Symp Quant Biol 1989;54(Pt 1):1–13.

[49] Palm NW, Medzhitov R. Pattern recognition receptors and control of adaptive immunity. Immunol Rev 2009;227(1):221–33.

[50] Iwasaki A, Medzhitov R. Regulation of adaptive immunity by the innate immune system. Science 2010;327(5963):291–5.

[51] Coffman RL, Sher A, Seder RA. Vaccine adjuvants: putting innate immunity to work. Immunity 2010;33(4):492–503.

[52] Desmet CJ, Ishii KJ. Nucleic acid sensing at the interface between innate and adaptive immunity in vaccination. Nat Rev Immunol 2012;12(7):479–91.

[53] Takeuchi O, Akira S. Pattern recognition receptors and inflammation. Cell 2010;140(6):805–20.

[54] Kumar H, Kawai T, Akira S. Pathogen recognition by the innate immune system. Int Rev Immunol 2011;30(1):16–34.

[55] Kawai T, Akira S. The role of pattern-recognition receptors in innate immunity: update on Toll-like receptors. Nat Immunol 2010;11(5):373–84.

[56] Kawai T, Akira S. Toll-like receptors and their crosstalk with other innate receptors in infection and immunity. Immunity 2011;34(5):637–50.

[57] Loo YM, Gale Jr. M. Immune signaling by RIG-I-like receptors. Immunity 2011;34(5):680–92.

[58] Elinav E, Strowig T, Henao-Mejia J, Flavell RA. Regulation of the antimicrobial response by NLR proteins. Immunity 2011;34(5):665–79.

[59] Osorio F, Reis e Sousa C. Myeloid C-type lectin receptors in pathogen recognition and host defense. Immunity 2011;34(5):651–64.

[60] Kadowaki N, et al. Subsets of human dendritic cell precursors express different toll-like receptors and respond to different microbial antigens. J Exp Med 2001;194(6):863–9.

[61] Iwasaki A, Medzhitov R. Toll-like receptor control of the adaptive immune responses. Nat Immunol 2004;5(10):987–95.

[62] Blasius AL, Beutler B. Intracellular toll-like receptors. Immunity 2010;32(3):305–15.

[63] Pichlmair A, Reis e Sousa C. Innate recognition of viruses. Immunity 2007;27(3):370–83.

[64] Barbalat R, Ewald SE, Mouchess ML, Barton GM. Nucleic acid recognition by the innate immune system. Annu Rev Immunol 2011;29:185–214.

[65] Querec T, et al. Yellow fever vaccine YF-17D activates multiple dendritic cell subsets via TLR2, 7, 8, and 9 to stimulate polyvalent immunity. J Exp Med 2006;203(2):413–24.

[66] Gaucher D, et al. Yellow fever vaccine induces integrated multilineage and polyfunctional immune responses. J Exp Med 2008;205(13):3119–31.

[67] Querec TD, et al. Systems biology approach predicts immunogenicity of the yellow fever vaccine in humans. Nat Immunol 2009;10(1):116–25.

[68] Coban C, et al. Novel strategies to improve DNA vaccine immunogenicity. Curr Gene Ther 2011;11(6):479–84.

[69] Schnare M, et al. Toll-like receptors control activation of adaptive immune responses. Nat Immunol 2001;2(10):947–50.

[70] Gavin AL, et al. Adjuvant-enhanced antibody responses in the absence of Toll-like receptor signaling. Science 2006;314(5807):1936–8.

[71] Matzinger P. Tolerance, danger, and the extended family. Annu Rev Immunol 1994;12:991–1045.

[72] Seong SY, Matzinger P. Hydrophobicity: an ancient damage-associated molecular pattern that initiates innate immune responses. Nat Rev Immunol 2004;4(6):469–78.

[73] Kono H, Rock KL. How dying cells alert the immune system to danger. Nat Rev Immunol 2008;8(4):279–89.

[74] Chen GY, Nunez G. Sterile inflammation: sensing and reacting to damage. Nat Rev Immunol 2010;10(12):826–37.

[75] Goto N, Akama K. Histopathological studies of reactions in mice injected with aluminum-adsorbed tetanus toxoid. Microbiol Immunol 1982;26(12):1121–32.

[76] Goto N, Kato H, Maeyama J-i, Eto K, Yoshihara S. Studies on the toxicities of aluminium hydroxide and calcium phosphate as immunological adjuvants for vaccines. Vaccine 1993;11(9):914–8.

[77] Goto N, et al. Local tissue irritating effects and adjuvant activities of calcium phosphate and aluminium hydroxide with different physical properties. Vaccine 1997;15(12–13):1364–71.

[78] Tang L, Hu W. Molecular determinants of biocompatibility. Exp Rev Med Dev 2005;2(4):493–500.

[79] Kool M, et al. An unexpected role for Aric acid as an inducer of T helper 2 cell immunity to inhaled antigens and inflammatory mediator of allergic asthma. Immunity 2011;34(4):527–40.

[80] Shi Y, Evans JE, Rock KL. Molecular identification of a danger signal that alerts the immune system to dying cells. Nature 2003;425(6957):516–21.

[81] Shi Y, Mucsi AD, Ng G. Monosodium urate crystals in inflammation and immunity. Immunol Rev 2010;233(1):203–17.

[82] Eisenbarth SC, Colegio OR, O'Connor W, Sutterwala FS, Flavell RA. Crucial role for the Nalp3 inflammasome in the immunostimulatory properties of aluminium adjuvants. Nature 2008;453(7198):1122–6.

[83] Franchi L, Nunez G. The Nlrp3 inflammasome is critical for aluminium hydroxide-mediated IL-1beta secretion but dispensable for adjuvant activity. Eur J Immunol 2008;38(8):2085–9.

[84] Hornung V, et al. Silica crystals and aluminum salts activate the NALP3 inflammasome through phagosomal destabilization. Nat Immunol 2008;9(8):847–56.

[85] Li H, Willingham SB, Ting JPY, Re F. Cutting edge: inflammasome activation by alum and alum's adjuvant effect are mediated by NLRP3. J Immunol 2008;181(1):17–21.

[86] Martinon F, Petrilli V, Mayor A, Tardivel A, Tschopp J. Gout-associated uric acid crystals activate the NALP3 inflammasome. Nature 2006;440(7081):237–41.

[87] Flach TL, et al. Alum interaction with dendritic cell membrane lipids is essential for its adjuvanticity. Nat Med 2011;17(4):479–87.

[88] Napirei M, et al. Features of systemic lupus erythematosus in Dnase1-deficient mice. Nat Genet 2000;25(2):177–81.

[89] Yasutomo K, et al. Mutation of DNASE1 in people with systemic lupus erythematosus. Nat Genet 2001;28(4):313–4.

[90] Al-Mayouf SM, et al. Loss-of-function variant in DNASE1L3 causes a familial form of systemic lupus erythematosus. Nat Genet 2011;43(12):1186–8.

[91] Hakkim A, et al. Impairment of neutrophil extracellular trap degradation is associated with lupus nephritis. Proc Natl Acad Sci USA 2010;107(21):9813–8.

[92] Okabe Y, Kawane K, Akira S, Taniguchi T, Nagata S. Toll-like receptor-independent gene induction program activated by mammalian DNA escaped from apoptotic DNA degradation. J Exp Med 2005;202(10):1333–9.

[93] Yoshida H, Okabe Y, Kawane K, Fukuyama H, Nagata S. Lethal anemia caused by interferon-beta produced in mouse embryos carrying undigested DNA. Nat Immunol 2005;6(1):49–56.

[94] Kawane K, et al. Chronic polyarthritis caused by mammalian DNA that escapes from degradation in macrophages. Nature 2006;443(7114):998–1002.

[95] Lande R, et al. Plasmacytoid dendritic cells sense self-DNA coupled with antimicrobial peptide. Nature 2007;449(7162):564–9.

[96] Kawasaki T, Kawai T, Akira S. Recognition of nucleic acids by pattern-recognition receptors and its relevance in autoimmunity. Immunol Rev 2011;243(1):61–73.

[97] Collins LV, Hajizadeh S, Holme E, Jonsson IM, Tarkowski A. Endogenously oxidized mitochondrial DNA induces in vivo and in vitro inflammatory responses. J Leukoc Biol 2004;75(6):995–1000.

[98] Zhang Q, et al. Circulating mitochondrial DAMPs cause inflammatory responses to injury. Nature 2010;464(7285):104–7.

[99] Oka T, et al. Mitochondrial DNA that escapes from autophagy causes inflammation and heart failure. Nature 2012;485(7397):251–5.

[100] McGill MR, et al. The mechanism underlying acetaminophen-induced hepatotoxicity in humans and mice involves mitochondrial damage and nuclear DNA fragmentation. J Clin Invest 2012;122(4):1574–83.

[101] Imaeda AB, et al. Acetaminophen-induced hepatotoxicity in mice is dependent on Tlr9 and the Nalp3 inflammasome. J Clin Invest 2009;119(2):305–14.

[102] Hemmi H, et al. A Toll-like receptor recognizes bacterial DNA. Nature 2000;408(6813):740–5.

[103] Martinon F. Dangerous liaisons: mitochondrial DNA meets the NLRP3 inflammasome. Immunity 2012;36(3):313–5.

[104] Shimada K, et al. Oxidized mitochondrial DNA activates the NLRP3 inflammasome during apoptosis. Immunity 2012;36(3):401–14.

[105] Viglianti GA, et al. Activation of autoreactive B cells by CpG dsDNA. Immunity 2003;19(6):837–47.

[106] Lamphier MS, Sirois CM, Verma A, Golenbock DT, Latz E. TLR9 and the recognition of self and non-self nucleic acids. Ann NY Acad Sci 2006;1082:31–43.

[107] Haas T, et al. The DNA sugar backbone 2' deoxyribose determines toll-like receptor 9 activation. Immunity 2008;28(3):315–23.

[108] Huck S, Deveaud E, Namane A, Zouali M. Abnormal DNA methylation and deoxycytosine-deoxyguanine content in nucleosomes from lymphocytes undergoing apoptosis. FASEB J 1999;13(11):1415–22.

[109] Lande R, et al. Neutrophils activate plasmacytoid dendritic cells by releasing self-DNA-peptide complexes in systemic lupus erythematosus. Sci Transl Med 2011;3(73):73ra19.

[110] Christensen SR, et al. Toll-like receptor 7 and TLR9 dictate autoantibody specificity and have opposing inflammatory and regulatory roles in a murine model of lupus. Immunity 2006;25(3):417–28.

[111] Yu P, et al. Toll-like receptor 9-independent aggravation of glomerulonephritis in a novel model of SLE. Int Immunol 2006;18(8):1211–9.

[112] Okabe Y, Kawane K, Nagata S. IFN regulatory factor (IRF) 3/7-dependent and -independent gene induction by mammalian DNA that escapes degradation. Eur J Immunol 2008;38(11):3150–8.

[113] Kim T, et al. Aspartate-glutamate-alanine-histidine box motif (DEAH)/RNA helicase A helicases sense microbial DNA in human plasmacytoid dendritic cells. Proc Natl Acad Sci USA 2010;107(34):15181–6.

[114] Miyashita M, Oshiumi H, Matsumoto M, Seya T. DDX60, a DEXD/H box helicase, is a novel antiviral factor promoting RIG-I-like receptor-mediated signaling. Mol Cell Biol 2011;31(18):3802–19.

[115] Zhang Z, et al. The helicase DDX41 senses intracellular DNA mediated by the adaptor STING in dendritic cells. Nat Immunol 2011;12(10):959–65.

[116] Burckstummer T, et al. An orthogonal proteomic-genomic screen identifies AIM2 as a cytoplasmic DNA sensor for the inflammasome. Nat Immunol 2009;10(3):266–72.

[117] Fernandes-Alnemri T, Yu J-W, Datta P, Wu J, Alnemri ES. AIM2 activates the inflammasome and cell death in response to cytoplasmic DNA. Nature 2009;458(7237):509–13.

[118] Hornung V, et al. AIM2 recognizes cytosolic dsDNA and forms a caspase-1-activating inflammasome with ASC. Nature 2009;458(7237):514–8.

[119] Roberts TL, et al. HIN-200 proteins regulate caspase activation in response to foreign cytoplasmic DNA. Science 2009;323(5917):1057–60.

[120] Unterholzner L, et al. IFI16 is an innate immune sensor for intracellular DNA. Nat Immunol 2010.

[121] Takaoka A, et al. DAI (DLM-1/ZBP1) is a cytosolic DNA sensor and an activator of innate immune response. Nature 2007;448(7152):501–5.

[122] Wang Z, et al. Regulation of innate immune responses by DAI (DLM-1/ZBP1) and other DNA-sensing molecules. Proc Natl Acad Sci USA 2008;105(14):5477–82.

[123] Yang P, et al. The cytosolic nucleic acid sensor LRRFIP1 mediates the production of type I interferon via a [beta]-catenin-dependent pathway. Nat Immunol 2010;11(6):487–94.

[124] Wegmann F, et al. Polyethyleneimine is a potent mucosal adjuvant for viral glycoprotein antigens. Nat Biotech 2013;30(9):883–8.

[125] Remijsen Q, et al. Dying for a cause: NETosis, mechanisms behind an antimicrobial cell death modality. Cell Death Differ 2011;18(4):581–8.

[126] Takeshita F, Ishii KJ. Intracellular DNA sensors in immunity. Curr Opin Immunol 2008;20(4):383–8.

[127] Fitzgerald KA, et al. IKKepsilon and TBK1 are essential components of the IRF3 signaling pathway. Nat Immunol 2003;4(5):491–6.

[128] Sharma S, et al. Triggering the interferon antiviral response through an IKK-related pathway. Science 2003;300(5622):1148–51.

[129] Marichal T, et al. Interferon response factor 3 is essential for house dust mite-induced airway allergy. J Allergy Clin Immunol 2010;126(4):836–44.

[130] Legutko A, et al. Sirtuin 1 promotes Th2 responses and airway allergy by repressing Peroxisome Proliferator-Activated Receptor-γ activity in dendritic cells. J Immunol 2011;187(9):4517–29.

[131] Khader SA, et al. Interleukin 12p40 is required for dendritic cell migration and T cell priming after *Mycobacterium tuberculosis* infection. J Exp Med 2006;203(7):1805–15.

[132] Robinson RT, et al. *Yersinia pestis* evades TLR4-dependent induction of IL-12(p40)2 by dendritic cells and subsequent cell migration. J Immunol 2008;181(8):5560–7.

[133] Brinkmann V, et al. Neutrophil extracellular traps kill bacteria. Science 2004;303(5663):1532–5.

[134] Brinkmann V, Zychlinsky A. Neutrophil extracellular traps: is immunity the second function of chromatin? J Cell Biol 2012;198(5):773–83.

[135] von Kockritz-Blickwede M, et al. Phagocytosis-independent antimicrobial activity of mast cells by means of extracellular trap formation. Blood 2008;111(6):3070–80.

[136] Schorn C, et al. Monosodium urate crystals induce extracellular DNA traps in neutrophils, eosinophils, and basophils but not in mononuclear cells. Front Immunol 2012;3:277.

[137] Yipp BG, et al. Infection-induced NETosis is a dynamic process involving neutrophil multitasking *in vivo*. Nat Med 2012;18(9):1386–93.

[138] Guiducci C, et al. Autoimmune skin inflammation is dependent on plasmacytoid dendritic cell activation by nucleic acids via TLR7 and TLR9. J Exp Med 2010;207(13):2931–42.

[139] Schorn C, et al. Bonding the foe – NETting neutrophils immobilize the proinflammatory monosodium urate crystals. Front Immunol 2012:3.

[140] Tillack K, Breiden P, Martin R, Sospedra M. T Lymphocyte priming by neutrophil extracellular traps links innate and adaptive immune responses. J Immunol 2013;188(7):3150–9.

[141] Keating SE, Baran M, Bowie AG. Cytosolic DNA sensors regulating type I interferon induction. Trends Immunol 2011;32(12):574–81.

[142] Shirahama T, Cohen AS. Ultrastructural evidence for leakage of lysosomal contents after phagocytosis of monosodium urate crystals. A mechanism of gouty inflammation. Am J Pathol 1974;76(3):501–20.

[143] Ishikawa H, Ma Z, Barber GN. STING regulates intracellular DNA-mediated, type I interferon-dependent innate immunity. Nature 2009;461(7265):788–92.

[144] Barber GN. STING-dependent signaling. Nat Immunol 2011;12(10):929–30.

[145] Barber GN. Cytoplasmic DNA innate immune pathways. Immunol Rev 2011;243(1):99–108.

[146] Ng G, et al. Receptor-independent, direct membrane binding leads to cell-surface lipid sorting and Syk kinase activation in dendritic cells. Immunity 2008;29(5):807–18.

[147] Dullaers M, et al. The who, where, and when of IgE in allergic airway disease. J Allergy Clin Immunol 2012;129(3):635–45.

[148] Talay O, et al. IgE+ memory B cells and plasma cells generated through a germinal-center pathway. Nat Immunol 2012;13(4):396–404.

[149] Xiong H, Dolpady J, Wabl M, Curotto de Lafaille MA, Lafaille JJ. Sequential class switching is required for the generation of high affinity IgE antibodies. J Exp Med 2013;209(2):353–64..

[150] Remijsen Q, et al. Neutrophil extracellular trap cell death requires both autophagy and superoxide generation. Cell Res. 2011;21(2):290–304.

[151] Ablasser A, et al. RIG-I-dependent sensing of poly(dA:dT) through the induction of an RNA polymerase III-transcribed RNA intermediate. Nat Immunol 2009;10(10):1065–72.

[152] Chiu Y-H, MacMillan JB, Chen ZJ. RNA polymerase III detects cytosolic DNA and induces type I interferons through the RIG-I pathway. Cell 2009;138(3):576–91.

Adjuvants Targeting the DNA Sensing Pathways – Cyclic-di-GMP and other Cyclic-Dinucleotides

Rebecca Schmidt and Laurel L. Lenz

Integrated Department of Immunology, National Jewish Health and University of Colorado School of Medicine, Denver, USA

INTRODUCTION

Vaccination remains one of the best ways to prevent or control infectious diseases. Current vaccine strategies range from attenuated live pathogens to killed pathogens, purified subunit antigens, or DNA coding for antigens [1,2]. These different vaccine formulations vary both in their immunogenicity and in their potential to elicit harmful or in some cases even deadly side effects. Live, attenuated pathogens contain numerous components that can stimulate immune responses and are thus often the most immunogenic vaccine formulations. However, such vaccines carry the risk of reversion to pathogenicity. By contrast, purified antigens can be safely administered as subunit vaccines as they completely lack infectivity. However, purified antigens are often poorly immunogenic and thus require the use of adjuvants to stimulate potent immune responses [1,2].

A diverse variety of immunostimulatory molecules have been tested or used as adjuvants for the development of vaccines. These range from microbial compounds such as CpG oligonucleotides and bacterial toxins to foreign particles such as oil emulsions and alum salts [1]. However, few of these adjuvants are approved for use in humans and for several of these the mechanistic basis for their adjuvanticity remains mysterious [3]. In addition, different arms of the immune response are required for resistance to different pathogens and in some cases the available approved adjuvants are not effective for vaccination against specific pathogens [2–4]. Hence, there is considerable interest and effort in testing and development of new adjuvants that stimulate different aspects of the immune response. Information from these studies may lead to approval of tailored adjuvants

Biological DNA Sensor.
DOI: http://dx.doi.org/10.1016/B978-0-12-404732-7.00013-7

that can be matched with appropriate antigens to generate more effective vaccine formulations.

Amongst the up-and-coming candidate vaccine adjuvants are the cyclic dinucleotide phosphates cyclic-di-guanosine-monophosphate (c-di-GMP) and cyclic-di-adenosine monophosphate (c-di-AMP). These c-di-NMPs play important roles in the regulation of bacterial functions such as virulence and gene expression during infection [5–7], and thus are necessary for bacterial physiology and pathogenesis. In addition, they can target host DNA-sensing innate immune pathways and thus elicit potent immuno-stimulatory effects [8,9]. In this chapter, we first introduce the cyclic-dinucleotides – including their production, regulation, and function in bacteria – then review the literature on their known effects on the immune system and their effectiveness in vaccine formulations.

PRODUCTION AND FUNCTIONS OF CYCLIC-DINUCLEOTIDES IN BACTERIA

Cyclic-Dinucleotides are Second Messengers in Bacteria

The use of modified nucleotides as messengers that relay and amplify signals induced by extracellular stimuli is a common theme in biology. For example, eukaryotes often use cyclic-guanosine monophospate (cGMP) to activate intracellular protein kinases and thus transduce signals from the cell membrane [6]. Similar functions are attributed to cyclic-adenosine mono-phospate (cAMP), though this messenger nucleotide is more common in single celled organisms. For example, cAMP regulates bacterial metabolic processes and is secreted by eukaryotic slime molds to coordinate colonial behaviors [6,10,11]. The alarmone guanosine tetraphosphate (ppGpp) is mainly used by bacteria, where it regulates the stringent response to inhibit replication and gene expression during limited nutrition conditions [12]. The cyclic-dinucleotides (c-di-NMPs) c-di-GMP and c-di-AMP are also involved in the regulation of numerous bacterial responses and were until recently considered to be exclusively produced by bacteria [6,13]. c-di-GMP and c-di-AMP are cyclic dimers of purine ribonucleotides [14]. These molecules consist of two NMP molecules joined by phosphodies-ter links at the 3′ and 5′ ribose carbons (Figure 13.1). Thus, the linkages forming the c-di-NMP dimers resemble those in the sugar–phosphate backbone of nucleic acids. However, the ribose and phosphate moieties are on the interior of the dimer with the nitrogenous bases facing out-ward and thus their conformation is 'inside-out' relative to Watson–Crick

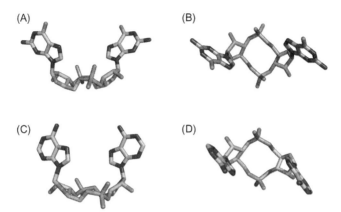

Figure 13.1 *Structure of c-di-NMPs in cyclase proteins.* **(A, B)** c-di-GMP in the cyclase domain of *Caulobacter cresentus* PleD [37]. Phosphodiester linkages between the 3' and 5' carbons of the ribose moiety form the cyclic basis of the dimeric GMP. **(C, D)** c-di-AMP crystallized with *Thermatoga maritime* DisA [17]. Please see color plate at the back of the book.

base-pairing in double-stranded nucleic acids. As discussed elsewhere in this book, mammalian cells possess a number of pattern recognition systems that can detect nucleic acids. The ability of c-di-NMPs to be recognized by and stimulate such systems gives them potent immunostimulatory properties and likely also contributes to their effectiveness as vaccine adjuvants in mammalian systems.

Production and Regulation of c-di-NMPs by Bacteria

Bacteria have a range of enzymes that regulate the synthesis and the degradation of c-di-NMPs. c-di-GMP is synthesized by di-guanylate cyclase (DGC) enzymes that contain a highly conserved GGDEF domain (Figure 13.2A). The GGDEF motif that gives rise to the domain's name is required for synthase activity [14]. Most GGDEFs are inhibited by allosteric binding of c-di-GMP to an inhibitory site, which is thought to prevent cellular GTP depletion in bacterial cells [5,6,14]. Another mechanism used by bacteria to regulate cellular c-di-GMP levels is the degradation of this messenger. The ribose–phosphate linkages of c-di-GMP are hydrolyzed to generate monomeric GMP molecules through two known pathways (Figure 13.2A). In the first, c-di-GMP is hydrolyzed by phosphodiesterase A (PDEA) to form a pGpG intermediate [5,6,14]. An EAL domain in PDEA mediates this hydrolysis, which requires Mg^{2+} and Mn^{2+}

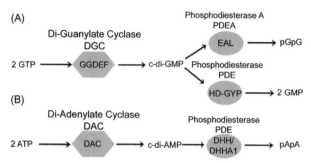

Figure 13.2 *Bacterial synthesis and degradation of c-di-GMP and c-di-AMP.* **(A)** 2 GTP are converted into c-di-GMP and 2pp by the conserved GGDEF domain found in di-guanylate cyclase (DGC) enzymes. C-di-GMP can be degraded by phosphodiesterase activity. The most common phosphodiesterase domain, EAL, degrades c-di-GMP into the intermediate pGpG, while the HD-GYP domain completely hydrolyzes c-di-GMP into 2 GMP molecules [5,7]. **(B)** Di-adenylate cyclase (DAC domains) synthesizes c-di-AMP from 2 ATP molecules. The described DAC domains were first identified in DisA and have since been found in other proteins. c-di-AMP can be degraded by phosphodiesterase activity from DHH/DHHA1 proteins, though it is unknown whether other phosphodiesterases also degrade c-di-AMP in bacteria [17,22].

but is inhibited by Ca^{2+} and Zn^{2+} An additional pathway for c-di-GMP hydrolysis also exists, and utilizes phosphodiesterase (PDE) enzymes with conserved HD-GYP domains [5,6,14]. The DGC and PDE enzymes that regulate c-di-GMP abundance are present in the genomes of both Gram-negative and Gram-positive bacteria, but these enzymes are more numerous in Gram-negative species [7]. In addition, there are usually multiple proteins of each type in a given bacterium. For example, *Pseudomonas aeruginosa* has 17 GGDEF proteins, 5 EAL proteins, 3 HD-GYP proteins, and 16 proteins containing both GGDEF and EAL domains [15]. The presence of multiple enzymes that serve ostensibly redundant functions is thought to permit bacteria to control bioavailability of c-di-GMP in a spatial and temporal manner. Indeed, the GGDEF and EAL proteins contain a variety of unique amino terminal sensory domains that appear to tailor the proteins for different uses by regulating their sub-cellular localization or their interactions with other regulatory proteins. In addition, it is surprisingly common to find the GGDEF and EAL domains in tandem in the same proteins. c-di-GMP binding to these domains may help regulate protein complex formation and activity and control spatial c-di-GMP availability [6,15]. Environmental signals known to regulate c-di-GMP levels include stimuli such as: blue light, oxygen, norspermidine, arginine, and nutrient starvation [5,16]. Host factors have also been shown to influence c-di-GMP regulation, including nitric oxide, bile salts, mucin, and metabolic chemical cues [5,16].

Synthesis of c-di-AMP involves cyclases that contain DAC domains (Figure 13.2B). The prototypical c-di-AMP cyclase is the *Bacillus subtilis* sporulation checkpoint protein DisA [17]. c-di-AMP production occurs in both non–pathogens and in human pathogens such as *Streptococcus pyogenes*, *Listeria monocytogenes*, *Staphylococcus aureus*, *Chlamydia*, and *Mycoplasma* [18–24]. The number of DisA domain proteins within a given species is typically low. For example, *L. monocytogenes* has a single DAC domain protein, *dacA* [21,24]. Indeed, only 330 DAC domain proteins have been identified in the genomes of 275 species [22]. It is unknown whether the EAL or HD-GYP-type PDE enzymes that degrade c-di-GMP can also hydrolyze c-di-AMP [18]. However, certain c-di-AMP-mediated bacterial processes are known to require PDE enzymes with DHH/DHHA1 domains that degrade c-di-AMP [25] (Figure 13.2B). Little is known about the regulation of c-di-AMP synthesis, but the fact that expression of two *Bacillus subtilis* DAC domain proteins is controlled by distinct bacterial sigma factors suggests that c-di-AMP production in this organism is differentially regulated by spatial and temporal signals [22]. In other species, such as *L. monocytogenes*, it is unknown whether c-di-AMP levels are regulated with the same sophistication as c-di-GMP levels. However, it remains a possibility that alternative DAC-functional enzymes may yet be discovered.

The DGC and DAC enzymes responsible for synthesis of c-di-NMPs show strong preferences, respectively, for GTP or ATP binding and there is little sequence similarity between the DAC domain of DisA and the GGDEF domain of DGC proteins [5,18]. It is thus likely that the enzymes responsible for synthesis of c-di-GMP and c-di-AMP evolved independently rather than by divergent evolution from a common ancestral domain [17]. A third class of bacterial nucleotide cyclase was also recently identified. This class is exemplified by the *Vibrio cholera* DncV enzyme, which synthesizes a hybrid cyclic dinucleotide, cyclic-GMP-AMP [26]. Thus, like c-di-GMP and c-di-AMP, cyclic-GMP-AMP is made and impacts bacterial cell signaling.

Biological Effects of c-di-NMPs in Bacteria

The importance of c-di-GMP in bacterial systems was first described in the 1980s. It was first shown that c-di-GMP acts as an allosteric regulator of the cellulose synthase enzyme, which is important for biofilm production by the bacterium *Gluconacetobacter xylinus* [27,28]. c-di-GMP can influence many bacterial processes, but most often inhibits motility and promotes biofilm formation. Biofilm formation can be a factor in virulence of several bacteria, though c-di-GMP can also influence bacterial virulence independent of its effects on bacterial locomotion. Other bacterial processes controlled by c-di-GMP include cell cycle progression, development, fimbrial synthesis, type III secretion, RNA modulation, stress response, bacterial predation, and virulence [5–7]. In certain pathogens, c-di-GMP production can also reduce virulence. For example, increased c-di-GMP concentrations lead to decreased replication of the intracellular bacterial pathogen *Francisella tularensis novicida* in macrophages and decreased virulence in mouse infections [29]. However, the DGC and PDE genes are absent in the more pathogenic *F. tularensis tularensis* strain [29]. It is also thought that c-di-GMP contributes to an increased sensitivity of *Salmonella typhimurium* to phagocyte oxidase killing and thus decreased virulence [30], and to regulate virulence in *Legionella*, *Brucella*, *Bordetella*, and *Vibrio* species [6,31–35]. Thus, c-di-GMP can increase or reduce virulence of various pathogens.

The first biological role for c-di-AMP in bacteria was in the regulation of *Bacillus subtilis* sporulation [22,36]. As mentioned above, the *B. subtilis* DNA integrity scanning protein A (DisA) contains a DAC cyclase domain that produces c-di-AMP. DisA scans chromosomal DNA and upon encountering branching Holliday junctions that are indicative of DNA damage repair, the cyclase activity of DisA is suppressed. The ensuing reduction in c-di-AMP concentration signals the bacterium to halt sporulation [17,20,36]. Subsequent work has shown that c-di-AMP helps to regulate cell wall characteristics including LTA incorporation and cell size in *Staphylococcus aureus* [18]. The function of c-di-AMP and its secretion by *Listeria monocytogenes* remain unclear [21]. Given the relative paucity of DAC enzymes identified to date, it is possible that c-di-AMP is a less ubiquitous bacterial second messenger when compared to c-di-GMP.

Bacterial 'Receptors' for c-di-NMPs

Bacterial responses to c-di-GMP often involve binding of this second messenger to proteins that act as molecular switches. Such binding can alter the ability of these switch proteins to transmit signals to downstream partners. The relevant switch proteins often contain PilZ domains [5], but GGDEF or EAL domains can also be involved. In addition, c-di-GMP

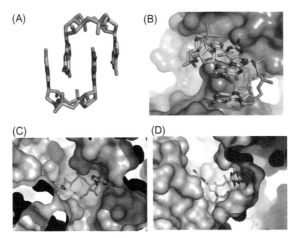

Figure 13.3 *c-di-GMP binding to receptor proteins.* **(A)** c-di-GMP can associate as a non-covalent dimer–dimer through base stacking interactions and hydrogen bonds between the nitrogenous base and the phosphate–sugar bonds of the opposite molecule [37]. **(B)** c-di-GMP binds to allosteric inhibition I sites, such as the one shown in *Caulobacter cresentus* PleD, as a dimer–dimer [37]. **(C)** The only mammalian c-di-NMP receptor to be crystallized with c-di-GMP is STING. c-di-GMP apparently does not bind STING as a dimer–dimer, though it is possible that if a c-di-GMP-mediated complex is formed with DDX441, interface interactions with c-di-GMP could allow for a base-stacking dimer of c-di-GMP to occur [71,72,81]. Please see color plate at the back of the book.

can bind allosteric sites on metabolic enzymes or transcription factors to alter their activity or DNA binding [5–7]. The c-di-GMP bound to these various target or effector proteins in at least some cases is a non-covalent dimer–dimer in which the bases are stacked and hydrogen bonds are formed between the guanine group and the backbone of the opposite molecule [14,37] (Figure 13.3). In addition to proteins, c-di-GMP interacts with riboswitches in messenger RNA to regulate gene expression [38,39]. Riboswitches are most common in bacteria but have been described in certain plants and fungi [40,41].

PRODUCTION AND MEASUREMENT OF c-di-NMPS

The study and use of c-di-NMPs requires their availability in a purified state. Traditional chemical synthesis methods require the use of costly and sometimes hazardous reagents and typically achieve low mg yields of impure material that must then be purified using high performance liquid chromatography [42,43]. In the case of c-di-GMP, an early approach

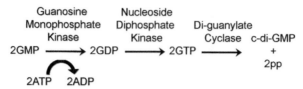

Figure 13.4 *One-pot synthesis of c-di-NMPs using recombinant enzymes.* An efficient method for synthesizing c-di-GMP without the use of hazardous and expensive chemicals relies on nucleotide precursors and three separate recombinant kinase and cyclase enzymes [43]. Using GMP and ATP rather than GTP as the reaction substrates significantly reduces the cost of production.

to improve upon chemical synthesis was the use of recombinant DGC enzymes to dimerize GTP. This method resulted in improved yields up to 100–200 mg, yet still suffered from expensive GTP as a precursor [43]. Subsequently, a 'one pot' synthesis method was developed to facilitate c-di-GMP production [43] (Figure 13.4). This method uses a cascade of three recombinant enzymes to first synthesize GTP from the less costly precursors GMP and ATP. Two GMP and two ATP molecules are converted by GMP kinases into two GDP molecules. These two GDP are converted into two GTP by nucleoside diphosphate kinase. Finally, the two GTP are converted to c-di-GMP by DGC. This method appears to be a more cost-effective and simple means of producing c-di-GMP [43]. Similar 'one-pot' procedures have also been explored for production of c-di-AMP by synthesis using recombinant DAC enzymes and ATP. For detection and quantification of c-di-NMPs in biological samples, most studies have relied on approaches involving mass spectrometry. While effective, mass spectrometry equipment and expertise are not universally available. An alternative approach to detect c-di-GMP is a bacterial reporter system that employs a fluorescent reporter driven by a promoter responsive to c-di-GMP [44]. Similar systems may be possible to develop for detecting c-di-AMP or other c-di-NMPs.

EFFECTS OF c-di-NMPS ON CANCER CELLS

Early Evidence that c-di-NMPs Act on Eukaryotic Cells

c-di-GMP was first characterized as an allosteric activator of bacterial cellulose synthase in *Gluconacetobacter xylinus* [22]. Given the similarity of bacterial and plant cellulose synthases, this finding suggested that c-di-GMP might have homologous functions in plants. Indeed, c-di-GMP was shown to bind a plant cellulose synthase enzyme [45]. It was initially postulated that c-di-GMP might be produced by plants to stimulate cellulose synthase enzymes [45]. However, it is now clear that c-di-GMP are unlikely to be endogenous regulators of these plant enzymes since the GGDEF domains required for cyclase activity in prokaryotic DGC enzymes are absent from the genomes of plants.

GGDEF domains are also absent from the genomes of most eukaryotes and Archaea, which gives support to the concept that c-di-GMP is an exclusive product of bacteria [5–7,46]. However, recent work identified GGDEF proteins in eukaryotic slime molds [47,48]. These *Dictyostelium* species secrete c-di-GMP for extracellular communication during the induction of stalk differentiation [47]. Additional recent evidence further suggests that cyclic dinucleotides participate in mammalian signaling. Specifically, the mammalian cGAS protein, which shares homology with the catalytic domain of oligoadenylate synthase (OAS1), was recently described to have nucleotide cyclase activity in mammalian cells [49]. Like the *Vibrio* cyclase DncV [26], cGAS synthesizes cyclic-GMP-AMP in response to the sensing of cytosolic DNA. As further discussed below, the presence of cyclic-GMP-AMP (like c-di-GMP and c-di-AMP) in the cytosol of mammalian cells contributes to the activation of pattern recognition systems that trigger production of type I interferons (IFNs).

Exogenous c-di-NMPs Target p21Ras

Much research in bacteria has focused on the role of c-di-NMPs as intracellular second messengers. However, there is evidence to suggest that c-di-AMP can be secreted from *Listeria monocytogenes* via multidrug resistance transporters [21,23,24]. It is unclear if c-di-GMP is also secreted from bacteria, but studies using *Dictyostelium* indicate that this eukaryote can secrete c-di-GMP and that this secreted c-di-NMP can elicit effects in neighboring cells [47]. It is unknown whether this or other secreted c-di-NMPs interact with specific eukaryotic cell surface receptors, but work by Amikam et al. indicated that radiolabeled c-di-GMP could enter lymphoblastoid Molt-4 cells without additional manipulation [50].

Analysis of lysates from the stimulated Molt4 cells further revealed that c-di-GMP bound to several cytosolic proteins, including the GTPase p21Ras [50]. p21Ras has an endogenous GDP/GTP binding site that was initially thought to bind c-di-GMP. However, c-di-GMP binding could not be blocked with cGMP, GTP, GDP, or GMP [50], and unlabeled c-di-GMP did not interfere with radiolabeled GTP binding [51]. Thus, it appears that c-di-GMP does not bind p21 via its active site. p21Ras integrates extracellular signals and regulates downstream intracellular effects in a variety of cell processes, including gene expression, cytoskeleton dynamics, cell growth/proliferation, and apoptosis. Consistent with its ability to bind p21Ras, c-di-GMP caused Molt4 cells to increase thymidine incorporation but blocked cell proliferation [50]. Subsequent studies by Steinberger et al. confirmed similar effects of c-di-GMP stimulation in immortalized Jurkat T cells [51]. As with the Molt4 cells, the radiolabeled c-di-GMP bound to p21Ras in Jurkat cells and increased thymidine incorporation but blocked cell proliferation. Cell cycle analysis confirmed that the Jurkat cells were arrested in S-phase. The stimulated Jurkat cells also exhibited a marked increase in CD4 expression [51], which could be due to targeting of p21Ras since, in thymocytes, p21Ras also influences cell differentiation into $CD4^+$ and $CD8^+$ T cell lineages [52].

Later work showed that H508 human cecal carcinoma cells treated with up to $50 \mu M$ c-di-GMP also showed decreased epidermal growth factor (EGF)-induced proliferation, with higher concentrations of c-di-GMP also decreasing basal proliferation [53]. Treatment of confluent normal rat kidney (NRK) cells did not impart cell toxicity, though some decreased proliferation was observed [53]. GMP and other nucleotide analogs did not exert anti-proliferative effects, suggesting that this activity was specific to c-di-GMP and not a metabolic breakdown product of c-di-GMP. MAPK effector proteins that mediate EGFR signaling include MAPK p44/42 (ERK1/2), but ERK1/2 activation was not affected in the H508 cells [53]. This finding is consistent with the studies showing that c-di-GMP inhibited proliferation in Jurkat and Molt4 cells via intranuclear effects, but the precise mechanism for this cell cycle arrest has not been determined.

EFFECTS OF c-di-NMPS ON MAMMALIAN INNATE IMMUNE RESPONSES

c-di-NMPs Activate Cultured Mouse and Human Myeloid Cells

As described in more detail elsewhere in this book, nucleic acids are potent immune stimulants due to their recognition by Toll-like receptors (TLRs), RIG-I-like receptors (RLRs), or other (some not yet known) pattern recognition receptors (PRRs). The detection of nucleic acids in endocytic compartments involves TLR 3, 7, 8, and 9, which can stimulate activate MyD88-dependent (TLR7,8,9) or MyD88-independent/Trif-dependent

(TLR3) signal transduction pathways [54]. In addition, nucleic acids present within the cytosol of immune cells can also stimulate immune responses independently of TLRs. Such detection can give rise to production of type I IFNs and other cytokines as well as the activation of IL-1B and IL-18 processing by enzymatic complexes known as inflammasomes.

Listeria monocytogenes and numerous other bacterial pathogens can stimulate inflammasomes and type I IFN production in infected cells in a manner that is independent of TLRs [54]. There is evidence that bacterial DNA can stimulate these responses when used to transfect host cells [55–57]. However, the extent to which DNA is released from the bacterium during infection remains unclear. Given the prevalence of c-di-GMP in bacterial responses and evidence discussed below suggesting that this messenger can act as a vaccine adjuvant, McWhirter et al. examined whether c-di-GMP could stimulate type I IFN responses [54]. They found that bone-marrow-derived macrophages (BMM) produced IFNβ when stimulated with c-di-GMP, but not with GTP, cGMP, and ppGpp. In addition, treatment of c-di-GMP with PDE enzymes ablated this activity. Similar responses were observed in peritoneal macrophages, dendritic cells, L929 fibroblasts, and Raw264.7 macrophage-like cells that were transfected with c-di-GMP, but not in MEFs or 293T human embryonic kidney (HEK) cells [54]. These findings and the fact that c-di-GMP, pA:T, and pI:C showed similar abilities to induce IFNβ and other genes suggested that c-di-GMP might be able to stimulate cytosolic nucleic acid recognition systems expressed selectively in the responsive cell types. In a separate study, c-di-GMP was also shown to enhance the induction of *il1b*, *tnf*, and *il6* mRNA in LPS stimulated J774A1 mouse macrophage cells [43], though it is unclear if these effects were independent of type I IFN production.

Like c-di-GMP, c-di-AMP was subsequently associated with the induction of IFNβ production by BMMs [21,23,24]. In these studies, it was found that mutant strains of *L. monocytogenes* which over-express the multidrug resistance transporter T (MdrT) secrete heightened amounts of c-di-AMP and trigger heightened type I IFN responses in cultured cells and in infected mice [21,23,24]. The fact that macrophages respond similarly to both c-di-NMPs suggests that secretion or 'accidental' release of these factors acts to boost immune stimulation during bacterial infections. However, given that bacteria produce a large number of products that are recognized by host pattern recognition systems, it remains unclear if stimulation by c-di-AMP and c-di-GMP play essential or non-redundant roles in eliciting host responses to infections by whole bacteria.

The use of transfection reagents to introduce c–di–GMP and c–di–AMP in several of the above studies and in other work [58,59] begs the question of whether these c–di–NMPs must be artificially introduced into cells in order to elicit responses. Initial work on stimulation of type I IFN production by c–di–GMP suggested that transfection increased this agent's potency in BMM [54]. However, the BMM in this study still responded in the absence of the transfection reagent. Transfection was also not required for the ability of c–di–GMP to stimulate J774A1 cells [43], and we have found that use of transfection reagents does not increase the responses of transformed RAW264.7 macrophages to c–di–GMP or c–di–AMP (K. Hill and LLL, unpublished result). As mentioned above, exogenous c–di–GMP was also suggested to enter MOLT4 cells and interact with p21Ras [50]. Thus, it appears there is some capacity for transport of these factors into the cytosol of at least some mammalian cell types. However, the mechanisms responsible for mammalian cell uptake of c–di–NMPs and their impact on the immunostimulatory effects of these factors during infections and in the context of vaccines are not presently clear.

Although c–di–GMP failed to induce IFNα from plasmacytoid DCs, monocyte-derived DCs cultured from PBMCs respond to exogenous c–di–GMP treatment with a maturation/activation program [60]. Treatment with 12–500 µM c–di–GMP (no detectable endotoxin) induced CD83 expression in a dose-dependent manner and use of 200 µM c–di–GMP also increased CD80, CD86, CCR7, and MHC II on treated immature DCs with an effectiveness similar to that of LPS treatment [60]. Other work found that c–di–AMP induced comparable activation of human DCs to c–di–GMP with regard to effects on CD80, CD83, CD86, and HLA-DR [61]. The human dendritic cells treated with c–di–GMP also increased transcription of IFNα, IL-1β, TNF, IFNγ, and IL-12p40, but not IL-4, IL-5, IL-10, and IL-13 [60]. Thus, c–di–NMP treatment appears to trigger production of Th1- versus Th2-type cytokines by human DCs. Similar to LPS-treatment, activation of human DCs with c–di–GMP was also shown to increase their ability to promote T cell proliferation. Cell signaling changes were induced in human DCs in response to c–di–GMP stimulation [60]. p38 MAPK was activated, based on an increase in phosphorylated p38 MAPK. However, NFκB, ERK, and JNK were not activated. In contrast, human macrophages stimulated with c–di–GMP did show pERK activation, but not p38 MAPK [60].

Maturation of mouse dendritic cells by c–di–NMPs has also been observed *in vitro* using bone-marrow-derived DCs [61,62]. c–di–GMP, ci-di-AMP, and the structurally related compound cyclic-di-inosine monophosphate (c-di-IMP) all induced increased cell surface MHC, CD54, CD40, CD80, and CD86 in cultured BMDCs [61].

In vivo Treatment with c-di-NMPs Activates Dendritic Cells and Macrophages

Karaolis et al. were the first to look at the specific responses of dendritic cells and macrophages after c-di-GMP administration *in vivo* [60]. Similar to studies with cultured DCs, dendritic (CD11c+) cells increased CD80 and CD86 expression and secretion of TNF and IL-8 following intraperitoneal (i.p.) treatment of naïve mice with c-di-GMP. In a later study, Ogunnyi et al. confirmed that dendritic cell activation in the lung (measured by CD40 and CD86 expression) was also higher in mice treated with c-di-GMP but not mice treated with cGMP [63], confirming that these *in vivo* responses were specific to c-di-GMP. Homing of activated DCs may also be affected by c-di-GMP treatment, since c-di-GMP increased CXCR4 and CCR7 expression while reducing CCR1 and CCR5 [60]. Treatment of cultured human DCs with c-di-GMP also affected their transcription of the chemokines IL-8/CxCL8, MIG/CXCl9, CXCl-10, CXCL11, MCP1/CCL-2, MIPiaCCL3, MIP-1b/CCL4, and RANTES [60]. The ability of c-di-GMP to alter chemokine and chemokine receptor expression likely impacts movement of DCs and other cell types within peripheral and lymphatic tissues.

In vivo treatment with c-di-GMP was also shown to affect pulmonary macrophages from mice pre-treated or not with c-di-GMP before infection with *Klebsiella pneumoniae* [64]. Nitric oxide (NO) predominantly plays a protective role in this infection model. The *in vivo* treatment with c-di-GMP increased production of iNOS protein and evidence of increased NO production in the lung macrophages [64]. However, it was not determined whether c-di-GMP alone impacted iNOS expression or NO production in the absence of a secondary bacterial stimulation [64]. Later studies found that c-di-GMP alone was unable to induce the release of NO from pre-stimulated macrophages *ex vivo* at concentrations up to $1\,\mu g/mL$, while c-di-IMP induced NO production at concentrations as low as 500 ng/mL [62]. In similar experiments, c-di-AMP was substantially more potent, inducing NO production from pre-activated macrophages at 1.6 ng/mL [61]. These data indicate that various c-di-NMPs have differing potencies or activities when administered *in vivo*. A promising approach in this regard was reported in the development and testing of chemically modified c–di–NMPs. Yan et al. made phosporothioate c–di–GMP analogs, replacing non-bridging oxygen at internucleotide linkages with sulfur [65]. The immune stimulatory effects of these synthetic analogs were comparable to unmodified c–di–GMP, although lung lavages showed a significant upregulation in IL–12(p40), IL–1β, IL–6, KC, MCP–1, MIP–1β, RANTES, and TNF–α compared to stimulation with unmodified c–di–GMP. The sulfur groups on these compounds reduced leukocyte infiltration of the lungs after intranasal administration by ~50% [65]. Further work with chemical variants of c–di–NMPs may facilitate more effective or tailored responses.

Effects of *in vivo* c-di-NMP Treatments on other Innate Immune Cell Types

Infections and stimuli that elicit type I IFN production are known to promote the activation of natural killer (NK) cells. Consequently, splenic NK cells harvested from mice given i.p. c–di–GMP injection were found to have increased responses to stimulation with a canonical NK cell target, YAC–1 tumor cells [54]. The NK cells primed by *in vivo* c–di–GMP treatment showed increased IFNγ secretion in response to the *in vitro* exposure to YAC–1 cells. Since NK cells from mice lacking both IRF3 and IRF7 failed to respond to the c–di–GMP and YAC–1 stimulation, this response was presumably dependent on the induction of type I IFNs by the c–di–GMP. Earlier studies also noted augmented NK cell activation in response to c–di–GMP pre-treatment prior to *Klebsiella pneumoniae* infection as measured by CD69$^+$ expression on NK cells [60].

Pre-treatment of mice with c–di–NMPs also appears to increase recruitment of NK and other immune cells to sites of bacterial infection. For example, a higher influx of immune cells was seen in the lungs of mice receiving c–di–GMP prior to *Klebsiella pneumoniae* infection [66]. The increased inflammatory response was specific to c–di–GMP and not cGMP pre-treatment and consisted of NK cells, neutrophils, and αβT cells. No changes were observed in the numbers of dendritic cells (plasmacytoid and myeloid), lung macrophages, NKT cells, or γδT cells. The intranasal pretreatment with c–di–GMP alone did not increase TNFα or IL–17 in the lungs, but increased MIP–2, IL–12p40, IFNγ, and IP–10 RNA levels were present both at the time of infection and 24 hours post-infection, compared to control mice. The increased production of these cytokines or other chemokines discussed above may account for the increased influx of neutrophils to the lung [64]. In another study, intraperitoneal inoculation of c–di–GMP was shown to recruit F4/80^{high} Ly6G^{med} monocytes and F4/80^{med} Ly6G^{high} granulocytes to the peritoneum as efficiently as 50 µg LPS [60].

In studies where the effects of intranasal administration of c-di-GMP alone were assessed in the absence of bacterial infection, the treatment failed to increase total immune cell numbers in lungs and draining lymph nodes of naïve mice [63]. Alveolar macrophage numbers were also unchanged, though there appeared to be an increased proportion of CD11b positive cells with high CD40 and CD86 staining. CD86 expression on DCs present in these tissues was also increased after c-di-GMP treatment. Numbers of neutrophils in the lung-draining lymph nodes were unaffected, but numbers of monocytes were 20-fold higher. In studies by another group, higher doses of intranasal c-di-GMP were reported to elicit a robust recruitment of lymphocytes and neutrophils to the lungs of the treated mice [65]. Together, these results suggested that pre-treatment of mice with c-di-GMP can elicit rapid inflammatory cytokine responses and more effective/prolonged phagocyte responses.

Mechanisms for Mammalian Cell Detection of c-di-NMPs

Treatment of cells with c-di-GMP activates a signaling cascade that culminates in activation of transcription factor IRF3 by the kinase TBK1 [54]. These responses are known to be downstream of several signaling pathways activated by nucleic acids, including those activated via TLRs and by the adaptors MAVS and STING. The ability of intact, live *L. monocytogenes* to stimulate IFNβ production had previously been shown to be independent of TLRs and MAVS when Barber and colleagues found that this response requires the endoplasmic reticulum–associated protein STING [67,68]. Subsequent work with an ENU mutant mouse harboring a point mutation in the *Sting* gene and with engineered STING$^{-/-}$ mice showed that STING was responsible for the production of type I IFNs in response to treatment with c-di-AMP, c-di-GMP or *Francisella tularensis* infection [21,59,69]. Later studies showed that STING directly binds radiolabeled c-di-GMP and that this binding is competed with by unlabeled c-di-NMPs but not other nucleic acid compounds [70]. The DEAD-box helicase DDX441 also binds both c-di-AMP and c-di-GMPs, and with higher affinities than does STING [71]. This binding to DDX441 requires the protein's DEAD domain rather than the helicase domain [72]. Given the higher affinity, it is possible that DDX441 rather than STING is the primary c-di-NMP detector in the STING-mediated type I IFN induction response [72]. However, signaling for

IFN production is thought to be driven by a complex of c-di-NMPs, DDX441, STING, and TBK1 [71].

Two recent studies revealed that mammalian cells themselves synthesize an endogenous c-di-NMP, the heteromeric cyclic di-GMP-AMP (c-GMP-AMP), as a step in the sensing of cytosolic DNA [49,73]. Like c-di-AMP and c-di-GMP, endogenous c-GMP-AMP is also detected by STING to activate type I IFN production. As mentioned above, an OAS1-like enzyme, cGAS, is responsible for the production of c-GMP-AMP. Whether cGAS evolved to mimic c-di-NMP detection by the existing STING pathway, or STING and DDX441 bind microbial c-di-NMPs due to their similarity to endogenous c-GMP-AMP is not clear. Nonetheless, these recent studies provide a mechanistic explanation for the sensing of both DNA and c-di-NMPs by the STING-dependent pathway (Figure 13.5).

c-di-NMPS AS VACCINE ADJUVANTS

Protective Effects of c-di-NMPs in Infection Models

Given the immune stimulatory effects of c-di-NMPs on cultured cells and in animals, it is perhaps not surprising that administration of c-di-GMP has been found to have protective effects in several animal infection models (Table 13.1). Amongst the early studies demonstrating therapeutic efficacy of c-di-GMP, Karaolis and colleagues used a murine mastitis model to show that c-di-GMP administration at the time of bacterial infection or 4 hours later inhibited *Staphylococcus aureus* colonization of mouse mammary glands in the absence of any direct bactericidal effects [9,60]. These results suggested an effect of the c-di-GMP treatment on biofilm formation and/or the host immune response. To distinguish between these possibilities, Karaolis and colleagues evaluated the ability of c-di-GMP to exert prophylactic effects when given 6 or 12 hours prior to infection. Such pre-treatments reduced bacterial burdens by up to 10,000-fold, arguing in favor of an immune-stimulatory basis for protection. Later work in a pneumonia model with the Gram-negative pathogen *Klebsiella pneumoniae* also showed prophylactic effects of c-di-GMP administration as much as 48 hours prior to infection [64]. Both subcutaneous and intranasal administrations of c-di-GMP were effective. The c-di-GMP significantly reduced bacterial burdens in the lungs and blood by 48 hours post-infection and increased mean survival from 5 to over 7 days. Given the lack of antigen and the short time between c-di-GMP administration and challenge, however, these effects were likely attributable to boosting of innate immune

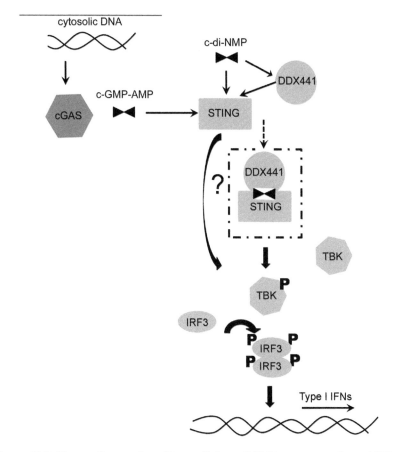

Figure 13.5 *Mammalian sensing of intracellular c-di-NMPs upstream of type I IFN signaling.* Two binding proteins have been identified in the sensing of c-di-NMPs leading to type I IFN signaling: STING and DDX441 [71,72]. These proteins are thought to form a c-di-NMP-dependent complex upstream of TBK1 phosphorylation. TBK1 phosphorylation leads to IRF3 activation and transcription of type I IFNs.

responses. Interestingly, co-administration or post–infection treatment with c–di–GMP did not increase survival in the *Klebsiella* model, but both pre- and post-treatments with c–di–GMP showed protective effects in a mouse model of pneumonia caused by *Streptococcus pneumoniae* [63]. It remains unclear how c–di–GMP pretreatment provides protection even in the absence of pathogen-specific antigens. Presumably this may be due to the ability of c–di–GMP to induce innate immune responses such as type I IFN production, and NK cell and myeloid cell activation. However, it is

Table 13.1 Summary of the literature on use of c-di-NMPs as adjuvants

Source (ref no.)	c-di-NMP	Pathogen	Antigen	Route of Vaccination	Protection?
[79]	c-di-GMP	N/A	β-Gal	Subcutaneous	N/A
[66]	c-di-GMP	N/A	β-Gal, Ova	Intranasal	N/A
[60]	c-di-GMP	Staphylococcus aureus	ClfA (clumping factor A)	Mammary gland injection, intramuscular	Dose–response reduction in mammary gland CFU
[64]	c-di-GMP	Klebsiella pneumoniae		Intranasal, intratracheal	Decreased CFU, increased survival
[63]	c-di-GMP	Streptococcus pneumoniae	PSA, PdB	Intraperitoneal, intranasal	Lower bacterial burden by bioluminescence measurement and CFUs, prolonged survival
[76]	c-di-GMP	Staphylococcus aureus (MRSA)	Clfa (clumping factor A), mSEC	Subcutaneous	Decreased CFUs in liver and spleen, Prolonged survival
[77]	c-di-GMP	Influenza	Plant-derived H5	Intranasal and intramusclular	Hemagluttanin aggregation, cross-protection against antigen-drifted virus
[78]	c-di-GMP	Influenza	H5N1	Sublingual, intranasal, and intramuscular	Hemagluttanin aggregation, cross-protection against antigen-drifted virus
[62]	c-di-IMP	N/A	β-Gal, Ova	Intranasal	N/A
[61]	c-di-AMP	N/A	β-Gal, Ova	Intranasal	N/A

alternatively conceivable that c-di-GMP is recognized as an innate antigen by T cells, analogous to the recognition of glycolipid antigens and vitamins by NKT and MAIT cells [74,75].

With regard to the use of c-di-GMP as a vaccine adjuvant an early study evaluated the effects of administering c-di-GMP versus alum on generation of immunity to purified *Staphylococcus aureus* mSEC or Clfa antigens and intact *S. aureus* [76]. Five weeks after the first immunization, mice were challenged with a lethal dose of a methicillin-resistant *S. aureus* (MRSA) strain, administered intravenously. The c-di-GMP administered on its own significantly reduced bacterial burdens in livers and spleens of the challenged mice, but further significant reductions were seen when c-di-GMP was used as an adjuvant with mSEC or Clfa antigen. Vaccination with the c-di-GMP adjuvant also prolonged survival during infection with lethal doses of MRSA. Thus, c-di-GMP functioned as adjuvant. The differences in protection induced by c-di-NMP and alum adjuvants in these studies were not significant, but the c-di-GMP elicited significantly stronger IgG2a, IgG2b, and IgG3 antibody responses to the mSEC and Clfa antigens.

Subsequently, c-di-GMP was tested as an adjuvant in the induction of protection against influenza virus infection. In one study, mice were immunized intramuscularly or intranasally with plant-derived H5N1 antigen (pH5N1) alone or with c-di-GMP and boosted 21 days later [77]. Sera were collected 21 days after the boost and tested for antibodies against the influenza hemagglutinin using hemagglutination inhibition (HI) assays. Increased HI titers correlate with resistance to seasonal influenza and are used as a measure of predicted protection. Use of the c-di-GMP adjuvant intranasally induced HI titers in the 1000s, well above the accepted protection level of 40, while pH5N1 alone elicited negligible responses. Administration of pH5N1 intramuscularly induced low HI titers that were still considered protective at geometric mean titer levels of 40, but c-di-GMP did not enhance the HI titer through this vaccination route. Sera were also tested for cross-reactive antibodies that indicate protection against antigenically drifted influenza strains, but little cross-reactivity was observed [77]. A separate study evaluated use of c-di-GMP as an adjuvant for virosomal influenza A H5N1 NIBRG-14 vaccine administered by sub-lingual, intranasal, and intramuscular routes [78]. A booster was given after 21 days, and serum antibody levels were assessed 21 days after the booster. The c-di-GMP adjuvant increased HI titers over antigen virosomes alone in all cases and geometric mean titers were considered protective at greater than 40 in c-di-GMP-adjuvanted vaccines through all administration routes [78]. These studies on the use of c-di-GMP as a vaccine adjuvant suggest it potently boosts antibody-dependent protective immunity as well as innate immune responses to pathogens.

There is also evidence that the use of c-di-GMP as an adjuvant boosts cellular immune responses. Ebensen and colleagues observed that vaccination of mice with c–di–GMP plus β-galactosidase versus β-gal alone elicited increased antigen-specific proliferation in cultures of splenocytes 42 days later [79]. The splenocyte proliferation was significantly higher in mice administered a higher dose of c–di–GMP (5 µg versus 1 µg). However, even the lower dose primed stronger responses than vaccination with β-gal alone [79]. Splenocytes from mice immunized with c–di–IMP plus β-gal also showed a higher proliferative response when stimulated *ex vivo* compared to both c–di–GMP and CTB as a control [61,62]. Splenocytes from mice vaccinated with c–di–GMP also secreted more IFNγ (580✕), IL-2 (5✕), IL-4 (1.6✕), IL-5 (327✕) and TNFα (48✕) compared to the non-adjuvanted vaccinated conditions [65]. Similar results were later seen after mucosal vaccination using c–di–GMP and c–di–IMP adjuvants [62,61,79].

Effects of c-di-NMPs on Antibody Responses and Lymphocyte Activation

B cell activation and subsequent antibody production are essential for protection elicited by many vaccines. In BALB/c mice immunized subcutaneously (s.c.) with c–di–GMP and the model antigen β-gal, IgG antibody titers specific for the β-gal antigen were found to be detectable after a single immunization with booster immunizations drastically increasing antibody titers [79]. Other investigators showed that c–di–GMP was an effective adjuvant for i.p. immunization to induce antibody responses to human serum albumin (HSA) [54]. c–di–GMP was also found to be effective as a mucosal adjuvant administered intranasally for immunization against β-gal and ovalbumin model antigens [66]. As with subcutaneous administration, c–di–GMP adjuvant significantly increased IgG secretion. The mucosal vaccine regimen with c–di–GMP also increased class-switching to IgA as revealed by increased β-gal-specific IgA in lung and vaginal lavages. As with subcutaneous studies, mucosal vaccination using c–di–GMP significantly increased titers of β-gal-specific IgG1 and IgG2a versus antigen alone. At higher doses (5 µg) of c–di–GMP, anti-β-gal IgG2a titers were markedly increased over IgG1. A disproportionate increase in IL-2 but not IL-4 secretion from splenocytes was correspondingly observed [66]. Comparison of c–di–GMP to c–di–AMP, c–di–IMP, and the commonly used mucosal adjuvant cholera toxin B (CTB) further revealed that the c–di–NMPs were similarly effective at inducing mucosal IgA in nasal and vaginal lavages while significantly more effective at inducing antigen-specific serum IgG [61,62].

Since Ig class-switching from IgM to IgG, IgA, or IgE is influenced by T helper cells through both co-stimulatory ligand/receptor engagement on the B cell surface and T cell-derived cytokines, the ability of c-di-NMPs to stimulate strong class-switched antibody responses suggests they also potently stimulate T helper cell responses. Indeed, use of c-di-GMP as an adjuvant stimulated both Th1 and Th2 responses to β-galactosidase, based on the cytokine production profiles of splenocytes from c-di-GMP/β-gal vaccinated mice [79]. Mucosal administration of c-di-GMP with OVA peptide also induced OVA-specific cytotoxic T cell responses, in contrast to peptide alone [66]. This is consistent with a predominantly Th1 cytokine profile from T helper cells based on the detection of TNFα, IFNγ, Il-5, and Il-2. Use of c-di-GMP as an adjuvant also induced increased numbers of total and CD69+ αβTCR+ T cells in the lungs of mice immunized against *S. pneumoniae* antigens [64]. Subsequent studies revealed that use of c-di-IMP and c-di-AMP as mucosal adjuvants also induced significant OVA peptide-specific cytotoxicity [61,62]. The c-di-IMP and c-di-AMP also induced Th1-type IFNγ and IL-2 cytokine profiles, as well as Th2-type IL-4 and Th17-type IL-17 production [61,62]. Many of these trends seen with model peptides were also observed in the context of vaccination with influenza virus. For example, intranasal immunization with c-di-GMP and pH5N1 significantly increased IL-2/IFNγ and IL-17 production by splenocytes activated *in vitro* with H5N1 peptide [77]. Sublingual H5N1 virosomes administered with c-di-GMP also induced responses dominated by production of the Th1-type cytokines IL-2, IFNγ and TNFα [78]. Administration of c-di-GMP with influenza virosomes also induced cross-protective T cells against heterologous influenza antigens and high frequencies of homologous H5N1 and heterologous pH1N1-specific CD4+Th1 cells [78]. The immunization also elicited high frequencies of multi-functional IL-2/IFNγ/TNFα CD4+ cells. Such multifunctional T cells are thought to correlate with increased protective immune responses. Therefore, c-di-GMP effectively stimulates humoral and cytotoxic vaccine effector responses in addition to enhancing innate immune activation (Figure 13.6).

Comparisons of c-di-NMPs and Conventional Adjuvants

Ideally, vaccine adjuvants should increase the efficacy of vaccination – presumably by increasing immune stimulation – while minimizing toxicity [80]. It appears that c-di-GMP is non-toxic, at least in mice. In addition, this compound is stable in human serum [60], suggesting that it may have prolonged immune stimulatory effects. However, no detailed pharmacokinetics have been published for systemically or mucosally administered

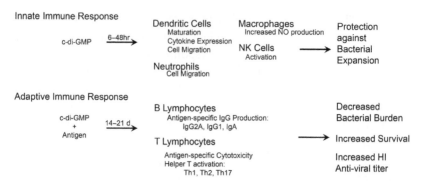

Figure 13.6 *Immunostimulatory effects of c-di-GMP on innate and adaptive immune responses.* Studies with c-di-GMP have explored the immunostimulatory effect of c-di-GMP administered up to 48 hours prior to infection, or with antigen administered 14–21 days prior to infection. Within 48 hours, c-di-GMP induces maturation and stimulation of dendritic cells and macrophages, migration of antigen-presenting cells and neutrophils, and activation of natural killer cells. After 14–21 days, c-di-GMP administered with antigen results in enhanced antigen-specific antibody production from B cells, increased antigen-specific cytotoxicity from CD8$^+$ T cells, and T helper stimulation of the Th1, Th2, and Th17 subtypes.

c-di–NMPs in rodents and/or humans. Future work in this area may provide insight on the differing *in vivo* efficacies of different NMPs and further guide development of vaccination regimens using these adjuvants.

Another major outstanding question is whether c-di–GMP or other c-di–NMPs are more safe and effective than adjuvants currently in use. Only a few studies thus far have directly compared the adjuvant properties of c-di–NMPs with adjuvants in clinical use, such as TLR agonists or alum. Thus, the jury is still out on this question. However, the available results are quite promising. One study, discussed in detail above, found that immunization of mice with pneumococcal antigens plus c-di–GMP was significantly better than antigens plus alum at increasing survival following subsequent challenge with live *S. pneumoniae* [63]. In another study, the efficacy of c-di–GMP for immunizing responses to *S. aureus* was found to be similar to that of alum with regard to reducing bacterial numbers [76]. However, the mice immunized with c-di–GMP showed substantially stronger antibody responses to the immunizing antigens. Thus, in these studies c-di–GMP appears to be at least as effective (and by some measures more effective) as alum at immunizing anti-bacterial immune responses.

Another potential advantage of c-di-GMP as an adjuvant is its demonstrated efficacy in stimulation of mucosal immune responses [61–64,66,76–78]. Presumably, the ability to vaccinate for strong mucosal immunity to a given pathogen is advantageous as mucosal tissues are typically the points of pathogen entry [3]. Currently available mucosal vaccines in humans are exclusively live attenuated pathogens [4]. Such attenuated pathogens often exert some toxicity and in addition have the potential for reversion to pathogenicity. Development of adjuvants that induce mucosal immune activation to purified antigens, rather than tolerance, could thus be a huge advance in vaccinology. As mentioned above, studies in mice thus far suggest that c-di-GMP is an effective mucosal adjuvant in conjunction with influenza vaccines [77,78]. Furthermore, the c-di-NMPs as a group induced greater mucosal adjuvant effects compared to the widely used model mucosal adjuvant CTB [61,62]. The finding that both c-di-IMP and c-di-AMP showed more potent adjuvant effects than c-di-GMP [61,62] suggest that these or other modified c-di-NMPs might prove to be even more efficacious mucosal adjuvants than c-di-GMP.

CONCLUSIONS

Since its discovery 30 years ago, c-di-GMP has revealed itself to be both a second messenger involved in regulation of bacterial behaviors and a potent immune stimulant. The more recently discovered c-di-AMP differs from c-di-GMP in its synthesis and some biological functions, but also seems to act as a second messenger. In addition, c-di-AMP may prove to be an even more potent immune stimulatory molecule than c-di-GMP. Recent findings that Archaea and eukaryotic cells can also produce c-di-NMPs, including c-GMP-AMP, suggest that c-di-NMPs impact signaling in all three kingdoms of life.

In mammals, endogenous c-di-NMPs appear to play an important role in the sensing of cytosolic DNA as a stage immediately proximal to IRF3 activation. Exogenous c-di-NMPs also readily enter immune cells, where they apparently mimic c-GMP-AMP to elicit the production of type I IFNs. Given their immune stimulatory effects and apparent stability *in vivo*, exogenous c-di-NMPs have considerable promise for development as immune stimulants and vaccine adjuvants. However, further studies to better define the unique and conserved effects of the various c-di-NMPs and their variants on immune cells and in animal models will be necessary if their full potential is to be realized.

REFERENCES

[1] Kuolee W, Yan R, Chen H. The potential of 3′,5′-cyclic diguanylic acid (c-di-GMP) as an effective vaccine adjuvant. Vaccine 2010;28(18):3080–5.

[2] Siegrist C-A. Vaccine immunology Stanley A, Plotkin M, Walter Orenstein MD, Paul A, Offit MD, editors. Vaccines (5th ed.). Philadelphia, PA: Elsevier; 2008. p. 17–36.

[3] Rueckert C, Guzman CA. Vaccines: from empirical development to rational design. PLoS Pathog 2012;8(11):e1003001.

[4] Lycke N. Recent progress in mucosal vaccine development: potential and limitations. Nat Rev Immunol 2012;12(8):592–605.

[5] Jenal U, Malone J. Mechanisms of cyclic-di-GMP signaling in bacteria. Annu Rev Genet 2006;40:385–407.

[6] Tamayo R, Pratt JT, Camilli A. Roles of cyclic diguanylate in the regulation of bacterial pathogenesis. Annu Rev Microbiol 2007;61:131–48.

[7] Hengge R. Principles of c-di-GMP signalling in bacteria. Nat Rev Microbiol 2009;7(4):263–73.

[8] Chen W, Kuolee R, Yan H. The potential of 3′,5′-cyclic diguanylic acid (c-di-GMP) as an effective vaccine adjuvant. Vaccine 2011;28(18):3080–5.

[9] Karaolis DK, Rashid MH, Chythanya R, Luo W, Hyodo M, Hayakawa Y. c-di-GMP (3′-5′-cyclic diguanylic acid) inhibits Staphylococcus aureus cell-cell interactions and biofilm formation. Antimicrob Agents Chemother 2005;49(3):1029–38.

[10] Siu CH, Wong LM, Lam TY, Kamboj RK, Choi A, Cho A. Molecular mechanisms of cell-cell interaction in dictyostelium discoideum. Biochem Cell Biol 1988;66(10):1089–99.

[11] Kalia D, Merey G, Nakayama S, Zheng Y, Zhou J, Luo Y, et al. Nucleotide, c-di-GMP, c-di-AMP, cGMP, cAMP, (p)ppGpp signaling in bacteria and implications in pathogenesis. Chem Soc Rev 2013;42(1):305–41.

[12] Traxler MF, Summers SM, Nguyen HT, Zacharia VM, Hightower GA, Smith JT, et al. The global, ppGpp-mediated stringent response to amino acid starvation in ssscherichia coli. Mol Microbiol 2008;68(5):1128–48.

[13] Sondermann H, Shikuma NJ, Yildiz FH. You've come a long way: c-di-GMP signaling. Curr Opin Microbiol 2012;15(2):140–6.

[14] Schirmer T, Jenal U. Structural and mechanistic determinants of c-di-GMP signalling. Nat Rev Microbiol 2009;7(10):724–35.

[15] Galperin MY, Nikolskaya AN, Koonin EV. Novel domains of the prokaryotic two-component signal transduction systems. FEMS Microbiol Lett 2001;203(1):11–21.

[16] Srivastava D, Waters CM. A tangled web: regulatory connections between quorum sensing and cyclic di-GMP. J Bacteriol 2012;194(17):4485–93.

[17] Witte G, Hartung S, Buttner K, Hopfner KP. Structural biochemistry of a bacterial checkpoint protein reveals diadenylate cyclase activity regulated by DNA recombination intermediates. Mol Cell 2008;30(2):167–78.

[18] Corrigan RM, Abbott JC, Burhenne H, Kaever V, Grundling A. c-di-AMP is a new second messenger in Staphylococcus aureus with a role in controlling cell size and envelope stress. PLoS Pathog 2011;7(9):e1002217.

[19] Kamegaya T, Kuroda K, Hayakawa Y. Identification of a Streptococcus pyogenes SF370 gene involved in production of c-di-AMP. Nagoya J Med Sci 2011;73(1-2):49–57.

[20] Oppenheimer-Shaanan Y, Wexselblatt E, Katzhendler J, Yavin E, Ben-Yehuda S. c-di-AMP reports DNA integrity during sporulation in Bacillus subtilis. EMBO Rep 2011;12(6):594–601.

[21] Woodward JJ, Iavarone AT, Portnoy DA. c-di-AMP secreted by intracellular Listeria monocytogenes activates a host type I interferon response. Science 2010;328(5986):1703–5.

[22] Romling U. Great times for small molecules: c-di-AMP, a second messenger candidate in Bacteria and Archaea. Sci Signal 2008;1(33):pe39.

[23] Schwartz KT, Carleton JD, Quillin SJ, Rollins SD, Portnoy DA, Leber JH. Hyperinduction of host beta interferon by a *Listeria monocytogenes* strain naturally overexpressing the multidrug efflux pump MdrT. Infect Immun 2012;80(4):1537–45.

[24] Yamamoto T, Hara H, Tsuchiya K, Sakar S, Fang R, Matsuura M, et al. *Listeria monocytogenes* strain-specific impairment of the TetR regulator underlies the drastic increase in cyclic di-AMP secretion and beta interferon-inducing ability. Infect Immun 2012;80(7):2323–32.

[25] Rao F, See RY, Zhang D, Toh DC, Ji Q, Liang ZX. YybT is a signaling protein that contains a cyclic dinucleotide phosphodiesterase domain and a GGDEF domain with ATPase activity. J Biol Chem 2011;285(1):473–82.

[26] Davies BW, Bogard RW, Young TS, Mekalanos JJ. Coordinated regulation of accessory genetic elements produces cyclic di-nucleotides for *V. cholerae* virulence. Cell 2012;149(2):358–70.

[27] Aloni Y, Cohen R, Benziman M, Delmer D. Solubilization of the UDP-glucose:1,4-beta-D-glucan 4-beta-D-glucosyltransferase (cellulose synthase) from Acetobacter xylinum. A comparison of regulatory properties with those of the membrane-bound form of the enzyme. J Biol Chem 1983;258(7):4419–23.

[28] Aloni Y, Delmer DP, Benziman M. Achievement of high rates of in vitro synthesis of 1,4-beta-D-glucan: activation by cooperative interaction of the *Acetobacter xylinum* enzyme system with GTP, polyethylene glycol, and a protein factor. Proc Natl Acad Sci USA 1982;79(21):6448–52.

[29] Zogaj X, Wyatt GC, Klose KE. cdGMP stimulates biofilm formation and inhibits virulence of *Francisella novicida*. Infect Immun 2012;80(12):4239–47.

[30] Hisert KB, MacCoss M, Shiloh MU, Darwin KH, Singh S, Jones RA, et al. A glutamate-alanine-leucine (EAL) domain protein of *Salmonella* controls bacterial survival in mice, antioxidant defence and killing of macrophages: role of cyclic diGMP. Mol Microbiol 2005;56(5):1234–45.

[31] Arico B, Miller JF, Roy C, Stibitz S, Monak D, Falkow S, et al. Sequences required for expression of *Bordetella pertussis* virulence factors share homology with prokaryotic signal transduction proteins. Proc Natl Acad Sci USA 1989;86(17):6671–5.

[32] Bruggemann H, Hagman A, Jules M, Sismeiro O, Dillies MA, Goyette C, et al. Virulence strategies for infecting phagocytes deduced from the *in vivo* transcriptional program of *Legionella pneumophila*. Cell Microbiol 2006;8(8):1228–40.

[33] Kim YR, Lee SE, Kim CM, Kim SY, Shin EK, Shin DH, et al. Characterization and pathogenic significance of *Vibrio vulnificus* antigens preferentially expressed in septicemic patients. Infect Immun 2003;71(10):5461–71.

[34] Lestrate P, Dricot A, Delrue RM, Lambert C, Martinelli V, De Bolle X, et al. Attenuated signature-tagged mutagenesis mutants of *Brucella melitensis* identified during the acute phase of infection in mice. Infect Immun 2003;71(12):7053–60.

[35] Tischler AD, Camilli A. Cyclic diguanylate regulates vibrio cholerae virulence gene expression. Infect Immun 2005;73(9):5873–82.

[36] Bejerano-Sagie M, Oppenheimer-Shaanan Y, Berlatzky I, Rouvinski A, Meyerovich M, Ben-Yehuda S. A checkpoint protein that scans the chromosome for damage at the start of sporulation in *Bacillus subtilis*. Cell 2006;125(4):679–90.

[37] Chan C, Paul R, Samoray D, Amiot NC, Giese B, Jenal U, et al. Structural basis of activity and allosteric control of diguanylate cyclase. Proc Natl Acad Sci USA 2004;101(49):17084–17089.

[38] Sudarsan N, Lee ER, Weinberg Z, Moy RH, Kim JN, Kink KH, et al. Riboswitches in eubacteria sense the second messenger cyclic di-GMP. Science 2008;321(5887):411–3.

[39] Smith KD, Strobel SA. Interactions of the c-di-GMP riboswitch with its second messenger ligand. Biochem Soc Trans 2011;39(2):647–51.

[40] Sudarsan N, Barrick JE, Breaker RR. Metabolite-binding RNA domains are present in the genes of eukaryotes. RNA 2003;9(6):644–7.

[41] Prasanth KV, Spector DL. Eukaryotic regulatory RNAs: an answer to the 'genome complexity' conundrum. Genes Dev 2007;21(1):11–42.

[42] Kawai R, Nagata R, Hirata A, Hayakawa Y. A new synthetic approach to cyclic bis(3'–>5')diguanylic acid. Nucleic Acids Res Suppl 2003;3:103–4.

[43] Spehr V, Warrass R, Hocherl K, Ilg T. Large-scale production of the immunomodulator c-di-GMP from GMP and ATP by an enzymatic cascade. Appl Biochem Biotechnol 2011;165(3–4):761–75.

[44] Rybtke MT, Borlee BR, Murakami K, Irie Y, Huntzer M, Nielsen TE, et al. Fluorescence-based reporter for gauging cyclic di-GMP levels in *Pseudomonas aeruginosa*. Appl Environ Microbiol 2012;78(15):5060–9.

[45] Amor Y, Mayer R, Benziman M, Delmer D. Evidence for a cyclic diguanylic acid-dependent cellulose synthase in plants. Plant Cell 1991;3(9):989–95.

[46] Romling U, Gomelsky M, Galperin MY. C-di-GMP: the dawning of a novel bacterial signalling system. Mol Microbiol 2005;57(3):629–39.

[47] Chen ZH, Schaap P. The prokaryote messenger c-di-GMP triggers stalk cell differentiation in *Dictyostelium*. Nature 2012;488(7413):680–3.

[48] Eichinger L, Pachebat JA, Glockner G, et al. The genome of the social amoeba *Dictyostelium discoideum*. Nature 2005;435(7038):43–57.

[49] Sun L, Wu J, Du F, Chen X, Chen ZJ. Cyclic GMP-AMP synthase is a cytosolic DNA sensor that activates the type I interferon pathway. Science 2013;339(6121):786–91.

[50] Amikam D, Steinberger O, Shkolnik T, Ben-Ishai Z. The novel cyclic dinucleotide 3'-5' cyclic diguanylic acid binds to p21ras and enhances DNA synthesis but not cell replication in the Molt 4 cell line. Biochem J 1995;311(Pt 3):921–7.

[51] Steinberger O, Lapidot Z, Ben-Ishai Z, Amikam D. Elevated expression of the CD4 receptor and cell cycle arrest are induced in Jurkat cells by treatment with the novel cyclic dinucleotide 3',5'-cyclic diguanylic acid. FEBS Lett 1999;444(1):125–9.

[52] Swat W, Shinkai Y, Cheng HL, Davidson L, Alt FW. Activated Ras signals differentiation and expansion of CD4 + 8+ thymocytes. Proc Natl Acad Sci USA 1996;93(10):4683–7.

[53] Karaolis DK, Cheng K, Lipsky M, Elnabawi A, Catalano S, Hoyodo M, et al. 3',5'-Cyclic diguanylic acid (c-di-GMP) inhibits basal and growth factor-stimulated human colon cancer cell proliferation. Biochem Biophys Res Commun 2005;329(1):40–5.

[54] McWhirter SM, Barbalat R, Monroe KM, Fontana MF, Hyodo M, Joncker NT, et al. A host type I interferon response is induced by cytosolic sensing of the bacterial second messenger cyclic-di-GMP. J Exp Med 2009;206(9):1899–911.

[55] Brunette RL, Young JM, Whitley DG, Brodsky IE, Malik HS. Stetson DB. Extensive evolutionary and functional diversity among mammalian AIM2-like receptors. J Exp Med 2012;209(11):1969–83.

[56] Hornung V, Ablasser A, Charrel-Dennis M, Bauerenfeind F, Horvath G, Caffrey DR, et al. AIM2 recognizes cytosolic dsDNA and forms a caspase-1-activating inflammasome with ASC. Nature 2010;458(7237):514–8.

[57] Unterholzner L, Keating SE, Baran M, Horan KA, Jensen SB, Sharma S, et al. IFI16 is an innate immune sensor for intracellular DNA. Nat Immunol 2010;11(11):997–1004.

[58] Jin L, Hill KK, Filak H, Mogan J, Knowles H, Zhang B, et al. MPYS is required for IFN response factor 3 activation and type I IFN production in the response of cultured phagocytes to bacterial second messengers cyclic-di-AMP and cyclic-di-GMP. J Immunol 2010;187(5):2595–601.

[59] Sauer JD, Sotelo-Troha K, von Moltke J, Monroe KM, Rae CS, Brubaker SW, et al. The N-ethyl-N-nitrosourea-induced Goldenticket mouse mutant reveals an essential function of Sting in the in vivo interferon response to *Listeria monocytogenes* and cyclic dinucleotides. Infect Immun 2010;79(2):688–94.

[60] Karaolis DK, Means TK, Yang D, Takahashi M, Yoshimura T, Muraille E, et al. Bacterial c-di-GMP is an immunostimulatory molecule. J Immunol 2007;178(4):2171–81.

[61] Ebensen T, Libanova R, Schulze K, Yevsa T, Morr M, Guzman CA. Bis-(3′,5′)-cyclic dimeric adenosine monophosphate: strong Th1/Th2/Th17 promoting mucosal adjuvant. Vaccine 2011;29(32):5210–20.

[62] Libanova R, Ebensen T, Schulze K, Bruhn D, Norder M, Yevsa T, et al. The member of the cyclic di-nucleotide family bis-(3′, 5′)-cyclic dimeric inosine monophosphate exerts potent activity as mucosal adjuvant. Vaccine 2010;28(10):2249–58.

[63] Ogunniyi AD, Paton JC, Kirby AC, McCullers JA, Cook J, Hyodo M, et al. c-di-GMP is an effective immunomodulator and vaccine adjuvant against pneumococcal infection. Vaccine 2008;26(36):4676–85.

[64] Karaolis DK, Newstead MW, Zeng X, Hyodo M, Hayakawa Y, Bhan U, et al. Cyclic di-GMP stimulates protective innate immunity in bacterial pneumonia. Infect Immun 2007;75(10):4942–50.

[65] Yan H, Wang X, KuoLee R, Chen W. Synthesis and immunostimulatory properties of the phosphorothioate analogues of cdiGMP. Bioorg Med Chem Lett 2008;18(20):5631–4.

[66] Ebensen T, Schulze K, Riese P, Morr M, Guzman CA. The bacterial second messenger cdiGMP exhibits promising activity as a mucosal adjuvant. Clin Vaccine Immunol 2007;14(8):952–8.

[67] Ishikawa H, Barber GN. STING is an endoplasmic reticulum adaptor that facilitates innate immune signalling. Nature 2008;455(7213):674–8.

[68] Ishikawa H, Ma Z, Barber GN. STING regulates intracellular DNA-mediated, type I interferon-dependent innate immunity. Nature 2009;461(7265):788–92.

[69] Jin L, Hill KK, Filak H, Mogan J, Knowles H, Zhang B, et al. MPYS is required for IFN response factor 3 activation and type I IFN production in the response of cultured phagocytes to bacterial second messengers cyclic-di-AMP and cyclic-di-GMP. J Immunol 2011;187(5):2595–601.

[70] Burdette DL, Monroe KM, Sotelo-Troha K, Iwig JS, Eckert B, Hyodo M, et al. STING is a direct innate immune sensor of cyclic di-GMP. Nature 2011;478(7370):515–8.

[71] Parvatiyar K, Zhang Z, Teles RM, Ouyang S, Jiang Y, Iyer SS, et al. The helicase DDX41 recognizes the bacterial secondary messengers cyclic di-GMP and cyclic di-AMP to activate a type I interferon immune response. Nat Immunol 2012;13(12):1155–61.

[72] Bowie AG. Innate sensing of bacterial cyclic dinucleotides: more than just STING. Nat Immunol 2012;13(12):1137–9.

[73] Wu J, Sun L, Chen X, Du F, Shi H, Chen C, et al. Cyclic GMP-AMP is an endogenous second messenger in innate immune signaling by cytosolic DNA. Science 2013;339(6121):826–30.

[74] Kjer-Nielsen L, Patel O, Corbett AJ, Le Nours J, Meehan B, Liu J, et al. MR1 presents microbial vitamin B metabolites to MAIT cells. Nature 2012;491(7426):717–23.

[75] Rossjohn J, Pellicci DG, Patel O, Gapin L, Godfrey DI. Recognition of CD1d-restricted antigens by natural killer T cells. Nat Rev Immunol 2012;12(12):845–57.

[76] Hu DL, Narita K, Hyodo M, Hayakawa Y, Nakane A, Karaolis DK. c-di-GMP as a vaccine adjuvant enhances protection against systemic methicillin-resistant *Staphylococcus aureus* (MRSA) infection. Vaccine 2009;27(35):4867–73.

[77] Madhun AS, Haaheim LR, Nostbakken JK, Ebensen T, Chichester J, Yusibov V, et al. Intranasal c-di-GMP-adjuvanted plant-derived H5 influenza vaccine induces multifunctional Th1 CD4+ cells and strong mucosal and systemic antibody responses in mice. Vaccine 2011;29(31):4973–82.

[78] Pedersen GK, Ebensen T, Gjeraker IH, Swindland S, Bredhold G, Guzman CA, et al. Evaluation of the sublingual route for administration of influenza H5N1

virosomes in combination with the bacterial second messenger c–di–GMP. PLoS One 2011;6(11):e26973.

[79] Ebensen T, Schulze K, Riese P, Link C, Morr M, Guzman CA. The bacterial second messenger cyclic diGMP exhibits potent adjuvant properties. Vaccine 2007;25(8):1464–9.

[80] Alving CR, Peachman KK, Rao M, Reed SG. Adjuvants for human vaccines. Curr Opin Immunol 2012;24(3):310–5.

[81] Shu C, Yi G, Watts T, Kao CC, Li P. Structure of sting bound to cyclic di–gmp reveals the mechanism of cyclic dinucleotide recognition by the immune system. Nat Struct Mol Biol 2012;19(7):722–4.

Note: Page numbers followed by "*f*" and "*t*" refers to figures and tables, respectively.

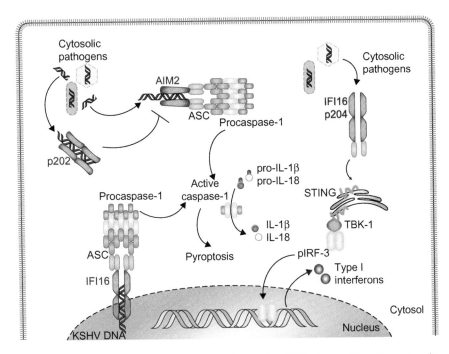

Figure 2.1 *Role of PYHIN proteins in innate immunity.* DNA released from the cytosolic bacteria and viruses is detected by AIM2, which leads to the formation of AIM2 inflammasome complexes comprising of AIM2, ASC, and procaspase-1. The recruitment of procaspase-1 to the inflammasome scaffold triggers the autoproteolytic processing and generation of active caspase-1. The active caspase-1 then catalyzes the proteolytic conversion of pro-IL-1β and pro-IL-18 to their active forms and ultimately leads to pyroptosis. IFI16 has also been reported to form an ASC-containing inflammasome complex only in response to Kaposi sarcoma associated herpes virus (KSHV). IFI16 and p204 also play a role in cytosolic DNA-driven type I interferon responses. In response to cytosolic DNA, IFI16 and p204 associate with STING, which then leads to the activation of TBK-1 and phosphorylation of IRF3. pIRF3 translocates to the nucleus to turn on the transcription of type I interferon genes.

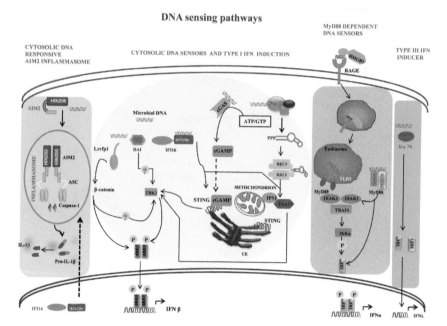

Figure 5.1 *DNA sensing pathways.* A number of DNA sensors have been identified in the cytosol. Five DNA sensors that induce type I IFN production are shown here. LRRFIP1 can bind both dsDNA and dsRNA by activating β-catenin, which migrates to the nucleus to increase IFN-β transcription in an IRF3-dependent and TBK1-independent manner. IFI16 can recognize viral DNA in the nucleus via its HIN200 domains and then migrate to the cytosol to induce IFN-β production mediated by the STING–TBK1 and IRF3 pathways. DAI can recognize both the B-form and left-handed Z-form of DNA and induce IFN-β in a TBK1/IRF3-dependent manner. Cytoplasmic AT-rich dsDNA is recognized by RNA Pol III and transcribed into 5′-triphosphate RNA, which is further recognized by RIG-I to induce IFN-β via IPS-1. The newly identified cyclic GMP-AMP (cGAMP) synthase functions as a cytosolic DNA sensor. In the presence of DNA, it converts cGTP and cATP to cGMP and cAMP, respectively, and the cyclic GMP-AMP interacts with STING and triggers IFN-β production dependent on IRF3. Self-DNA can also gain access to TLR9 with the aid of HMGB1 and RAGE, which can bind to extracellular DNA and facilitate DNA uptake and TLR9 engagement, whereby TLR9 is activated and signals through MyD88 to induce IFN-α production. Another class of DNA sensors dependent on MyD88 are the cytosolic DExDH box helicases DHX9 and DHX36. The AIM2-like receptors AIM2 and IFI16 can form inflammasomes in association with ASC in response to bacterial and viral DNAs, and these inflammasomes mediate caspase-1-dependent cleavage of pro-IL-1β and pro-IL-18 and secretion of active IL-1β and IL-18. Ku-70 induces IRF1/IRF7-dependent IFN-λ production.

Different modes of self nucleic acid recognition

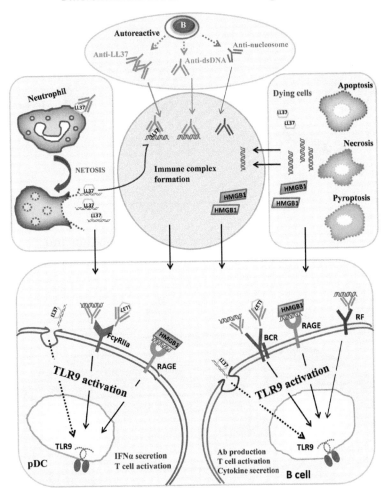

Figure 5.2 *Different modes of self-DNA recognition by TLR9.* Autoreactive B cells can produce ANAs in the form of anti-dsDNA, anti-LL37, and anti-nucleosomes antibodies, and these autoantibodies can form immune complexes with subcellular components, nuclear autoantigens (DNA, chromatin, nucleosomes), and proteins such as LL37 and HMGB1 derived from inefficient clearance of dead cells arising from different forms of cell death, namely apoptosis, necroptosis, and pyroptosis. LL37–DNA complexes can also come from neutrophil death due to NETosis, induced by anti–LL37 antibodies binding to LL37 on the neutrophil surface. The various ICs can promote delivery of self-DNA to endosomal compartments of pDCs or B cells to activate TLR9 through engagement of a variety of receptors. RAGE binds to HMGB1–DNA complexes, and FcγR binds to the Fc portion of autoantibody complexes with LL37–DNA or DNA on pDCs. On B cells, B cell receptor (BCR) binds to the Fc portion of autoantibody complexes with LL37–DNA or dsDNA, RAGE binds to HMGB1–DNA complexes, and RF to anti-dsDNA antibodies.

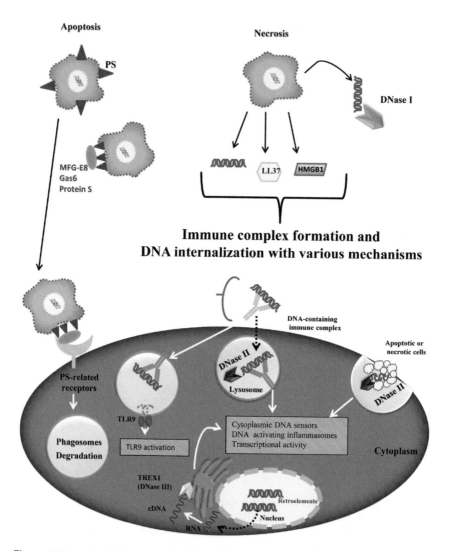

Figure 5.3 *Impaired clearance of self nucleic acids.* Dead cell debris arising from apoptosis and necrosis is the source of self-DNA that can activate various DNA sensors. Several molecules including receptors that bind and clear cell debris and secreted molecules such as MFG-8, GAS6, and Protein S act in concert to effectively clear this self-DNA containing dead cells. DNase I can degrade extracellular DNA and prevent its internalization. DNase II present in lysosomes digests DNA from engulfed apoptotic and necrotic cells or DNA that has escaped from DNase I and is internalized with the help of various receptors (BCR, FcγR, RAGE) and accessory proteins (LL37, HMGB1). DNase III can degrade cell-intrinsic DNA derived from the reverse transcription of endogenous retroelements.

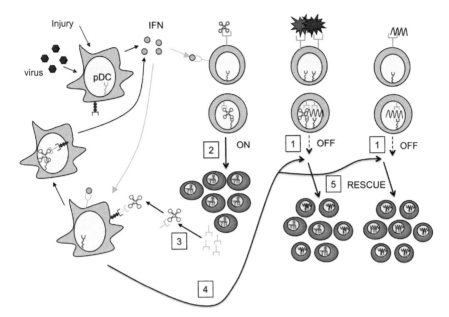

Figure 6.1 *Biphasic role of TLR9 in the etiology of SLE* [1]. TLR9 negatively regulates resting B cells that are stimulated either by BCR/TLR9 engagement, or by a combination of BCR/TLR9 and BCR/TLR7 engagement [2]. TLR7 promotes the activation of resting B cells stimulated by BCR/TLR7 engagement, especially in the presence of type I IFN, and these cells divide and differentiate into antibody producing cells [3]. Autoantibodies produced by BCR/TLR7 activated cells form ICs that stimulate dendritic cells to produce cytokines or survival factors or upregulate co-stimulatory factors [4]. Cytokines/survival factors produced by DCs or activated T cells reprogram resting B cells so that they can no longer be negatively regulated by BCR/TLR9 engagement [5]. Reprogrammed B cells are now activated through a BCR/TLR9 dependent mechanism.

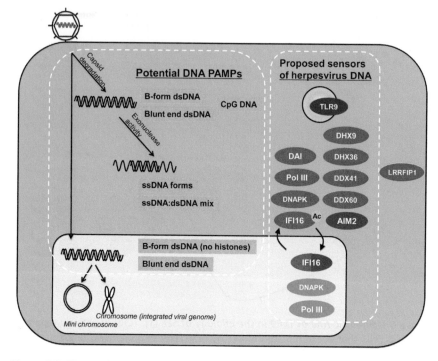

Figure 8.2 *Herpesvirus DNA PAMPs and PRRs.* The herpesvirus genome is in the form of dsDNA, which is believed to be the prime herpesvirus PAMP in the cytoplasm, and potentially also in the nucleus. The incoming genomic DNA is linear, and free DNA ends can potentially also be sensed by host PRRs. Finally, genomic DNA partially digested by exonucleases, and hence containing stretches of ssDNA, could also contribute to stimulation of innate responses during herpesvirus infection. It is currently not known if nuclear herpesvirus DNA in the form of episomes or integrated into the host genome is detected by the innate immune system. DNA sensors are well-established to detect DNA in the cytoplasm, and there is also evidence for innate DNA recognition in the nucleus. All proposed DNA sensors – except LRRFIP1 – have been suggested to recognize herpesvirus DNA. Purple, DNA sensors stimulating IFN expression upon DNA recognition in endosomes. Red, DNA sensors stimulating IFN expression upon DNA recognition in the cytoplasm. Green, DNA sensors potentially stimulating IFN expression upon DNA recognition in the nucleus. Blue, DNA sensors stimulating inflammasome activation.

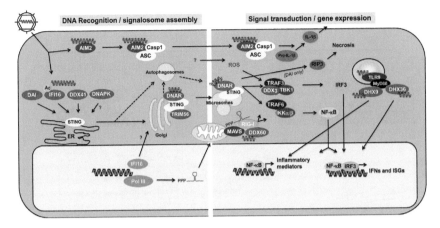

Figure 8.3 *Activation of signaling pathways by DNA.* Left: DNA recognition and signalosome assembly. Cytosolic DNA is recognized by DNA sensors and stimulates assembly of a STING signalsome and an AIM2 inflammasome. STING resides in the ER and proceeds through the Golgi to vesicular structures isolated together with mitochondria-associated ER membranes. The subcellular localization of the AIM2 inflammasome is not well described. RNA polymerase III can transcribe AT-rich dsDNA to induce immunostimulatory RNA species. Right: The STING signalsome and RIG-I-driven MAVS signaling stimulate induction of TBK1–IRF3-dependent IFN responses. DAI specifically signals through RIP3 to induce necrotic cell death. The AIM2–ASC–caspase1 inflammasome stimulates proteolytic maturation of pro-IL-1β and pro-IL-18.

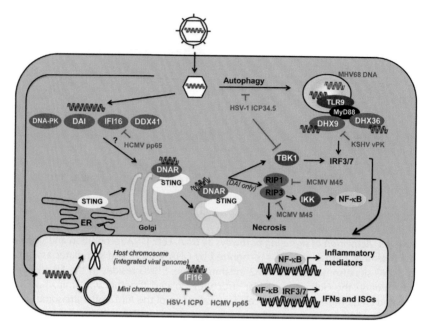

Figure 8.4 *Evasion of DNA sensing and signaling by herpesviruses.* Herpesviruses evade DNA-driven innate immune responses by inhibiting exposure of viral DNA to DNA sensors (e.g. HSV-1 ICP34.5), modulating the genome composition to become less immunostimulatory (MHV68), and by encoding proteins that specifically target DNA sensors or DNA-activated signaling molecules (e.g. MCMV, M45). The figure only includes herpesvirus evasion strategies targeting the upstream steps in DNA recognition and signaling. Viral strategies to evade innate response at the more downstream level are described elsewhere [27].

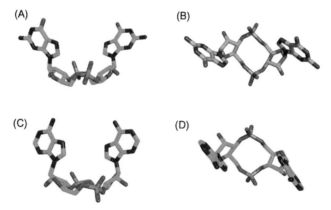

Figure 13.1 *Structure of c-di-NMPs in cyclase proteins.* **(A, B)** c-di-GMP in the cyclase domain of *Caulobacter cresentus* PleD [37]. Phosphodiester linkages between the 3′ and 5′ carbons of the ribose moiety form the cyclic basis of the dimeric GMP. **(C, D)** c-di-AMP crystallized with *Thermatoga maritime* DisA [17].

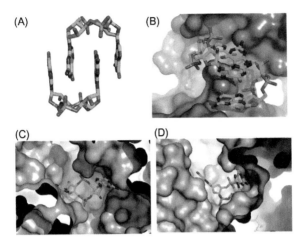

Figure 13.3 *c-di-GMP binding to receptor proteins.* **(A)** c-di-GMP can associate as a non-covalent dimer–dimer through base stacking interactions and hydrogen bonds between the nitrogenous base and the phosphate–sugar bonds of the opposite molecule [37]. **(B)** c-di-GMP binds to allosteric inhibition I sites, such as the one shown in *Caulobacter cresentus* PleD, as a dimer–dimer [37]. **(C)** The only mammalian c-di-NMP receptor to be crystallized with c-di-GMP is STING. c-di-GMP apparently does not bind STING as a dimer–dimer, though it is possible that if a c-di-GMP-mediated complex is formed with DDX441, interface interactions with c-di-GMP could allow for a base-stacking dimer of c-di-GMP to occur [71,72,81].

Printed and bound by CPI Group (UK) Ltd, Croydon, CR0 4YY

08/05/2025

01864982-0001